Nobel Symposium 21

From Plasma to Planet

Proceedings of the Twenty-First Nobel Symposium held September 6–10, 1971, at Saltsjöbaden, near Stockholm, Sweden

Edited by

AINA ELVIUS

Stockholm Observatory

WILEY INTERSCIENCE DIVISION

John Wiley & Sons, Inc. *New York, London, Sydney*

ALMQVIST & WIKSELL *Stockholm*

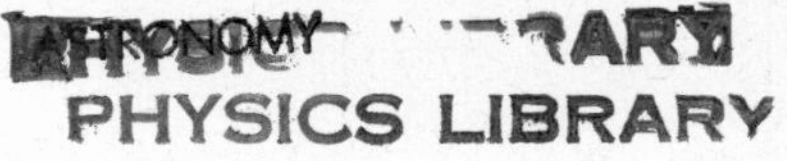

Library of Congress Catalog Card Number 79–39645

Wiley ISBN 0-471-23875-9

Almqvist & Wiksell ISBN 91-20-05045-3

Printed in Sweden by
Almqvist & Wiksells Boktryckeri AB, Uppsala 1972

Preface

The Nobel Symposium 21 "From Plasma to Planet" was held at Saltsjöbaden, Southeast of Stockholm, Sweden, September 6–10, 1971. During the five days nine sessions were held, viz.

1. Chemical state of plasma and gas in space. Chairman: F. L. Whipple.
2. Irradiation of grains in space. Chairman: Y. Öhman.
3. Interaction of plasma and gas with solids. Chairman: P. M. Millman.
4. Chemical studies of meteorites and Moon. Chairman: Z. Kopal.
5. Small bodies in the Solar system. Chairman: H. E. Newell.
6. Formation of comets. Chairman: G. I. Petrov.
7. Space research and the origin of the Solar system. Chairman: Sir Harrie Massey.
8. Plasma dynamics. Chairman: D. Lal.
9. Future space exploration in relation to research on the origin of the Solar system–general discussion. Chairman: H. Alfvén.

Session number 7, on Thursday September 9, was held as an open session in a lecture hall at the Royal Institute of Technology, Stockholm, and was well attended by students and scientists from various fields as well as other persons interested in the exploration of the Solar system by means of space research. After the two lectures a twenty minutes long film produced by NASA was shown to illustrate the planned Grand Tours to the outer planets.

The remaining eight sessions were all held in the conference room "Erstavik" at Grand Hotel Saltsjöbaden and were open only to invited participants and observers.

Professor Hannes Alfvén opened the meeting on Monday morning by presenting the plans for the symposium. His introductory talk as well as all the papers presented at the symposium are contained in the present volume together with an almost complete record of the discussions following the various communications.

Although the exploration of the Solar system presents many more problems than can be solved during a one week conference, the contacts and discussions during this symposium between scientists from many different fields like astronomy, chemistry, meteoritics, plasma physics, and space research, may prove to be of great importance in the future work, not only because of the interesting new information presented there but also because of the better mutual understanding created in discussions on different ways of approach to the same problems and the clarification of some questions of semantics.

The last session was devoted to a general discussion and summing up of results and impressions. Future plans were outlined and suggestions offered for work in various fields which may contribute to a better understanding of the origin and evolution of our planetary system. On behalf of the organizing committee I take this opportunity to express our sincere thanks to the participants of Nobel Symposium 21 for their valuable contributions to the symposium and to this volume which should provide a useful survey of the most important new information and remaining problems in this highly active field of research.

On behalf of the organizing committee I wish to express our deep gratitude to the Nobel Foundation and the Royal Academy of Sciences for the grant that made it possible to arrange this symposium. We also want to thank the Academy for the reception at the Stockholm Observatory, Saltsjöbaden, where the President of the Academy, professor C. G. Bernhard and the Director of the Observatory, professor P. O. Lindblad, welcomed the guests, and for the luncheon given in connection with a visit to the Royal Academy of Sciences where we were received by professor E. Rudberg, Permanent Secretary of the Academy. Our thanks are also due to Dr Marcus Wallenberg for his great hospitality in connection with the dinner given for the symposium participants at "Täcka Udden" and for the favourable conditions offered at Grand Hotel Saltsjöbaden.

We had the pleasure of having with us the Executive Director of the Nobel Foundation, N. Ståhle, who welcomed the guests at the Monday evening banquet.

To the success of the symposium contributed also a visit to the Laboratory for Plasma Physics at the Royal Institute of Technology. Excursions were made to the sculptural garden "Millesgården" and to Drottningholm where participants enjoyed the performance of the Gluck opera "Orfeus and Euridice" at the 18th century Court Theatre.

Finally I wish to express my personal thanks to the two organizers for a pleasant collaboration and to all who contributed to the success of the symposium for their valuable help.

Saltsjöbaden, October 21, 1971

Aina Elvius

Contents

Nobel Symposium Committee 11

Organizing Committee 11

List of participants 12

H. ALFVÉN
From plasma to planet. Plans for the symposium 15

Chemical state of plasma and gas in space

H. MASSEY
Atomic and molecular reactions in space 17
Discussion: Alfvén, Arrhenius, Massey, Vanýsek and Whipple

Z. KOPAL
Origin of the planetary systems: Astronomical evidence in other stars . . 39
Discussion: Anders, Kopal, Lal, Millman, Pellas, Vanýsek and Whipple

Irradiation of grains in space

D. LAL
Accretion processes leading to formation of meteorite parent bodies . . 49
Discussion: Alfvén, Anders, Arrhenius, Lal, Pellas and Whipple

P. PELLAS
Irradiation history of grain aggregates in ordinary chondrites. Possible clues to the advanced stages of accretion 65
Discussion: Anders, Arrhenius, Lal, Pellas and Whipple

Interaction of plasma and gas with solids

H. SATO
A gasdynamical view on the motion, heating and accretion of solid bodies in the solar system 93
Discussion: Alfvén, Anders, Kopal, Lehnert, Pellas, Runcorn, Sato and Whipple

B. LEHNERT
On the conditions for cosmic grain formation 111
Discussion: Lal, Lehnert, Pellas and Runcorn

H. E. NEWELL
A computer-generated motion picture showing a rendezvous mission with comet Encke 115
Discussion: Alfvén, Gehrels, Lindblad, Millman, Newell and Whipple

G. ARRHENIUS
Chemical effects in plasma condensation 117
Discussion: Arrhenius, Gehrels, Kopal, Lodén, Öhman and Whipple

Chemical studies of meteorites and moon

E. ANDERS
Conditions in the early solar system, as inferred from meteorites 133
Discussion: Alfvén, Anders, Arrhenius, Lal, Lindblad, Pellas, Sherman and Whipple

Small bodies in the solar system

P. M. MILLMAN
Cometary meteoroids 157
Discussion: Anders, Arrhenius, Massey, Millman, Pellas, Vanýsek and Wallis

T. GEHRELS
Physical parameters of asteroids and interrelations with comets. 169
Discussion: Anders, Gehrels, Runcorn and Vanýsek

J. TRULSEN
Theory of jet streams 179
Discussion: Alfvén, Anders, Gehrels, Kopal and Trulsen

B. A. LINDBLAD
Meteor and asteroid streams 195
Discussion: Lehnert, Lindblad, Millman and Whipple

Formation of comets

F. L. WHIPPLE
On certain aerodynamic processes for asteroids and comets 211
Discussion: Alfvén, Anders, Arrhenius, Pellas, Sato and Whipple

V. VANÝSEK
The structure and formation of comets 233
Discussion: Elvius, Gehrels, Öhman, Vanýsek and Wallis

A. MRKOS
Observation and feature variations of comet 1969e before and during the Perihelion passage 261
Discussion: Alfvén, Anders, Danielsson, Lal, Lindblad, Trulsen, Vanýsek, Wallis and Whipple

Space research and the origin of the solar system

G. I. PETROV
Investigation of solar system evolution by automatic vehicles on the Moon 273

H. E. NEWELL, D. H. HERMAN AND P. TARVER
Potential contributions of the United States space program to exploration of the solar system 285

Plasma dynamics

J. C. SHERMAN
The critical velocity of gas-plasma interaction and its possible hetegonic relevance 315
Discussion: Alfvén, Lehnert, Sherman and Whipple

Short communications

Y. ÖHMAN
On some plasma rotation phenomena on the Sun 343
Discussion: Lehnert, Lindblad, Öhman and Runcorn

D. LAL
A »cometary» suggestion 349
Discussion: Gehrels, Lal, Vanýsek, Wallis and Whipple

L. DANIELSSON AND W.-H. IP
On the existence of a resonance-captured "quasi-satellite" of the Earth . 353
Discussion: Alfvén, Anders, Danielsson, Gehrels, Lal, Lindblad and Whipple

B. A. LINDBLAD
The possibility of a trans-Saturnian belt of particulate matter 359

H. ALFVÉN
Semantics 361
Discussion: Alfvén, Danielsson, Gehrels, Kopal, Lal, Lindblad, Millman, Newell, Persson and Vanýsek

Future space exploration in relation to research on the origin of the solar system — General discussion

H. ALFVÉN

Introduction to summary papers 367

P. M. MILLMAN

The meteoritic complex 367

H. S. W. MASSEY

Some suggested research activities 368

Discussion: Alfvén, Anders, Arrhenius, Lal, Lindblad, Millman, Newell, Pellas and Runcorn

Discussion after a Summary Paper by G. Arrhenius and E. Anders . . . 372

S. K. RUNCORN

Fossil magnetic fields 373

Discussion: Alfvén and Arrhenius

T. GEHRELS

Remarks on techniques for future studies 377

Discussion: Alfvén, Anders, Kopal and Whipple

H. E. NEWELL

Planning of space experiments 381

Discussion: Alfvén

Z. KOPAL

Studies of other planetary systems with space techniques 381

Discussion: Kopal and Newell

G. I. PETROV

Comments on space programs 384

General discussion 385

Alfvén, Arrhenius, Gehrels, Lal, Lindblad, Millman, Newell, Petrov and Whipple

Nobel Foundation Symposium Committee

Ståhle, Nils K., Chairman, Executive Director of the Nobel Foundation
Hulthén, Lamek, Professor, Member of the Nobel Committee for Physics
Tiselius, Arne, Professor, Chairman of the Nobel Committee for Chemistry
Gustafsson, Bengt, Professor, Secretary of the Nobel Committee for Medicine
Gierow, Karl Ragnar, Ph.D., Permanent Secretary of the Swedish Academy and Chairman of the Nobel Committe for Literature
Schou, August, Director of the Norwegian Nobel Institute (peace)

Organizing Committee

Hannes Alfvén
Gustaf Arrhenius
Aina Elvius
Hans Rickman (assistant editor)

List of Participants

Hannes Alfvén, Professor, Dept. of Applied Physics, University of California, San Diego, *La Jolla*, California, 92 037, USA and Division of Plasma Physics, Royal Institute of Technology, *S-100 44 Stockholm 70*, Sweden

Edward Anders, Professor, Enrico Fermi Institute, University of Chicago, 5640 Ellis Avenue, *Chicago*, Illinois 60637, USA

Gustaf Arrhenius, Professor, Scripps Institution of Oceanography, University of California, San Diego; *La Jolla*, California 92037, USA

Lars Danielsson, Dr., Division of Plasma Physics, Royal Institute of Technology, *S-100 44 Stockholm 70*, Sweden

Aina Elvius, Professor, Stockholm Observatory, *S-133 00 Saltsjöbaden*, Sweden

Carl-Gunne Fälthammar, Professor, Division of Plasma Physics, Royal Institute of Technology, *S-100 44 Stockholm 70*, Sweden

T. Gehrels, Professor, Lunar and Planetary Laboratory, University of Arizona, *Tucson*, Arizona 85721, USA

Torsten Gustafson, Professor, Inst. of Theoretical Physics, *S-223 62 Lund*, Sweden

Øivind Hauge, Dr., Institute of Theoretical Astrophysics, University of Oslo, Blindern, *Oslo 3*, Norway

Nicolai Herlofson, Professor, Royal Institute of Technology, *S-100 44 Stockholm 70*, Sweden

Zdeněk Kopal, Professor, Department of Astronomy, University of Manchester, *Manchester M13 9PL*, England

D. Lal, Professor, Tata Institute of Fundamental Research, Homi Bhabha Road, *Bombay-5*, India

Bo Lehnert, Professor, Division of Plasma Physics and Fusion Research, Royal Institute of Technology, *S-100 44 Stockholm 70*, Sweden

B. A. Lindblad, Dr., Astronomical Observatory, *S-222 24 Lund*, Sweden

Per Olof Lindblad, Professor, Stockholm Observatory, *S-133 00 Saltsjöbaden*, Sweden

Lars Olof Lodén, Professor, Astronomical Observatory, Box 515, *S-751 20 Uppsala 1*, Sweden

Sir Harrie Massey, Professor, Physics Department, University College, Gower Street, *London W.C. 1 E 6 BT*, England

Peter M. Millman, Dr., National Research Council of Canada, *Ottawa 7*, K 1A OR8 Ontario, Canada

A. MRKOS, Dr., Charles University, Astronomical Institute, Svédská 8, *Praha 5-Smichov, Czechoslovakia*

HOMER E. NEWELL, Dr., Associate Administrator, National Aeronautics and Space Administration, *Washington,* D.C. 20546, USA

YNGVE ÖHMAN, Professor, Baltzar von Platens Gata 1, 5tr., *S-112 42 Stockholm,* Sweden

PAUL PELLAS, Dr., Muséum National d'Histoire Naturelle, 61, Rue de Buffon, *Paris 5e,* France

HANS PERSSON, Dr., Division of Plasma Physics, Royal Institute of Technology, *S-100 44 Stockholm 70,* Sweden

G. I. PETROV, Academician, Space Research Institute, Academy of Sciences USSR, 88, Profsojuznaja st., *Moscow, V-485,* USSR

S. KEITH RUNCORN, Professor, School of Physics, The University, *Newcastle upon Tyne NE1 7RU,* England

HIROSHI SATO, Professor, Institute of Space and Aeronautical Science, University of Tokyo, Komaba, Meguro-Ku, *Tokyo,* Japan

JOHN SHERMAN, Dr., Dept. of Electrical Engineering, Liverpool University, *Liverpool,* England

JAN TRULSEN, Dr., Auroral Observatory, P.O. Box 387, *9001 Tromsö,* Norway

V. VANÝSEK, Professor, Charles University, Astronomical Institute, Svédská 8, *Praha 5-Smichov,* Czechoslovakia

MAX WALLIS, Dr., Division of Plasma Physics, Royal Institute of Technology, *S-100 44 Stockholm 70,* Sweden

FRED L. WHIPPLE, Dr., Smithsonian Institution, Astrophysical Observatory, 60, Garden Street, *Cambridge,* Massachusetts 02138, USA

From Plasma to Planet
Plans for the Symposium

By H. Alfvén

The symposium "From plasma to planet" will be devoted to a discussion of a number of physical and chemical processes which are likely to have been essential to the formation and evolution of the Solar System. As the clarification of these events or series of events is one of the main goals of space research, it is important to establish contacts between the scientific leaders of the space efforts on one hand and the scientists working in cosmogony on the other. We are very glad to have prominent representatives of both the USSR and the USA space reasearch here. The scientific result of the space missions have of course already been of great importance to clarifying the evolution of the Solar System and we expect still more of the coming missions. In fact mostly due to space research this field is now slowly changing from a patchwork of hypotheses to a respectable branch of science. We hope that the interaction will go both ways: that the theoretical results will help to formulate the space research priorities.

Because of the character of our field of research it is obviously very difficult to formulate a program and to select the very few invitees which a Nobel symposium traditionally restricts itself to. As the origin and evolution of the Solar System is one of the key problems in science, it has for a long time attracted much interest. The number of hypotheses which have been published, are probably closer to 100 than to 10. It is obvious that it has no sense to discuss all these.

The first thing to do is to exclude all such suggestions which are obviously against the laws of nature. This seems to be a trivial remark, but in this field it is not. Many new suggestions are published by people who seem not to know how celestial mechanics works. Still worse is the knowledge in plasma physics and hydromagnetics. Although there are rather few people today who explicitly deny that plasma phenomena were of essential importance in the early state of the Solar System, there are many who neglect them or apply a plasma physics which was commonly used 10 or 20 years ago but is obsolete today. In fact, the progress in this field has been revolutionary. Partly due to what we have learned about cosmic plasmas from space research, some of the fundamental concepts have changed. But the majority of people working in cosmogony seem not yet to be aware of this new knowledge.

This is regrettable but it has no sense to discuss theories which employ plasma mechanisms which we know do not work. It will be an important task for this symposium to clarify whether there also are other fields of science which are erroneously applied to our problem.

Furthermore there are a number of theories which pretend to explain the origin of planets but which obviously are not applicable to the formation of satellite systems. As has been pointed out many times by several authors the general structure of the satellite systems are so similar to that of the planetary system that only theories (by Arrhenius called "hetegonic theories")—which are applicable to both satellite and planetary formation should be taken seriously. This is another principle which I think justifies the exclusion of a number of theories from our discussions.

Finally I should like to point out that the present state of hetegony seems to be such that we should not aim too much at the formulation of detailed theories but more to a mapping what phenomena should attract most interest and by what methods we should approach them theoretically and experimentally. This includes an attempt to select what space missions are most desirable for clarifying the processes by which a primeval plasma was transformed into planets—and, not to forget, into satellites and all the smaller bodies.

Atomic and Molecular Reactions in Space

By H. S. W. Massey

Physics Department, University College, London

1. *Introduction*

Since the war and particularly during the last ten years there has been a great increase in research activity in atomic and molecular physics especially in the study of the rates of collision processes in the gas phase (Massey et al. 1969, 1971). As a result much more reliable and detailed information is available and there are few basic processes for which the rates are not known to within an order of magnitude, at least under normal temperature conditions. For applications to the interpretation of phenomena in planetary atmospheres and in interplanetary and interstellar space, rates are required often for very exotic reactions, as judged from normal laboratory experience, under unusual conditions. Even if it has not proved possible to reproduce these conditions in the laboratory the progress in understanding the factors which determine the rates of different kinds of reactions has been such that extrapolation of observed results to other circumstances can usually be done without much risk of serious error.

For astronomical and cosmological applications it is also necessary to have information about the rates of reactions occurring at the surfaces of dust grains of different surface composition. In fact it is only by invoking such processes that it seems possible to produce a sufficiently rapid means for formation of molecules in interstellar space. Considerable attention has been paid to the problem of analysing the factors which determine the effectiveness of solid grains in producing molecule formation as well as condensation to larger granules. It is important to realise, however, that the origin of the grains themselves needs to be understood in terms of processes which do not require their previous existence. At the present time it is not at all clear that the processes of production proposed, occurring in the atmospheres of cool stars, satisfy this condition. For this to be so it would be necessary that the various processes giving rise to the stellar condensation in the first place did not involve the previous presence of solid nucleation particles.

To give some idea of the problems and possibilities in relation to the early stages of condensation in planetary formation, a brief preliminary summary will first be given of the information available, or potentially available, about reaction rates which might be relevant. By way of illustration we will then discuss two applications which are sufficiently close in nature to the planetary problem to be of interest—molecule formation in interstellar space and the properties of the Earth's outer atmosphere. The former subject has become of greatly increased topical interest because of the remarkable discoveries in the last few years, through application of mm wave radio techniques, of an unexpected variety of interstellar molecules, some quite complex.

The first direct observation of interstellar molecular hydrogen has also been made through the use of rocket-borne absorption spectrometry. Space techniques, using instrumentation aboard rockets and satellites, has greatly improved and expended our knowledge of the earth's outer atmosphere. Combining these data with the new information now available on atomic reaction rates we are not far from possessing a fairly complete picture of many aspects of the earth's ionosphere.

2. *Basic Processes and Their Reaction Rates*

We now consider briefly the relevant processes which occur in a gaseous, partly-ionized medium at low concentration containing a dispersed system of fine solid grains.

2.1. Ion Recombination Processes.

In a purely gaseous atomic plasma at an ion concentration less than 10^8 cm^{-3} the only effective recombination process is the radiative one

$$A^+ + e \rightarrow A + h\nu \text{ (radiative recombination).} \tag{1}$$

The rate at which this proceeds does not depend very much on the nature of the atom A. It may be calculated accurately when A^+ is a proton (Bates et al. 1939) giving a recombination coefficient which increases from 10^{-13} cm^3 s^{-1} at an electron temperature of 64 000 K to 4.8×10^{-12} at 250 K. Thus in a plasma containing 1 ion cm^{-3} the mean time before recombination will take place, at an electron temperature of 250 K, is 2×10^{11} s or about 10 000 years. For recombination to O^+ the rate coefficient is only slightly different.

If the gas contains molecular ions a much faster process, dissociative recombination, is possible (Massey, Burhop and Gilbody 1972)

$$A_2^+ + e \rightarrow A + A \text{ (dissociative recombination).} \tag{2}$$

For this process the rate tends to increase with the complexity of the molecular ion involved. Thus at ordinary temperatures for both the ions and electrons, it is less than 10^{-9} cm^3 s^{-1} for He_2^+ (Berlande et al 1970) but increases from 1.7 to 7 to 12×10^{-7} cm^3 s^{-1} in going from Ne_2^+ to Ar_2^+ to Kr_2^+. For O_2^+ it is 3×10^{-7} cm^3 s^{-1} while for O^+ it is as high as 2.3×10^{-6} cm^3 s^{-1} (Kasner and Biondi, 1970). The variation of the rate with electron temperature T_e and ion vibrational temperature T has not been investigated over a very wide range. Present evidence indicates a variation with T_e for fixed T as $T_e^{-\alpha}$ where α is between 0 and 1.

Recombination will also occur on the surface of solid grains, probably at each impact of an ion with the surface.

2.2. Rearrangement Collisions

These include charge-transfer reactions such as

$$A^+ + B \rightarrow A + B^+, \text{ (charge transfer)} \tag{3}$$

and chemical rearrangements of the type

$$AB + C \rightarrow AC + B, \text{ (chemical rearrangement).} \tag{4}$$

A great deal of information is available about charge transfer reactions (Massey, Burhop and Gilbody 1971, 1972). In many cases the effective cross section is small when

$$\lambda = a|\Delta E|/\hbar v \gg 1, \tag{5}$$

ΔE being the change of internal energy in the collision, v the relative velocity of impact and a a length of about 8 Å. The maximum cross section, which occurs in these cases when $\lambda \simeq 1$, is usually of order of gas-kinetic. This rule is a rough guide but there are many exceptions.

The most important feature to note about the chemical rearrangement collisions is that an activation energy is usually associated with them. This is found semi-empirically (Glasstone, Laidler and Eyring 1941) to be around $E_i/20$ where E_i is the binding energy of AB. Once the impact energy exceeds the activation energy the effective collision area may well be about gas-kinetic though there will be exceptional cases in which steric factors may reduce it well below this value.

2.3. Diatomic Molecule Formation

2.3.1. *In the gas phase*

The only two-body process in which a diatomic molecule may be formed from separate atoms is the radiative one

$$A + B \rightarrow AB + h\nu \text{ (radiative association).} \tag{6}$$

This process is so slow that it is very difficult to measure its rate. For the simplest reaction of this kind

$$H + H^{+} \rightarrow H_2^{+} + h\nu, \tag{7}$$

it is possible to calculate the rate quite accurately. This was done by Bates (1951) who found a rate, at 500K, of 1.3×10^{-18} cm^3 s^{-1}. Other cases must be discussed individually.

To analyse the process in more detail we note that in general the colliding atoms or ions, both in their ground states, will interact in one of a number of different ways corresponding to the different molecular states which can arise. In general some of these states will be repulsive and others attractive. The most favourable situation which can arise is when an allowed radiative transition can take place from one state to another, lower, state which is attractive. This is the case for the H_2^+ reaction and indeed will always be so for reactions between an atom and a positive ion of the same kind. For H_2 on the other hand it does not occur. The two molecular states arising from the interaction of ground state H atoms differ in total electron spin so that a radiative transition between them is strongly forbidden and the rate of the association reaction will be less than 10^{-24} cm^3 s^{-1}. However, for collisions involving more complex atoms there will certainly be many cases in which allowed radiative transitions to attractive states are possible. For many of the simpler cases the nature of the molecular states which arise can be determined so that cases in which the radiative association rate is likely to be comparable with that for H_2^+ can be selected. These include CH, but NH, OH and MgH are unlikely (Herbig 1963). CH^+ is a special case. If the C^+ is in its ground $^2P_{\frac{1}{2}}$ state no allowed transition is possible but the $^2P_{\frac{3}{2}}$ state requires only 0.008 eV excitation. C^+ atoms in this state may associate radiatively to H atoms through allowed transitions.

Even if the process may be classified as allowed or forbidden there still remains a considerable uncertainty in the reaction rate. It depends not only on the radiative transition probability between the molecular states concerned as a function for nuclear separation but also on the dynamics of the collision. Full allowance for the latter requires a more detailed knowledge of the interaction energy as a function of the nuclear separation in the relevant states than is usually available Thus the interaction, while attractive at small separations, may possess a subsidiary small repulsive maximum at larger separations which would drastically effect the chance of the systems approaching closely enough, at low temperatures, for the radiative transition to be at all likely. A recent discussion of these matters has been given by Bain and Bardsley (1971).

The only other associative process in the gas phase is the three-body one

$$A+B+C \rightarrow AB+C \text{ (three-body association).} \tag{8}$$

It is possible to make an estimate of the rate of such reactions by the following simple argument. We can suppose that the effective cross section for association of A and B will be almost geometrical if it is very probable that, when they collide, an atom C will be found close enough to receive the surplus energy released by molecule formation. As a rough estimate we can take it that this condition will be satisfied if the mean distance between atoms C is of order 10^{-7} cm. The rate for the process (8) can therefore be written

$$\frac{dn_{AB}}{dt} = R_3 = 10^{-21} n_A n_B n_C Q v \tag{9}$$

where n_A, n_B and n_C are the concentrations in cm^{-3} of A, B and C respectively, Q is a cross section of order 10^{-15} cm^2 and v is the mean relative velocity of A and B which may be taken as 10^5 cm s^{-1}. This gives

$$R_3 \simeq 10^{-31} n_A n_B n_C \text{ cm}^6 \text{ s}^{-1}. \tag{10}$$

To show that this gives the correct order of magnitude we may refer to some measured rate such as

$$He^+ + He + He \rightarrow He_2^+ + He \tag{11}$$

which is 0.8×10^{-31} cm^6 s^{-1} at ordinary temperatures (Oskam and Mittlestadt 1963).

It is of interest to note that, according to (10), the three-body reaction rate will certainly be comparable with that for radiative association when 10^{-31} $n_C \simeq 10^{-18}$ cm^{-3}, i.e. $n_C \simeq 10^{13}$ cm^{-3} (corresponding to a pressure of 2×10^{-4} torr at ordinary temperatures). If the radiative process is forbidden then the three-body rate will be comparable for much smaller values of n_C. On the other hand, in terms of time scale, for a three-body process to be effective in a time comparable with the lifetime of the solar system, say 10^9 years, the concentration of neutral atoms must exceed 10^7 cm^{-3}.

It is this low rate which has drawn attention to the important role of dust grains in promoting molecule formation. At the same time it points to the difficulty of understanding how nucleation centres are first built up in astronomical evolutionary processes.

2.3.2. *By physical adsorption on grains*

Formation of molecules on dust grains may take place through physical adsorption or through chemical bond formation. The former can be effective at low gas and grain temperatures but at high temperatures is likely to be unimportant.

The rate at which molecules A_2 are formed through collisions of atoms A with a grain can be written

$$n_A^2 \pi \varrho^2 v S \gamma \tag{12}$$

where ϱ is the effective radius of a grain, v the mean velocity of an atom relative to the grain, S the chance that an atom will stick when it collides with the grain and γ the chance that once stuck the atom will remain long enough for a second atom to collide with the grain and combine with the first to produce a molecule which evaporates from the surface.

Most attention has been concentrated on the special case of the recombination of H atoms (Hollenbach and Salpeter 1970, 1971). Usually when the conditions are such that S and γ in (12) are of order unity for this case they will also be so for heavier atoms.

The sticking probability is determined by the strength of the van der Waals attraction exerted by the surface atoms on the impinging atom and by the rate at which this latter atom loses energy through surface collisions. For light atoms such as hydrogen, allowance must also be made for the zero point energy which reduces the effective binding strength of the van der Waals attraction. Hollenbach and Salpeter (1970) find that, provided

$$kT < kT_g \ll D \tag{13}$$

where T, T_g are the grain and gas temperatures respectively and D is the van der Waals binding energy, the sticking factor is given for H atoms on a perfect ice crystal grain by

$$S(T_g) = \frac{\Gamma^2 + 0.8\Gamma^3}{1 + 2 \cdot 4\Gamma + \Gamma^2 + 0.8\Gamma^3} \tag{14}$$

where $\Gamma = (D\Delta E_s)^{\frac{1}{2}}/kT_g$, ΔE_s being the mean energy transferred per collision with the surface. For H atoms on ice D was calculated to be about 300 k and ΔE_s about 43 k so that $\Gamma \simeq 1$ for $T_g = 72$K. It follows from (14) that $S \simeq 1$ for $T_g \ll 110$K and $\simeq (72/T_g)^2$ for $T_g \gg 72$K.

The mobility of an atom on a perfect crystal surface is very high so that the recombination probability γ is determined by the chance that a second atom will collide and stick to the surface while the first is still adsorbed. The time t_s between sticking collisions is

$$t_s = (\pi S n \bar{v} \varrho^2)^{-1} \tag{15}$$

and the mean time an atom will remain on the surface before evaporation, t_e, is given by

$$t_e \simeq \nu^{-1} \exp(D/kT) \tag{16}$$

where ν is the characteristic lattice frequency of the solid grain. γ will be nearly unity when $t_s < t_e$. A critical temperature T_c for the grains may be defined for which $t_s = t_e$. This is given by

$$kT_c = D/\ln(t_s \nu) \tag{17}$$

For grain temperatures $T \gg T_c$, $\gamma \ll 1$.

For H atoms the problem is further complicated by the possibility that a molecular monolayer will be formed over the grains. If such a monolayer is formed the binding energy for an incident H atom falls so reducing T_c below (17). The condition for formation of a monolayer is that the time t_{s2} for a molecule to stick should be small compared to the time t_{e2}/N for a molecule on any of the N surface sites in the monolayer to evaporate. The binding energy D_2 for a molecule is estimated to be about 550°. The critical temperature T_m for formation of an H_2 monolayer is given by

$$kT_m = D_2/\ln(N t_{s2} \nu) \tag{18}$$

As we shall see later, for typical grains the range of temperatures within which recombination could take place comes out to be very small. However, Hollenbach and Salpeter (1970) pointed out that an actual grain would certainly not have an ideal surface crystal structure. There will be surface irregularities due to the nature of the grain growth process and to radiation bombardment, while impurity molecules will be present. On some of these defect sites the binding energies may be greater than for the pure crystal surface and also may be greater for atoms rather than molecules so effects corresponding to H_2 monolayer formation may not occur. Hollenbach and Salpeter (1971) show that if D', D_2' are the binding energies for H and H_2 attracted to a defect site with enhanced attraction, D' must be $> D_2'$ by 50 k or more. An atom will normally come to rest on a patch of the regular surface and there will be an average time t_{tr} required before the atom becomes bound to the nearest enhanced site. For these sites to help in recombination we must have

$$t_{tr} \ll t_e \tag{19}$$

where t_e is the evaporation time given by (16). The equality $t_{tr} = t_e$ defines a new critical temperature T_{tr} which was estimated by Salpeter and Hollenbach as about 40–50 K. For low grain temperatures $T \simeq 15°$ it is only necessary to have $700\ k < D' < 3\,000\ k$ and $D_2' < 1\,000\ k$ on a few sites per grain surface. At higher grain temperatures there must be a few sites per grain with $50\ k < D' \leqslant 5\,000\ k$.

2.3.3. *By chemical reactions at grain surfaces* (Stecher and Williams 1966)

As mentioned earlier the conditions for formation of H_2 molecules are in general more restrictive than for formation of other molecules because of the relatively low rate of energy loss per collision with the surface and the high zero point energy. However, for all atoms the sticking probability falls rapidly as the gas temperature T_g, which determines the mean impact velocity, increases. Certainly for $T_g > 1\,000$ K physical adsorption effects will be ineffective. Under these conditions molecule formation can still occur through chemical rearrangement collisions involving atoms G of the grain surface

$$GX + Y \rightarrow G + XY. \tag{20}$$

Such reactions will only be effective if they are exothermic so that the GX bond energy is less than that of the molecule XY.

It is necessary (as mentioned in 2.2) to take into account in considering these processes that an exothermic chemical rearrangement reaction involving an initial molecule with binding energy E_i, is usually associated with an activation energy $A \simeq E_i/20$. Because of this the process is only likely to be important when the gas temperature is in excess of the grain temperature by an amount $(E_i/20\,k)$. It is therefore quite unimportant for reactions at low gas temperatures T_g for which the physical adsorption mechanism outlined above is required. For $T_g > 1\,000$ K on the other hand the reverse is the case and, as pointed out by Stecher and Williams (1966), must be invoked as the major source of molecule formation.

The rate at which a reaction of this kind will proceed at a gas temperature T is given by

$$\tfrac{1}{2} n_y Q (8\pi k T/m_y)^{\frac{1}{2}} N f n_g \exp(-A/kT) \tag{21}$$

n_y, n_g being the respective concentrations of molecules Y and of grains, Q the mean effective area for collisions between atoms X and Y at temperature T, m_y the mass of atom Y, N the total number of reaction sites on the grain surface and f the fraction available for the chemical reaction. The factor $\frac{1}{2}$ allows for the fact that the atom Y may approach only from outside the grain.

Stecher and Williams (1966) considered in particular graphite and ice surfaces. In the former case the edges of the basal planes contain free carbon valences so are strongly reactive and in the course of time will capture atoms from the surrounding space. In this process the bulk of the grain acts as the third body so that the capture process can be regarded as an associative reaction

$$G' + \text{C} + X \rightarrow G' + \overline{\text{C}}X \tag{22}$$

where G' is the residual grain and $\overline{\mathrm{C}}$ is a carbon atom bound to the edge of a basal plane. For such reactions there is no activation energy but for an atom X to be captured in this way the energy released must be insufficient to break up the bond of strength 5.5 eV holding the C atom in the basal plane i.e. $\overline{\mathrm{C}}$. This process we can regard as the formation on the surface of diatomic molecules $\overline{\mathrm{C}}X$. We can then write the reactions corresponding to (18) as

$$\overline{\mathrm{C}}X + Y \rightarrow \overline{\mathrm{C}} + XY. \qquad (23)$$

The most likely possibility for X in interstellar space is an H atom. The binding energy CH will be less than in free CH (3.47 eV) and was estimated as 3.2 eV, giving an activation energy near 0.17 eV (2 040 k). Exothermic reactions are then possible when Y is an H, C, O and N atom. If X is a C atom the activation energy A is about 0.24 eV (2 870 k) and Y may be N or O. Again X may be an N atom in which case A is about 0.4 eV (4 780 k).

For pure ice grains there are fewer possibilities, the only exothermic reactions being those which lead to H_2 and CO production.

In an ionized medium the grains will be charged, the sign and magnitude of the charge being determined by the need to balance the net rates of collection of negative and positive charge. Unless the loss of electrons from the grain by the photoelectric effect is high the balance is achieved only when there is a retarding potential applied to the incoming electrons which will be of the order of the electron temperature. A negative potential of 1 V on a grain of radius 0.1 μm acquires a surface charge of about 60 electrons. The chance of permanent capture of positive atomic ions at vacant valence sites will be affected by the energy released through recombination with surface electrons. As in dissociative recombination this may break the bond and release the neutralized atom once again. Whether these effects are important it is not yet clear.

2.4. Photoionization

The cross section for photoionization of H has been calculated accurately as a function of frequency (Massey et al. 1969). Although experimental measurement of photoionization cross sections for other, chemically unstable atoms, such as O, N and C is difficult, calculated values are available which give good average results over an appreciable frequency range. In fact the true cross sections exhibit resonance effects due to autoionization at least out to frequencies of 2 to 3 times the threshold value. Typical resonance widths are around 0.01 eV. For many purposes they are blurred out by averaging but it is important to remember that they exist.

Photoionization cross sections have been measured with quite high frequency resolution for most diatomic gases but little reliable information, except as to order of magnitude, is available for chemically unstable molecules such as CH, NH and CN. The cross-section frequency curves exhibit even more detailed structure at frequencies out to a few times the threshold, than for atoms.

2.5. Photodissociation

Optical absorption by molecules which results in dissociation is a very important process in determining the equilibrium between atomic and molecular forms. One obvious way in which this can occur is by the inverse of the photoassociation processes discussed earlier—absorption of a photon in a suitable frequency range produces a transition from a bound to an unstable molecular state. It is not necessary, however, that this latter state should dissociate into ground state atoms only. A transition to an upper unstable state leading to dissociation into one or both excited atoms will be equally effective.

In earlier considerations, particularly those concerning the photodissociation of H_2, account was taken only of these possibilities in which case the estimated dissociation rates by interstellar radiation came out to be very slow indeed. This is because a transition to the nearest unstable state from the ground state in H_2 requires an energy as high as 14.7 eV (840 Å) which is beyond the Lyman limit. However, it was pointed out by Stecher and Williams (1967) that dissociation may occur at considerably lower quantum energies through excitation of the upper, B, state of the Lyman bands. Although this state is bound optical transitions back to the ground state may occur to vibrational levels of that state lying in the continuum and so lead to dissociation. Radiation with wavelength between 960 and 1 010 Å is the most effective. Assuming the intensity of radiation to have the constant value G_0 in photons cm^{-2} s^{-1} Hz^{-1} the rate of photodissociation due to the process is approximately

$$3 \times 10^{-3}\, G_0 \text{ s}^{-1}. \tag{24}$$

Because they depend very much on the detailed distribution of the interaction energy curves for the different molecular states photodissociation cross sections vary very much from molecule to molecule. For CH they are large (Herzberg and Johns 1969) and are such that almost every photon of wavelength 3 200 Å absorbed leads to dissociation. On the other hand for N_2 and CO, which have large binding energies, photodissociation is difficult but it is not possible as yet to give any reliable quantitative data.

2.6. Ionization and Dissociation by Impact of Energetic Particles

Although the flux of cosmic ray particles in interstellar space is very low there may be circumstances in which it will cause significant ionization or dissociation. Experimental and theoretical information is available about the rates at which these processes will occur due to impact of energetic charged particles (Massey, Burhop and Gilbody 1972).

3. *Molecules in the Interstellar Medium*

3.1. Molecular Hydrogen

Until very recently there was no direct evidence available about the concentration of H_2 in interstellar space. However it was found (Garzoli and Varsavsky 1966; Varsavsky 1968; Mészáros 1968) that the great majority of dark nebulae, or dust clouds, gave little or no evidence of emission or absorption of the 21 cm of atomic hydrogen indicating a much greater ratio of dust grain to H atom concentration than for 'normal' regions. This suggested the possibility that, in dust clouds, the hydrogen is mainly molecular.

To establish the presence of H_2 and measure its concentration directly it is necessary to observe in the ultraviolet, particularly between 1 000–1 150 Å within which lies the Lyman band resonance absorption of H_2. Such observations only became possible with the development of space observing techniques. The first observations using rocket borne spectrographs (Carruthers 1967; Smith 1969, 1970) showed that H_2 is not present in appreciable concentration in the general interstellar medium. However in 1970 Carruthers observed the absorption bands of interstellar H_2 in the far ultraviolet spectrum of ξ Persei. Along the line of sight the total visual extinction by dust was about 1 magnitude which corresponds to at least a dark grey cloud. The column density of H_2 (1.3×10^{20} cm^{-2}) found was comparable with that of interstellar H (4×10^{20} cm^{-2}) determined from the Ly α absorption line in the same spectrum, indicating that nearly half the total hydrogen may be molecular.

The latest theoretical analyses (Hollenbach, et al. 1971; Solomon and Wickramasinghe 1969) of the H–H_2 equilibrium in interstellar space are in general agreement with the observations. In an H region we suppose that the ratio n_g/n of the concentration of grains to gas atoms and molecules is constant. In low density clouds the mass ratio of dust to gas, which is in these cases, as we shall see, almost exclusively atomic hydrogen, is $\simeq 10^{-2}$. We therefore take

$$n_g/n \simeq 10^{-2}\, m_{\mathrm{H}}/m_g. \qquad (25)$$

A typical grain (Spitzer 1968; Wickramasinghe 1967) is of density 2 g cm^{-3}, mean radius 0.17 μm and mass 4×10^{-14} g. This gives

$$n_g/n \simeq 4 \times 10^{-13} \tag{26}$$

The gas temperature T_g is around 100 K and $n \simeq 10\ \text{cm}^{-3}$ in typical clouds at low concentration. A further important quantity is the grain temperature T. This arises as a balance between the rate of absorption of ambient radiation and that of emission from the grain, mainly in the infrared. The interstellar radiation field is usually represented as that corresponding to a Planck distribution at a temperature of 10^4 K diluted by a factor 10^{14}. In such a field a black body would take up a temperature of about 3 K. However the emissivity of an actual grain at long wavelengths will fall below that of the black body and so T will be substantially above 3 K but probably not greater than 15 K. This assumes that the radiation field is not appreciably reduced by absorption within the cloud. In dust clouds this will be strong and the grain temperature will be much lower.

The H–H_2 equilibrium is essentially determined by a balance between the rate of formation of H_2 molecules by physical adsorption on grains and destruction by photodissociation. Following the analysis of the formation process by Hollenbach and Salpeter (1970, 1971), which allows for the effect of surface imperfections, the sticking and recombination probabilities S and γ in (12) will be close to unity under the conditions we have assumed. This gives for the rate of production

$$R_2 = 10^{-17}\, n_{\text{H}} n\ \text{cm}^{-3}\ \text{s}^{-1} \tag{27}$$

where n_H is the concentration of H atoms. For the radiation field assumed, the rate of photodissociation is given by

$$R_d \simeq 10^{-10}\, n_2\ \text{s}^{-1} \tag{28}$$

where n_2 is the concentration of H_2 molecules, so that, for equilibrium,

$$n_2/n_{\text{H}} \simeq 10^{-7} \tag{29}$$

This low relative concentration is well below the upper limits assigned from rocket observations.

In dense clouds the situation is changed because the strong absorption reduces the radiation field within the cloud. Hollenbach, Werner and Salpeter (1970) have carried out detailed calculations for representative cases of spherical clouds under these conditions. If the optical depth to the centre of the cloud due to absorption by the grains is τ_ν and that in the ultraviolet near 1 000 Å is taken as 2.5 τ_ν then their results can be represented roughly as follows. If $f(\xi)$ is the fractional abundance $2\,n_2(\xi)/n$ of the hydrogen in molecular form at a fractional depth ξ (=actual depth /radius of the cloud) then, when (29) applies at the surface of the cloud,

$$f(\xi) \simeq 4\times 10^{-7}\, n \exp(10n\xi\tau_\nu), \quad n\xi\tau_\nu < 0.4, \tag{30}$$

$$\simeq X(1+X^{\frac{1}{2}}+X)^{-1}, \quad n\xi\tau_\nu \gg 1, \tag{31}$$

where $X = 10^{-4}n^2\xi\tau_\nu \exp(5\xi\tau_\nu)$. In (30) the dominating exponential factor arises from molecular line absorption and in (31) absorption by dust. If $X(1) \gg 1$ the cloud is mostly molecular.

As an example, if $\tau_\nu = 0.33$ and the cloud mass is 500 $M_\odot$, corresponding to a gas atom concentration of 50 cm^{-3}, f rises from about 10^{-3} for $\xi = 0.1$ to nearly 0.1 at the centre of the cloud. These results are not inconsistent with the rocket observations in which the H_2 concentration was found to be comparable with that of H in a cloud for which $\tau_\nu \simeq \frac{1}{2}$.

An estimate may be made of the fraction of interstellar gas in molecular form as about 0.4.

The atomic hydrogen concentration in the cores of dense clouds will depend on the cosmic ray flux which penetrates. It is difficult to estimate how strongly the effective particle radiation (energies a few MeV) is absorbed. Estimates of the rate of production of H through dissociation by impact of the general flux of cosmic radiation have been made by Solomon and Werner (1971). They find a rate which appears to be too large, indicating that substantial absorption of cosmic rays must occur within dark clouds.

3.2. Other Molecules in Interstellar Space

The first molecular lines were observed in absorption around 4 000 Å. These were identified as lines of CH^+ and CH (near 4 300 Å) and of CN (near 4 000 Å) (Swings and Rosenfeld 1937; McKellar 1940; Douglas and Herzberg 1941). The abundances of these molecules were estimated at about 10^{-7} to 10^{-8} that of hydrogen.

The next major discovery was of the 18 cm line of OH (Weinreb et al. 1963) both in emission and absorption. With the development of millimetre wave astronomy the first observations of spectra due to polyatomic molecules were made—the emission lines of H_2O (Cheung et al. 1968) and of NH_3 (Cheung et al. 1969) at 13 mm, as well as absorption by NH_3 at the same wavelength. In the last 2 years a remarkable series of discoveries have been made. Observations of the 2.6 mm line of CO both in emission and absorption (Wilson et al. 1970; Penzias et al. 1971) have shown that this molecule is remarkably abundant (in some directions up to 10^{-3} of that of hydrogen). Great interest was attached to the observation of absorption lines at 6.2 and 1 cm of formaldehyde H_2CO (Snyder et al. 1969; Buhl et al. 1969; Palmer et al., Zuckerman et al., 1969, 1970) a molecule also observed in emission at 2 mm. This is one of the simplest organic molecules. Since then CH_3CN and CH_3CH_2 have been observed in absorption at 2.7 and 3.5 mm respectively. The other molecules which have been detected up till the time of writing are HCN (Snyder and Buhl 1971) (in emission and absorption) CS, COS and SiO (all in absorption) (Solomon 1971).

The richest source of these molecular lines is near the direction of the galactic centre but regions of star formation such as H II regions, infrared objects and the Orion nebulae are usually strong sources also. At the time of writing OH is the only molecule detected in an external galaxy but relatively little effort has yet been devoted to searches of this kind.

From a study of the brightness temperatures, line shapes etc. it is possible to derive information about the kinetic temperature, the radiation field, the velocities of mass flow and other properties of the interstellar medium. The ubiquity and high abundance of CO make it particularly useful, as for example in the study of very dense clouds.

In a number of cases the observed emission or absorption line intensities are not consistent with thermal equilibrium with the 2.7° isotropic radiation field. This is true for example of the OH and H_2O emission and of the NH_3 and HCHO absorption. Various suggestions have been made about mechanisms for producting the anomalous level populations but the situation is still far from clear. Eventually, when these anomalies are understood, they should provide much more detailed information than hitherto about conditions in the regions concerned.

The problem of determining the origin of these molecules is a difficult one and, at the present time, when new information is still being accumulated rapidly, only relatively little can be said.

The formation of CH and CH^+ was the first case considered, in the now classic paper by Bates and Spitzer (1951). Account was taken as quantitatively as possible of the processes of formation and of destruction of these molecules. It was found that the latter processes, mainly photodissociation, proceeded at such a rate as to require a balancing rate of production one or two orders of magnitude faster than that estimated for radiative association. The suggestion was then made that the molecules were most probably formed on the surface of dust grains. Later work, while indicating that the rate of radiative association was probably underestimated also showed that this applied a fortiori to the rate of photodissociation so that it seems even more necessary to invoke surface recombination.

The case of CO is very interesting. Very intense emission lines due to the rotational transition $J=1\rightarrow0$ have been observed along directions passing through H II regions (W 51, Sag A, Sag B for example) (Penzias et al. 1971). These include not only the line from the main isotope $^{12}C^{16}O$ but also lines from $^{13}C^{16}O$ and $^{12}C^{18}O$. In fact the emission is so strong in many cases that the $^{12}C^{16}O$ line is probably strongly saturated. Analysis of the data is consistent with the assumption that the isotopic abundance ratios are not very different from those on earth ($^{12}C/^{13}C = 89$, $^{16}O/^{18}O = 488$), although this cannot be regarded as finally established, and column densities as high as 10^{19} molecules/cm^2 have been derived along directions passing through W 51, assuming that the rotational distribution is thermal. This corresponds to concentrations of as much as 1 molecule/cm^3 throughout the emitting region.

It seems likely that, in very general terms, the relative abundance of CO is high because not only does the molecule possess an exceptionally high binding energy but is also very difficult to dissociate by absorption of radiation.

The fact that the CO emission is especially strong along directions passing through H II regions appears inconsistent with the relatively high rate with which electrons can excite rotation in a dipolar molecule. If the electron concentration is > 3 cm^{-3}, which would be expected in H II regions, the excitation rate would be so high that the brightness temperature of the emission would be considerably higher than observed. Excitation in neutral gas collisions would be quite adequate to produce a thermal rotational distribution in equilibrium with the kinetic temperature of the surrounding interstellar gas.

In contrast to CO, the CN distribution (Jeffert, Penzias and Wilson 1970) is certainly optically thin and the column density of CN is $< 10^{-4}$ of that for CO. Estimates of collision rates, taking account of the fact that the dipole moment of CN is greater than 10 times that of CO, suggests that the rotational distribution in CN should be thermal in HI regions but not in HII regions, so that the CN emission may be used as a radiation thermometer under the former conditions.

It is remarkable that the triatomic molecule HCN is considerably more abundant (Snyder and Buhl 1971) than CN. This may mean that the molecules have arisen from fragmentation of more complex aggregates rather than from associative processes.

Zuckerman et al. (1970) found that in about 50% of the clouds in which HI absorption was observed H_2CO was detectable with a projected density $\simeq 10^{-8}$ of that of H. Again it was found by comparison with the OH absorption survey (Goss 1968) that there were six sources in which OH but not H_2CO was detected but none in which H_2CO was detected and not OH. Typically the OH projected densities were about 30 times those of H_2CO. The origin of the latter is obscure.

4. *The Earth's Upper Atmosphere—the Ionosphere*

Ever since the discovery of the reflecting layers for radio waves in the Earth's upper atmosphere a great deal of research has been carried out to study the properties of the layers and investigate the nature of the processes which determine these properties. Very great advances have been made in recent years because of the rapid growth of knowledge in relevant reaction rates from laboratory research and of ionospheric properties and the short wave solar spectrum from the use of space techniques.

The ionized regions are largely produced by photoionization of upper atmospheric atoms and molecules by solar ultraviolet and X-radiation. Particle radiation from the sun makes only a small contribution except under special conditions such as at polar latitudes or following solar disturbances. At any time the equilibrium concentration of electrons is determined by a balance between the rate of production in this way and the net rate of loss by various reactions with atmospheric ionized and neutral constituents. During the daytime the greatest concentration n_e of electrons (5×10^5 cm^{-3}) occurs at altitudes between 250 and 400 km in the so called F_2 region. Smaller daytime maxima are found at lower altitudes near 160 km ($\simeq 2 \times 10^5$ cm^{-3}) in the F region and 110 km ($\simeq 10^5$ cm^{-3}) in the E region.

A considerable amount of information is now available about the electron concentration n_e, the composition of the positive ions, the kinetic temperatures of the electrons, positive ions and neutral gas molecules and the motions of the ionization and of the neutral gas. These quantities vary with altitude, time of day, latitude and longitude, magnetic activity and epoch in the solar cycle so it is important to make observations of correlated quantities at the same time and place. Such sets of observations are just now becoming available but already a good deal can be done by considering the conditions at altitudes and times where variability is a minimum.

For a proper theoretical understanding of the ionosphere it is necessary to have as basic information, (*a*) the intensity and composition of short wave solar radiation (*b*) the composition of the neutral atmosphere (*c*) the absorption and photoionization cross sections of atmospheric atoms and molecules and (*d*) the rates of recombination and of ionic reactions involving atmospheric ions and neutral molecules. From (*a*), (*b*) and (*c*) the rate of production and from (*b*) and (*d*) the rates of loss of free electrons and of different positive ions may be calculated. In addition account must be taken of the redistribution of charged particles through diffusion and drift. In general transport effects are not important in the E and F_1 regions but are of dominant importance in F_2.

A great deal of information is now available about (*a*) from rocket and satellite borne spectrographs, about (*b*) from rocket and satellite studies including day air glow observations and about (*c*) and (*d*) from laboratory work so that an effective start can be made towards a detailed interpretation of ionospheric properties.

One interesting feature of atmospheric composition is that the ionosphere occurs in a dissociative region in which the solar radiation as well as producing ionization is also effective in photodissociation in a highly selective way. Thus above about 100 km the rate of photodissociation of O_2 by solar radiation near 1 760 Å dominates the rate of three-body recombination so that the oxygen is very largely in the atomic form. On the other hand N_2, like CO, is very resistant to photodissociation and remains predominantly as N_2 out to very high altitudes. Diffusive separation begins to be very significant above about 150 km so that the O atoms float above the N_2 and the O/N_2 ratio rises.

During the daytime the main positive ions are NO^+ and O_2^+ in comparable concentrations up to about 140 km. Above this altitude O^+ becomes very rapidly the only important ion until altitudes of 800 km or so. Depending on conditions, the dominant ion at higher altitudes is then H^+, or at first He^+ and then at even higher altitudes H^+. N_2^+ and N^+ are unimportant at all altitudes, the concentration of N_2^+ at 150 km being about 10^{-2} of that of O^+ and that of N^+ even smaller.

We now consider in more detail the situation in the E and F_1, regions which are less responsive not only to dynamical effects but also to other variables than the higher regions.

Using typical data (Donahue 1966) on the composition of solar radiation and of the neutral atmosphere together with laboratory information on photoionization rates, the rate of ionization at 3 altitudes within the region comes out to be as shown in Table 1.

The equilibrium concentration of electrons is given by

$$\frac{dn_e}{dt} = q - \alpha n_e^2 = 0, \tag{32}$$

where q is the total rate of production of electron-ion pairs and α is the effective recombination coefficient given by

$$n_e \alpha = \alpha_2[N_2^+] + \alpha_3[O_2^+] + \alpha_4[NO^+]. \tag{33}$$

α_2, α_3 and α_4 are the coefficients for dissociative recombination, (2), of electrons to N_2^+, O_2^+ and NO^+ respectively. $[A^+]$ denotes the concentration

Table 1.

Altitude (km)	Temp. (T°K)	Concentration of neutral particles (cm^{-3})				Ionization rates (ion pairs cm^{-3} s^{-1})		
		O	N_2	O_2	NO	O^+	N_2^+	O_2^+
130	335	2.2×10^{10}	1.1×10^{11}	1.1×10^{10}	1.6×10^{7}	7.3×10^{2}	2.1×10^{3}	810
160	620	4.2×10^{9}	9.3×10^{9}	7×10^{8}	—	7.7×10^{2}	1.5×10^{3}	300
220	937	7.5×10^{8}	6.1×10^{8}	3.5×10^{7}	—	2.5×10^{2}	2.3×10^{2}	26

of A^+. $[N_2^+]$ is so small as to be ignored. At 300K, corresponding to 150 km altitude, using the data given in the Table together with data on $[O_2^+]$ and $[NO^+]$, we find

$$\alpha_3 + 0.7\,\alpha_4 = 5 \times 10^{-7}\ \text{cm}^3\ \text{s}^{-1}. \tag{34}$$

This is consistent with laboratory data which gives $\alpha_3 = 3 \times 10^{-7}$ (Kasner and Biondi 1968) and $\alpha_4 = 4 \times 10^{-7}$ cm^3 s^{-1} (Weller and Biondi 1967). At 220 km where the gas temperature is approaching 1 000K (34) is replaced by

$$0.7\,\alpha_3 + \alpha_4 = 7.5 \times 10^{-8}\ \text{cm}^3\ \text{s}^{-1}. \tag{35}$$

Because of the high gas and presumably also electron temperatures α_3 and α_4 are not so well known as at 300K and complications due to atmospheric motions are likely to be more important as the altitude increases. It is therefore uncertain whether any inconsistency arises.

The very low concentration of N_2^+ at all altitudes appears surprising at first sight because these ions are produced at a faster rate than any other ions during the daytime (see Table 1). Laboratory measurements (Massey, Burhop and Gilbody 1971) show that the fast reaction which depletes N_2^+ is

$$N_2^+ + O \rightarrow NO^+ + N \tag{36}$$

with a rate coefficient of 2.5×10^{-10} cm^3 s^{-1} at 300K.

The equilibrium concentration of O^+ is determined by the two reactions

$$O^+ + N_2 \rightarrow NO^+ + N, \tag{37}$$

$$O^+ + O_2 \rightarrow O_2^+ + O, \tag{38}$$

with rate coefficients at 300K of 4×10^{-12} and 4×10^{-11} cm^3 s^{-1} respectively.

Special interest attaches to NO^+ because, although it is a dominant ion, it is not formed appreciably by primary photoionization. At the higher altitudes it is mainly formed through the reaction (36) but at lower altitudes (37) and also

$$O_2^+ + N_2 \rightarrow NO^+ + NO \tag{39}$$

become of comparable importance. In general NO^+ tends to dominate because NO has the lowest ionization potential of all the species which can be formed from N and O atoms.

Turning now to the F_2 region transport processes cannot be ignored. In addition the effective loss coefficient for electrons changes for the following reasons. Dissociative recombination can only take place to molecular ions. In F_2 in which O is the dominant neutral species, and O^+ the dominant ion, the effective loss rate is determined by the rate at which O^+ ions can be converted into molecular ions (Bates and Massey 1947). We may therefore write the loss rate as βn_e where

$$\beta = \gamma_2[N_2] + \gamma_3[O_2], \tag{40}$$

where γ_2 and γ_3 are the rate coefficients for the reactions (37) and (38) respectively. Since $[O_2] \ll [N_2]$ at these high altitudes $\beta \simeq \gamma[N_2]$. The equilibrium condition for the electron concentration at height h is that

$$q - \beta n_e - \frac{d}{dh}\{n_e(w_a + w)\} = 0, \tag{41}$$

where w_d and w are respectively the vertical diffusion and drift velocities of the ions. The peak ionization will be located at a height where the loss rate βn_e is equal to the rate of vertical diffusion. For an exponential lapse rate of atmospheric concentration with scale height H this will occur when

$$\beta = \gamma_2[N_2] \simeq D/H^2 \tag{42}$$

where D is the ambipolar diffusion coefficient for O^+ions. Using γ_2 derived from laboratory data, $[N_2]$ and H derived from observations with rocket-borne equipment and a theoretical estimate for D, good general agreement is obtained with the observed location of the F_2 maximum of n_e (Rishbeth 1966).

The situation is less satisfactory as regards the behaviour of He and He^+ in the extreme outer atmosphere and inadequate laboratory data is available for the discussion of the H^+ equilibrium.

Unlike interstellar space, dust grains seem to play no important role in the ionosphere (Bates and Massey 1947) which can be regarded as a region in which dissociation rather than association predominates. It is of interest to note that the particle concentrations involved are comparable with those assumed in discussing planetary accretion.

I am grateful to Mr Spitzer for his assistance in obtaining reference material and for discussion of the recent work on molecules in interstellar space.

References

Bain, R. A. and Bardsley, J. N. 1971, Preprint Univ. of Manchester, Phys. Lab.

Bates, D. R., 1951, M.N.R.A.S., *111*, 303.

Bates, D. R. Buckingham, R. A., Massey, H. S. W. and Unwin, J. J., 1939, Proc. Roy. Soc. A., *170*, 322.

Bates, D. R. and Massey, H S. W., 1947, Proc. Roy. Soc. A., *192*, 1.

Bates, D. R. and Spitzer, L., 1951, Ap. J., *113*, 441.

Berlande, J., Cheret, M., Deloche, R., Gonfalone, A. and Manus, C., 1970, Phys. Rev. *A1*, 887.

Buhl, D., Snyder, L. E., Zuckerman, B. and Palmer, P., 1969, 12th meeting of American Astron Soc. Honolulu, March 30–April 2.

Carruthers, G. R., 1967, Ap. J., (Letters) *148*, L 141. 1970, Ap. J., *161*, L 81.
Cheung, A. C., Rank, D. M., Townes, C. H., Thornton, D, D. and Welch, W. J., 1968, Phys. Rev. Letters, *21*, 1701.
—— 1969. Nature, 221, 626.
Donahue, T. M., 1966, Planetary and Space Science, *14*, 33.
Douglas, A. E. and Herzberg, G., 1941, Ap. J., *94*, 381.
Garzoli, S. L. and Varsavsky, C. M., 1966, Ap. J., *145*, 79.
Glasstone, S., Laidler, K. J. and Eyring, H., 1941. The Theory of Rate Processes, New York, McGraw Hill.
Goss, W. M., 1968, Ap. J. Suppl., *15*, 131.
Herbig, G. H., 1963, J. Quant. Spectry. Radioactive Transfer, *3*, 529.
Herzberg, G. and Johns, J. W. C., 1969, Ap. J., *158*, 399.
Hollenbach, D. and Salpeter, E. E., 1970, J. Chem. Phys., *53*, 79.
Hollenbach, D., Werner, M. W. and Salpeter, E. E., 1971, Ap. J., *163*, 165.
Kasner, W. H. and Biondi, M. A., 1968, Phys. Rev., *174*, 139
McKellar, A., 1940, Pub. A. S. P., *52*, 187.
Massey, H. S. W., Burhop, E. H. S. and Gilbody, H. S., 1969, Electronic Impact Phenomena, Clarendon Press, Oxford, 2nd edn. Vols. I & II.
Massey, H. S. W., Burhop, E. H. S. and Gilbody, H. S., 1971, ibid Vol. III.
Massey, H. S. W., Burhop, E. H. S. and Gilbody, H. S., 1972, Vol. IV in preparation.
Mészáros, P., 1968, Astrophys. Space Sci., *2*, 510.
Oskam, H. J. and Mittlestadt, V. R., 1963. Phys. Rev., *132*, 1435.
Palmer, P., Zuckerman, B., Buhl, D. and Snyder, L. E., 1969, Ap. J. (Letters), *156*, L 147.
Penzias, A. A., Jefferts, K. B. and Wilson, R. W., 1971, Ap. J., *165*, 229.
Rishbeth, H., 1966, J. Atoms and Terrestrial Phys., *28*, 911.
Smith, A. M., 1969, Ap. J., *156*, 93.
Smith, A. M., 1970, ibid, *160*, 595.
Snyder, L. E. and Buhl, D., 1971, Ap. J. (Letters), *163*, L 47.
Snyder, L. E., Buhl, D., Zuckerman, B. and Palmer, P., 1969. Phys. Rev. Lett., *22*, 679.
Solomon, P. M., 1971, Discussion Meeting, Cambridge, Aug. 3.
Solomon, P. M. and Werner, M. W., 1971, Ap. J., *165*, 41.
Solomon, P. M. and Wickramasinghe, N. C., 1969, Ap. J., *158*, 449.
Stecher, T. P. and Williams, D. A., 1966, Ap. J., *146*, 88, 1967, ibid *149*, L 29.
Swings, P. M. and Rosenfeld, L., 1937, Ap. J., *86*, 483.
Varsavsky, C. M. 1968, Ap. J., *153*, 627.
Weinreb, S., Barrett, A. H., Meeks, M. L. and Henry, J. C., 1963, Nature, *200*, 829.
Weller, C. S. and Biondi, M. A., 1967, Phys. Rev. Lett., *19*, 59.
Wilson, R. W., Jefferts, K. B. and Penzias, A. A., 1970, Ap. J. (Letters), *161*, L 43.
Zuckerman, B., Buhl, D., Palmer, P. and Snyder, L. E., 1970, Ap. J., *160*, 485.
Zuckerman, B., Palmer, P., Snyder, L. E. and Buhl, D., 1969, Ap. J. (Letters), *157*, L 167.

Discussion

H. Alfvén

When discussing the space density which is required for a certain process, it is essential to distinguish between average density and local density.

Homogeneous models of cosmical plasmas are commonly used in spite of the fact that we are not sure whether a cosmic plasma ever is homogeneous over large volumes. Solar prominences, coronal streamers and filamentary structures in interstellar clouds are examples of inhomogeneous plasmas. In all these cases the average density may differ by orders of magnitude from the local density.

H. Massey I agree entirely.

V. Vanýsek
You mentioned the maximum grain temperature of 15 K. In a recent paper in Ap.J. a critical temperature limit of 10 K or 8 K was given. How do you calculate the critical temperature?

Augason, G. C., 1970, Ap.J., *162*, 463.
Hollenbach, D. et al., 1971, Ap.J., *163*, 155.

H. Massey
In my paper I have described what might happen. But the situation is very complex due to the imperfect surfaces of the actual grains.

G. Arrhenius
As pointed out by Sir Harrie, very low temperatures (<10 K) are required by general theory to permit sufficient dwelling time of hydrogen atoms on crystal surfaces for production of H_2 at any appreciable rate. This has caused concern among astrophysicists who occasionally observe substantial concentrations of H_2 in association with dust at much too high temperatures to permit the necessary production by the process considered. Resort to imperfections and edge effects seems to raise the upper temperature limit not more than another ten degrees. We have recently called attention to the possibility of non-activated adsorption of hydrogen on nickel-iron grains as a likely mediating process in the molecule production, remaining efficient up to several hundred degrees and hence providing an explanation both for the production of H_2 and other molecules observed.

Brecher, A. and Arrhenius, G., 1971. Nature, *230*, 107.

F. L. Whipple
What is your opinion regarding interstellar grain temperatures as low as 3 K so that H_2 ice could form on them?

H. Massey
I think it is clear that if you work out the theory for perfect crystal surfaces, you find conditions under which an H_2 layer can be formed. But I am not sure how well the theory applies to imperfect surfaces and what the temperature conditions are for them.

Origin of the Planetary Systems Astronomical Evidence in Other Stars

By Zdeněk Kopal

Department of Astronomy, University of Manchester, England

One of the principal aims of the astronomical science is to extrapolate, from observations that can be made at the present time, the past as well as the future of the celestial bodies and their systems—including the solar system—as far as this can be done within the framework of acceptable physical theories. The present state of our solar system is indeed amply documented by extensive evidence obtained by telescopic means as well as (more recently) with the aid of spacecraft. Yet—perhaps because of the size of this evidence—an extrapolation of the present state to its initial conditions by theoretical means alone has proved well-nigh impossible—both because of the obvious complexity of the problem as well as because of the long time-span (some 4.5 billion years) separating us from the origin of the Sun and its planets. On the other hand, in the present state of our knowledge, the origin of the solar system is manifestly not an "initial-value" problem which could be developed deductively from the first principles; so that an empirical strategy still remains our best avenue of approach.

In quest of such an empirical approach, the best strategy would seem to be to inquire into possible circumstantial evidence on astronomical processes by which planetary systems can be found, offered by other stars in our proximity. We have no reasons to conclude that our solar system is unique in the sky; moreover, it was probably the result of processes that are generally operative under similar circumstances; and if so, can we hope to see any one of them being born in our neighbourhood at the present time?

Recent work on the astrometry of nearby stars, which we know best, has disclosed a total of 55 stars (including the Sun) within a distance of 5 parsecs (16 light years) from us (cf. van de Kamp 1955), of which 31 appear to be single, and 18 are grouped in 9 binaries while 6 form two triple systems (i.e., α Cen and o^2 Eri). Moreover, within 20–25 parsecs from us more than 1 700 stars have been identified so far (cf. Gliese 1969; Woolley et al. 1790); and their total census (incomplete as it still is due to observational selection for objects of low luminosity) may well be larger. The most important result of such surveys has been the realization that single stars—far from being the dominant species in stellar population—are actually less numerous than systems consisting of two or more stars located so close together as to revolve around their common centre of gravity and share common translational motion through space.

In most such double or multiple stars, the mass-ratios of their constituents remains in the range of 10: 1, rendering both components luminous stars in their own right. A careful search for binaries in which the mass-ratios exceed this range has, however, led in recent years to the discovery of components whose masses are less than one-hundredth that of our Sun—i.e., of planetary rather than stellar order of magnitude. One of them is the well-known "Barnard's star"—a red dwarf of spectral type M 5 and apparent visual magnitude 9.5. Its annual parallax of $0''.545$ (corresponding to a distance of 5.98 light years) renders Barnard's star the nearest known star to us with the exception of the system of α Centauri. Moreover, a continuous tracking of its proper motion (which amounts to $10''.3$ per annum) in the sky led, in recent years, to the discovery of a 25-year periodic perturbation (cf. van de Kamp, and Dennison 1950)—indicating the presence of an invisible companion whose mass has been estimated by van de Kamp (1968) to 0.0015 $M_\odot$, or $1\frac{1}{2}$ times that of Jupiter.

Barnard's star is, moreover, not the only one of our stellar neighbours whose proper motion has disclosed the presence of unseen companions of small mass. Thus the brighter component of 61 Cygni (whose parallax of $0''.292$ corresponds to a distance of 11.1 light years) has recently been shown by Strand (1956) to possess a companion of mass close to 0.008 $M_\odot$ revolving around its primary in a period of only 4.8 years. Another nearby star—Lalande 21185, our third nearest stellar neighbour at a distance of 8.2 light years—was shown by similar astrometric methods to possess a companion of mass less than 0.01 $M_\odot$ revolving around its central star (a red dwarf of spectral class M 2) in a period close to 8 years.

These are the three most conspicuous cases of our evidence for celestial bodies in our neighbourhood with masses of planetary rather than stellar order of magnitude. All three are no more than 12 light years away—at a distance within which we find a total of 17 stars including our Sun. If, therefore, 4 out of 17 stars in our immediate stellar neighbourhood have proved to possess companions of masses comparable with that of Jupiter (and less massive companions can easily escape detection), planetary systems would seem to attend some 25 % of nearby stars—possibly more. In other words, astronomical evidence already on hand tends to indicate that planetary systems —far from being rare—may be actually very common in the Universe around us.

If such is indeed the case, the question immediately arises as to the number of planetary systems that may be in the process of formation at the present time. The duration of the formative process of our own solar system can be at least estimated from the dispersion of the measured radioactive ages of the oldest meteorites, as well as of the rocks brought back in the past two years from the lunar surface by different Apollo missions; and it appears to be of the order of magnitude of 10^8 years or less (as experimental errors are likely to exaggerate this difference). On the other hand, the present age of the solar system (i.e., the time which elapsed since the solidification of its oldest remains) is now known to lie between 4.5 to 4.6×10^9 years. Therefore, the formative stage of the solar system apparently did not last any longer than 1–2 % of its present age. If, moreover, not more than one star in ten were to have acquired such a system (and indications exist in our close neighbourhood that this may be an under-estimate), it should follow that approximately one star in a thousand should be passing through this formative act at any one time. As more than 1 700 individual stars have now been identified within a distance of less than 25 parsecs from us, one or two of these should now be in the throes of such a formative process.

The foregoing argument is, to be sure, over-simplified in one essential respect: namely, it assumes the formation of the stars to be a continuing process. In actual fact, most stars in our close neighbourhood are considerably evolved, and belong to the "disc-type" population contemporary with our Sun. Since, moreover, we know that the birth of the central star occurs simultaneously with the formation of its planetary system—the two events cannot be consecutive—it follows that the probability of an identification of this process could be greatly increased by confining our attention to a sample of stars of more recent origin. Such stars are rare in our close proximity; but can be spotted at a greater distance if they are sufficiently bright. In point of fact, supergiant stars of most spectral types are likely to be quite young (as their available supplies of nuclear fuel cannot maintain their excessive luminosities for very long time)—the younger, the greater the absolute luminosity of the respective object. It is among such stars that our search for planetary systems in the process of formation should primarily be directed with any hope of success.

The present writer pointed out recently (Kopal 1970) that the peculiar eclipsing system ε Aurigae may be one case in point. The astronomical aspects of the case having already been discussed in a previous paper just referred to, in what follows we shall limit ourselves only to a summary of the relevant part of the evidence.

The star ε Aurigae constitutes a single-spectrum eclipsing system, the visible component of which is an F2I supergiant of mass close to 35 solar masses and a luminosity about 250,000 times as large as that of the Sun. If this luminosity is derived from a conversion of hydrogen into helium, the star must be less than 10^6 years old—a very young object in the celestial time scale. This star is attended by a companion whose mass must lie between 20–25 $M_\odot$, but which shows no recognizable light in the visible domain of the spectrum; and infrared measures performed in recent years (Mitchell 1964) have restricted its absolute temperature to 500 ± 50K.

Once every 27.1 years, this "invisible" component eclipses its F2I mate for over 700 days, during some 330 of which the light of the system—diminished to approximately one-half of its full intensity—remains sensibly constant. If these minima of light were due to mutual eclipses of opaque stellar discs, they are much too deep for their observed durations (or, conversely, their constant phase lasts much too long for its observed amplitude)—facts which render ε Aurigae entirely unique among all eclipsing variables that have been discovered so far.

Several consequences follow immediately from these facts. Inasmuch as the principal F2I component remains spectroscopically visible throughout the entire cycle, the eclipsing body must obviously be semi-transparent. Secondly, inasmuch as the light of the system during the 330-day constant phase remains sensibly constant, it follows that the eclipsing body must be flat; its distribution of its mass cannot be characterized by spherical symmetry (in which case its optical depth would be bound to vary with the phase). Third, inasmuch as the depth of the light minima is quite independent of the wavelength of light in which they are observed, the particles of matter which causes obscuration must be large in comparison with the wavelength.[1]

That mass particles constituting it should indeed be relatively large follows also from the following consideration. Spectroscopic observations disclose the mass of the secondary component of ε Aurigae to be between 20–25 $M_\odot$; and radiometric observations show its effective temperature to be close to 500 K. *How could, however, so large a semi-transparent mass remain so cold—unless the mean free path of its constituent particles is so large that collisions between individual particles are unimportant?* This leads us to a conclusion that the particles in question—of masses ranging (say) from meteoritic to asteroidal or cometary—describe orbits around the centre of mass of the respective configuration, governed by the laws of celestial mechanics rather than of the kinetic theory of gases; and that the measured temperature of 500 ± 50K refers to the mean surface temperature of solid particles rather than that of any gas. In point of fact, this temperature coincides essentially with one at

[1] The scattering of light on free electrons, once considered in this connection (cf. Kuiper, Struve, and Strömgren 1937) cannot be invoked for this purpose because of the lack of a sufficient ionizing source.

which chondritic meteorites solidified in our solar system (cf. Anders, these Proceedings, p. 133)—a coincidence which may, but need not, be accidental.

It is, perhaps, becoming apparent by now that what we are describing under the term of the "secondary component of ε Aurigae" may bear more than a superficial resemblance to current pictures of a primordial "solar nebula" from which our own planetary system is supposed to have evolved. Just as in our solar system the solar equator is considerably inclined to the planes of planetary orbits (7° to the invariable plane of the system), the eclipses of the F2I component of ε Aurigae by its disc-like companion require the latter to be inclined to the orbital plane of the binary system—possibly for the same reasons. We propose, therefore, to identify tentatively the configuration which we call the secondary component of ε Aurigae with an early stage of a "planetary system in the making"—at a distance of some 1 350 parsecs from us, and very much more massive than our own. It is, in fact, possible that the particles (or bodies) responsible for periodic observations of the F2I component of ε Aurigae every 27.1 years may be identical with what Alfvén and Arrhenius (1970) have recently termed "stellesimals"—i.e., bodies whose eventual coalescence may result in the formation of the central star, while the left-overs continue to revolve around it as planets.

How far could such a process have progressed in ε Aurigae by this time? The very high luminosity of the F2I component (-8.5 abs. magn.) leaves but little room for doubt that we are dealing with an extremely young object. A star of -8.5 absolute bolometric magnitude radiates energy at a rate close to 10^{39} ergs/sec; and, in spite of its large mass, it could not have been doing so at this (or comparable) rate for more than a few hundred thousand years. While the principal (more massive) component of ε Aurigae could not thus have become a star until quite recently, its less massive (and, therefore, less evolved) companion has probably not become a star as yet. Moreover, for reasons which may be connected with an excess of angular momentum, this secondary component failed to assume spherical form but developed into a flattened disc, inclined to the orbital plane of the two stars. The eventual product of its evolution may indeed be a planetary system revolving around a central star which—as is attested by the present low temperature of the object—has so far failed to switch on any effective source of nuclear power.

Since the writer presented these views last year (Kopal 1970), an interesting alternative as to the nature of the core of the secondary component was put forward by Cameron (1971), who considered the possibility that the system of ε Aurigae has already attained the post-Main Sequence stage; and the secondary component has reached the ultimate stage of a "black hole" (or, as Cameron termed it, a "collapsar"). Without wishing to enter into a discussion of whether or not such "black holes" may actually exist in the Universe,[1] it should be pointed out that its existence in the system of ε Aurigae seems to be very unlikely on several grounds.

First, the geometry of the present system lends no support to a view that it has evolved past the Main Sequence stage. The present radius of its F2I component should be close to 290 $R_\odot$ or 1.4 AU; but large as it is, it constitutes only 4% of the semi-major axis of the system which should be close to 35 AU (cf. Kopal 1970); while fractional dimensions of the primary's Roche limit (corresponding to a mass-ratio $m_2/m_1=0.65$) should be 0.42—in contrast with the actual value of 0.04 which is ten times smaller. In other words, the presently observed characteristics of the system give no hint of secular expansion which follows the Main-Sequence stage. Besides, if the present secondary were to represent an erstwhile primary which evolved faster than its mate to reach the ultimate stage of a "black hole", its initial mass would have to have been very much larger than the present mass of 35 $M_\odot$ for the F2I star—probably 50 $M_\odot$ or greater; and such stars do not exist in the sky (probably on account of their instability).

It should also be kept in mind that the spectrum of ε Aurigae shows also evidence of some gas being present in the secondary's obscuring material (cf., e.g., Hack 1959, 1962)—though not in amount sufficient to cause obscuration at continuous (as distinct from discrete) frequencies; and the presence of such gas would be difficult to account for in the proximity of a "collapsar". The electron density of some 10^{11} electrons per cm^3, inferred by Hack (op. cit.) from the dilution factors of the observed spectral lines, very probably originates by ionization of molecules constituting solid particles of the obscuring disc rather than by ionization of a gas; for the ionization potentials of the latter would require a more energetic ionizing source than the radiation of the F2I star (which the recent observations by OAO—2 from space deprived of any hypothetical UV excess); while the weakly-bound electrons of solid state may just about produce the observed density of free electrons in the climate prevalent in the system ε Aurigae.

References

Alfvén, H. and Arrhenius, G., 1970, Astrophys. Space Sci., *8*, 338; *9*, 3.
Cameron, A. G. W., 1971, Nature, *229*, 178.
Gliese, W., 1969, Veröff d. Astr. Rechen-Institut. Heidelberg, Nr. 22.
Hack, M., 1959, Astrophys. J., *129*, 291.
Hack, M., 1962, Mem. Soc. Astro. Ital., *32*, 351.
Kopal, Z., 1970, Astrophys. Space Sci., *10*, 332.

[1] The possibility that such collapsed configurations may be found among components of binary systems was previously considered by Zeldovich.

Kuiper, G. P., Struve, O., and Strömgren, B., 1937, Astrophys. J., *86*, 570.
Mitchell, R. I., 1964, Astrophys. J., *140*, 1607.
Strand, K., Aa., 1956. Astron. J., *61*. 319.
Van de Kamp, P., 1955, Sky and Telescope, *14*, No. 12.
Van de Kamp, P., 1968, A.S.P. Leaflet No. 470.
Van de Kamp, P., and Dennison, E W., 1950, Sky and Telescope, *9*, No. 9.
Woolley, R. v. d. R., Epps, E. A., Penston, M. J. and Pocock, S. B., 1970, Ann. Roy. Obs. Greenwich, No. 5.

Discussion

F. L. Whipple

Your colleague, A. G. W. Cameron and an associate have suggested that the companion of ε Aurigae is a "Black hole". Have you any comments on this subject?

Z. Kopal

A similar suggestion for other binaries has earlier been made by academician Zeldovich. For astronomical as well as physical reasons it is difficult to believe in this interpretation. It was based on the assumption that the F2 star is not a young object. If, however, the present F2 star with its 35 solar masses is in the post-mainsequence stage, then its pre-main-sequence stage would have contained 50–70 solar masses and been of an entirely unknown type. It has also been assumed that the star is expanding towards the Roche limit. This is also improbable. The size of the star amounts to only 4 per cent of the dimension of the system. It is a factor of 10 less than its Roche limit. Moreover, a recent Dutch investigation by Blokland of the optics of space in the proximity of a black hole seems to indicate that the observed light curve for ε Aurigae cannot be fitted by the assumption that the star has a black-hole companion.

E. Anders

I am afraid the fraction of stars passing through the formation of a planetary system must be less than 1 in 1 000. Your calculation is based on a dispersion of meteorite ages of 10^8 years; actually Podosek's I^{129}/I^{127} ages of meteorites show a maximum spread of 1.5×10^7 years. Some new data, which I shall mention in my talk suggest that the time scale was even shorter, no more than 2×10^6 years.

Z. Kopal

If so, my estimate will have to be changed by the same order of magnitude. Even so, however, we should be able to find several interesting cases in the sample of stars available for discussion within 25 light years. The Orion nebula will be the main place to look for solar systems in the process of formation.

V. Vanýsek
Have polarization effects been observed during the eclipse?

Z. Kopal
No. The bodies are too large to polarize the light.

P. M. Millman
You stated that no polarization effects were detected. To what accuracy was the polarization measured?

Z. Kopal
I understand that the accuracy is better than one part in a thousand, and this is true for a wide range of wavelengths.

D. Lal
Can you comment on the time scales?

Z. Kopal
For lunar rocks you find that the oldest solidified 4.6×10^9 years ago, with a dispersion of only 1 or 2 per cent, and this dispersion may partly be due to observational errors. I assume that the solidification of different materials (including meteorites) may have a spread over a certain period, and stopped more or less after the main planets had come into being. The actual period of formation may have been very short.

E. Anders
The dating method based on extinct I^{129} has the greatest resolving power. There the total spread is only 15 million years, and part of this represents the cooling time of the meteorite parent bodies, which should not be included in the formation time of the solar system.

D. Lal
The interval of time which Dr Kopal referred to in this probability calculation is most probably not the same which you just talked about.

The probability of observing a solar system in making is dependent on the total time elapsed between formation of the solar nebula and the planetesimals or planets (in any case, at least the total time required for crystallisation of different meteorites) and not the spread in time between crystallisation of different meteorites.

P. Pellas
The Iodine-Xenon correlation is quite precise. The first results indicate that the material was formed at a high temperature and within a (relative) age in the range of 2 million years. Later, other experiments have enlarged this time due to the fact that a mixture of material tends to indicate too long times. The important fact is that the time elapsed between the isolation of the proto-solar system from the galactic nucleosynthetic source on one hand, and the retention of xenon by solids on the other, is estimated to be 50 to 100 million years. Our knowledge about the processes acting in this time interval is very poor.

Accretion Processes Leading to Formation of Meteorite Parent Bodies

By D. Lal

Tata Institute of Fundamental Research, Bombay, India and
Scripps Institution of Oceanography, La Jolla, California

1. *Introduction*

In all forms of matter, physical changes occur continuously. These changes form the basis for the study of evolutionary history of matter. During the last fifteen years, studies of cosmic ray produced cosmogenic alterations in the meteorites, and more recently in lunar materials, have provided important clues to a variety of collisional and accretional processes.

There are two distinct cosmogenic effects in matter exposed to corpuscular radiation. The first one is a change in the isotopic composition due to nuclear interactions (and nuclear stopping) of energetic particles. The second is alteration of the crystalline structure of matter due to ionisation produced by charged particles. In the case of the latter when the solid state damage is appreciable, a very easily observable effect is the chemical revelation of tracks along trails of ionising particles. The fossil track technique (Fleischer et al. 1970) has been extensively applied to terrestrial (Fleischer and Hart 1971) and extraterrestrial samples (Crozaz et al. 1970; Fleischer et al. 1970; Lal et al. 1970; Price and O'Sullivan 1970). To date most of the studies of extraterrestrial samples refer to small objects exposed to multiply charged cosmic ray particles in free space or on the lunar surface.

A very interesting recent observation is that individual grains now present in the matrix of certain meteorites were exposed to low-energy particles—shielded by $\lesssim 10^{-3}$ gm. cm^{-2} of matter (Lal and Rajan 1969; Pellas et al. 1969) for time periods of the order of 10^3–10^4 years. In this paper I will discuss the status of our present knowledge of the accretion history of meteorites, based on a study of the fossil track record of charged cosmic ray nuclei. This survey includes the unpublished results of a recent detailed analysis and intercomparison of fossil track record in lunar and meteoritic samples (Arrhenius et al. 1971 *a*). Before discussing this evidence, it seems useful to briefly summarize (i) the types of particles and the corresponding velocity regions where solid state damage becomes sufficient to result in a fossil track record and (ii) the reason why it is possible to characterize an irradiation with an essentially zero shielding, which is the basis for using the fossil track record to delineate pre- and post- accretional history of meteorites.

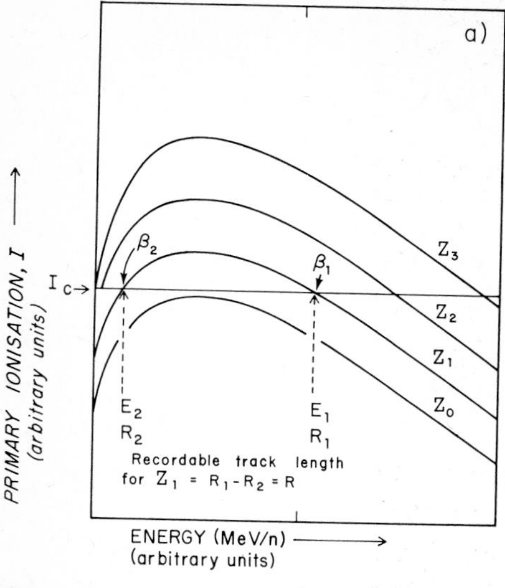

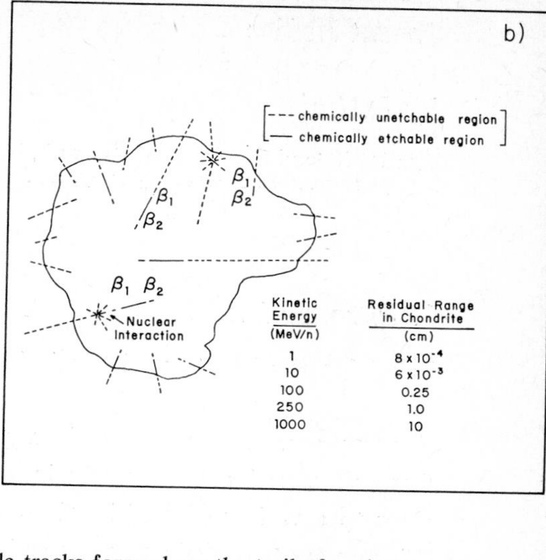

Fig. 1. In an insulating solid, chemically etchable tracks form along the trail of a charged particle of atomic number, Z, if its rate of primary ionization exceeds the critical value, I_c (fig. a). The maximum chemically etchable range of a particle is R, corresponding to the distance travelled between E_1 and E_2. In a solid object exposed to cosmic radiation in space, tracks revealed within correspond to particles of different *primary* kinetic energies.

Solid state damage due to ionization along the trail of a charged particle, if extensive enough, can be chemically enlarged and seen with an optical microscope. The chemically developed holes, cylindrical or conical, are termed as tracks. Extensive studies of formation of tracks in a variety of insulating solids by the Fleischer–Price–Walker team led them to the conclusion that if primary ionization I, exceeds a certain value I_c, tracks can be made visible chemically (cf. Price and Fleischer 1971). In Fig. 1(*a*), we show schematically the atomic number (Z)–velocity (v), combinations for formation of tracks: tracks along the trail of a charged particle are formed if the (Z, v) combination corresponds to $I \geqslant I_c$. The rate of formation of tracks within an object, say a grain within a large body, depends on several factors: flux and energy spectrum of cosmic ray nuclei, size of the object, nuclear interaction mean free path, and fragmentation parameters (Fleischer et al. 1967; Lal 1968); in cosmic radiation, iron is the most abundant nuclear species above $Z = 20$–22, the minimum atomic number to lead to chemically etchable tracks in rock silicates. In the inset of Fig. 1(*b*), the residual range of iron nuclei in chondrites is given for a few selected values of kinetic energies. Since the maximum recordable range for iron nuclei in silicates is of the order of 10 microns (Lal 1968), a track due to an iron nucleus will become revealable when its kinetic energy drops below 2 MeV/n. Most cosmic ray tracks within a solid body correspond to iron nuclei of different *primary*

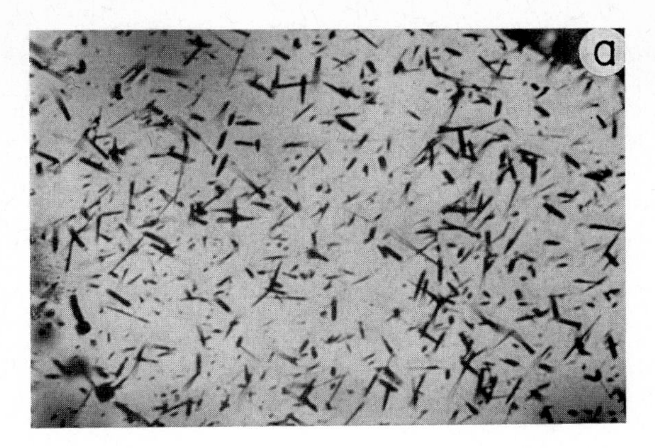

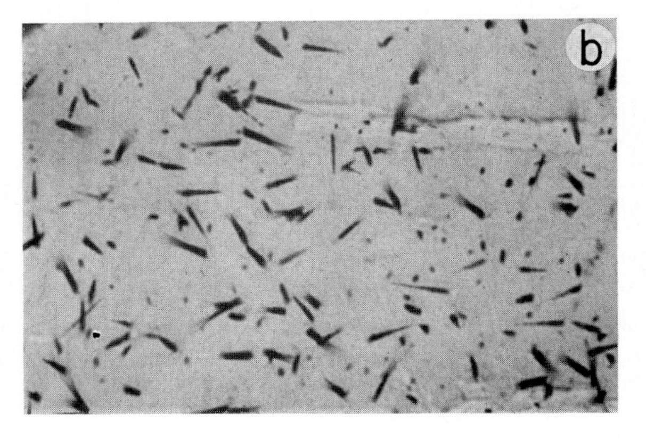

Fig. 2. Photomicrographs of fossil tracks in silicate crystals from the meteorite, Patwar (fig. a) and Apollo 14 rock 14310 (fig. b). Track holes are chemically plated with silver to increase optical contrast. The fields of view are 140×90 and 130×75 μm respectively for figs. (a) and (b).

energies (Fig. 1 (*b*)); as an illustration, tracks formed at the centre of a meteorite of radius 10 cm will correspond to iron nuclei of primary kinetic energy 1 BeV/n.

In Fig. 2, we show photomicrographs of cosmic ray tracks seen in an Apollo rock and in the meteorite, Patwar. These tracks correspond to iron nuclei of primary kinetic energies around 300–500 MeV/n. In Fig. 3 we show photomicrographs of cosmic ray tracks in Kapoeta meteorite and an Apollo 12 lunar grain from the soil sample 120; these tracks correspond to low energy iron nuclei of 0.5–20 MeV/n (Lal and Rajan 1969).

The fact that large gradients in track densities are observed in the case of low energy irradiation (Fig. 3) can be easily seen from the following approximate theoretical relation for the rate of formation of etchable tracks. In the case of iron nuclei this corresponds essentially to a rate at which they are brought to rest in the matrix, as discussed above. The track formation rate, $\dot{\varrho}(x) = d\varrho(x)/dt$ is then related to the range (dN/dR) or energy (dN/dE) spectrum of cosmic ray nuclei which lead to tracks:

$$\dot{\varrho}(x) \propto \left(\frac{dN}{dR}\right)_x \qquad (1)$$

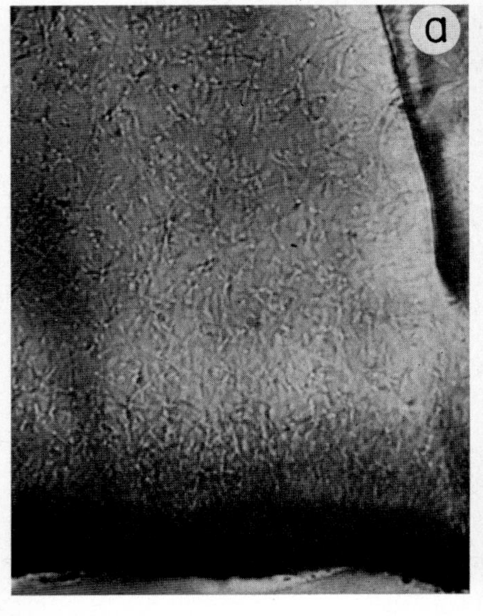

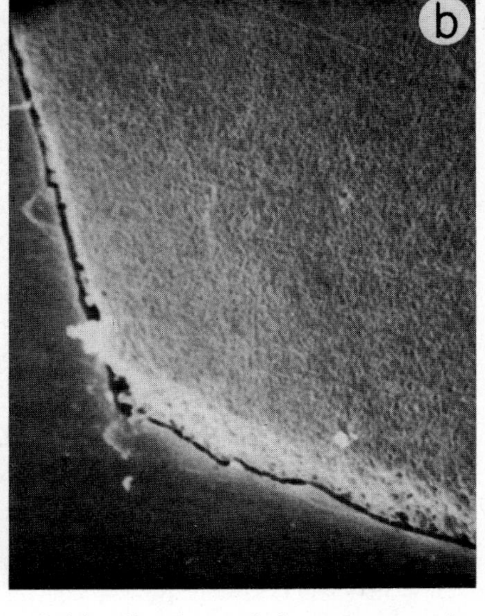

Fig. 3. Photomicrographs of fossil tracks in so-called irradiated crystals from the meteorite, Kapoeta (fig. a) and the lunar-regolith soil sample: 12070,15 (fig. b); tracks were photographed using optical and scanning electron mocroscopes respectively. In figs. (a) and (b) respectively, the crystals are pyroxene and olivine and the fields of view are 70×50 and 100×75 μm.

$$\dot{\varrho}(x) \propto \left(\frac{dN}{dE} \cdot \frac{dE}{dR}\right)_x \qquad (2)$$

Relation (1) is approximate since it neglects the loss of cosmic ray nuclei due to nuclear interactions and assumes a constant solid angle. With the following empirically observed relations for the functions in relation (2):

$$dN \propto E^{-\alpha} dE \qquad (3)$$

$$R \propto E^{\beta} \qquad (4)$$

one obtains (cf. Crozaz et al. 1971) a power law relation for $\dot{\varrho}(x)$:

$$d\varrho(x)/dt = \dot{\varrho}(x) = x^{-\eta} \qquad (5)$$

where $\eta = (\alpha + \beta - 1)/\beta$

Over a wide range of energies, the slope η will be expected to be of the order (1.5–2) since, assuming values of (1–1.5) and (1–2) respectively for β and α (cf. Henke and Benton 1966; Fan et al. 1968 and Hsieh and Simpson 1970).

For a detailed treatment, reference is made to Fleischer et al. (1967) and Lal et al. (1969). The expected power law dependence in the track production rate with depth has been experimentally verified by examination of fossil tracks in meteorites and in lunar samples, particularly in the latter (Lal and Rajan 1969; Pellas et al. 1969; Comstock et al. 1971; Crozaz et al. 1971) where material irradiated with low energy particles, i.e. corresponding to tracks formed at small values of x, is plentiful. We will not consider here complications arising due to erosion during irradiation, which result in a flattening of the slope, η (Bhandari et al. 1971; Crozaz et al. 1971; Fleischer et al. 1971). The essential point in deriving the approximate relation (5) is to show that the track formation rate due to cosmic rays is expected to be a power law function of the effective depth from the surface of the grain or rock exposed in space. Therefore, in the case of a grain, say of 1 mm size, one would expect to see marked variations in track densities within the grain only if it is exposed freely to low energy cosmic rays in space, without any shielding. Once the grain is shielded (or if it forms part of a rock of several mm–cm size), the gradients in track densities reduce markedly. The fossil track photomicrographs in Figs. 1 and 2 visually demonstrate this point. In Fig. 1 which refers to a crystal in Patwar meteorite shielded by at least 1 cm of meteorite material, no track gradients are seen. On the other hand, the crystal in Fig. 2 which is from an interior region of the meteorite, Kapoeta, shows marked track density gradients, ϱ decreasing rapidly as one goes away from the border of the grain. As pointed out above, the only way such gradients can occur is if the tracks were formed when the grain was exposed to cosmic rays as an individual in space (cf. eq. 5), not shielded by more than a few microns of meteorite material. Such a theoretical deduction now finds ample support from the direct observations of tracks in grains from the lunar regolith (Arrhenius et al. 1971 *b*).

For the sake of completeness, we present in Fig. 4, the theoretically expected gradient in track densities for meteorites of chondritic composition, based on our present day knowledge of the contemporary and long term averaged energy spectrum of cosmic radiation (Bhandari et al. 1971).

2. *Experimental Techniques*

We have studied fossil track records in a variety of silicate crystals:

Olivines ($(Mg, Fe)_2SiO_4$)

Pyroxenes ($(Mg, Fe)\ SiO_3$)

Plagioclase feldspars[1] ($(Ca\ Al_2\ Si_2O_8)_f + (Na\ Al\ Si_3O_8)_{1-f}$)

[1] A solid solution series with values of f ranging from 0 to 1.

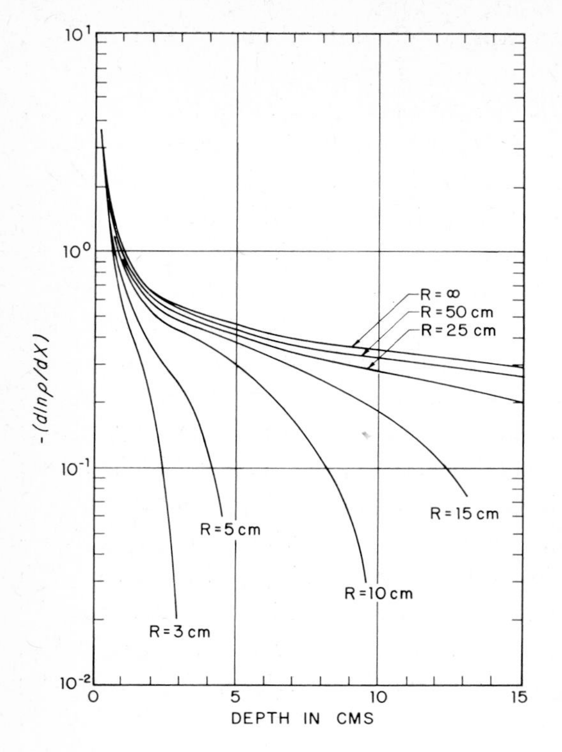

Fig. 4. Theoretically computed gradients in the track production rate as a function of depth (radially inwards) for chondritic meteorites of different radii, exposed in space. The primary energy spectrum of cosmic ray iron group nuclei is based on both contemporary cosmic ray and fossil track meteorite data (Bhandari et al. 1971).

The extraterrestrial samples studied are meteorites and lunar regolith components. Although several meteorites were included in our study (e.g. Allende, Bjurböle, Bununu, Chainpur, Fayetteville, Holbrook, Kapoeta, Khor-Temiki, Mokoia, Pesaynoe, Rupota, Weston and Vigarano), the studies have been most extensive in the case of Kapoeta.

In the case of lunar soil, most of the studies refer to Apollo 12 soil samples (12 001, 52; 12 041, 11; 12 057, 49; 12 070, 15; 12 070, 149); one Apollo 14 sample (14 259, 8) was also analysed. Gently segregated grains and cut sections of 200–500 micron thickness were chemically etched for an optical study of fossil tracks.

Techniques used for revealing fossil tracks in meteorites (and in lunar materials) will not be discussed here: reference is made to Lal (1968), Bhandari et al. (1971) and Macdougall et al. (1971). It should, however, be mentioned here that since the development of the olivine etch (Krishnaswami et al. 1971), it has become relatively easy to study fossil tracks in olivine rich materals. Firstly, the cone angle of tracks in olivine is small and it becomes possible to see high track densities optically. Secondly, the revelation of fossil tracks in olivine is carried out at a lower temperature (105°C) in a solution at pH 8 which does not attack any of the other principal minerals.

Measurements of track densities up to values of $(2–3) \times 10^8$ cm^{-2} were made using an optical microscope. Higher track densities were studied by the replica technique (Macdougall et al. 1971). The electron microscope observation of plastic replicas has been found to be satisfactory if track densities do not exceed 3×10^9 cm^{-2}. In Fig. 5 we show micrographs of tracks in lunar olivines, as examples. For semi-quantitative studies of the high track density regions, we examined the etched crystals directly with a scanning electron microscope (see Fig. 3). For track densities exceeding values of 3×10^9 cm^{-2}, it becomes necessary to use a high voltage (0.6–1.0 MeV) electron microscope. Using this technique, the French and Berkeley groups (Borg et al. 1970; Dran et al. 1970; Barber et al. 1971) have recently reported the presence of extremely high track densities ($>10^{10}$–10^{11} tracks per cm^2) in lunar and meteoritic materials. Such track densities occur in the outermost regions of grains, presumably at depths <2 μm. These measurements, however, are time-consuming and preclude any extensive examination of irradiated samples. The observed track density gradients at the grain borders are usually very large so that at distances of $\geqslant 10$ μm, tracks can be studied with an optical microscope; see for example, Figs. 3 and 5. Some lunar grains, even those of size >100 μm, are found to have been irradiated for appreciably longer periods of time (compared to grains in meteorites) so that their studies can be made using electron microscope only.

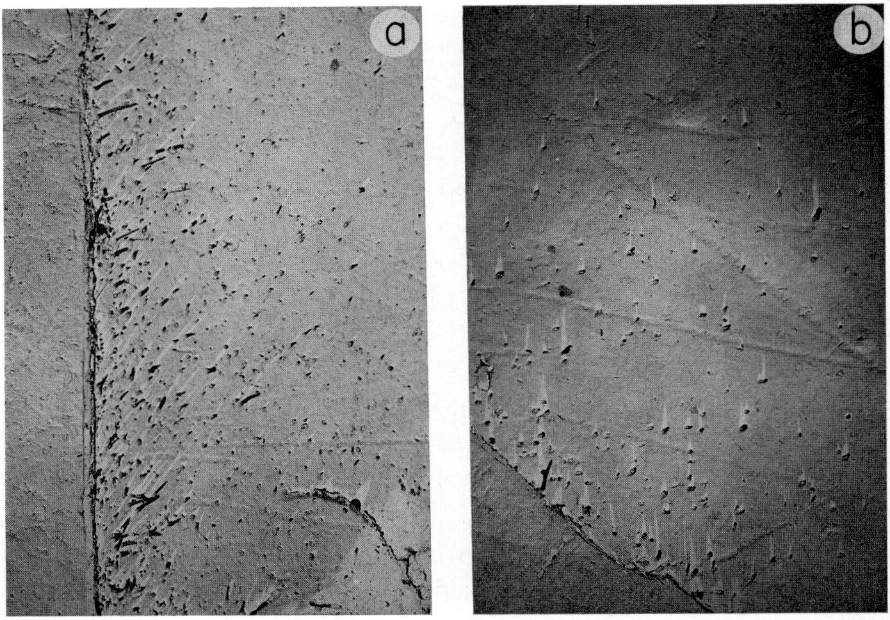

Fig. 5. Electron micrograph of palladium shadowed replicas of etched grain surfaces of olivines from Apollo 12 lunar fines, 12070, 15 (fig. a) and 12070, 149 (fig. b). The fields of view are 20×12 and 35×25 μm respectively in figs. (a) and (b).

3. *Results and Discussions*

Although the early observations of Lal and Rajan (1969) and Pellas et al. (1969) had already thrown considerable light on the nature of irradiation of grains with respect to (i) shielding, (ii) relative times of irradiation of different grain faces and (iii) the extent of alteration/fragmentation of crystals after the irradiation, they were not extensive enough to allow a reliable generalisation of the irradiation features. The work of Wilkening et al. (1971) sought to answer many of the key questions in the case of the meteorite Kapoeta. These authors confirmed the earlier observation of Lal and Rajan (1969) that in Kapoeta, the feldspar grains were also irradiated in a manner similar to the pyroxenes for which the evidence was abundant (Lal and Rajan 1969; Pellas et al. 1969). Furthermore the chemical compositions of irradiated grains i.e. the track-rich ones, was found to be similar to those which were not track-rich, i.e. those shielded from the low energy cosmic ray particles. The spatial distribution of track-rich grains in the meteorite matrix was found to be random. The studies of Wilkening et al. (1971) gave strong support to the earlier observation of Lal and Rajan (1969) that the track-rich borders of grains in the Kapoeta meteorite were, as a general rule, continuous for different types of crystals.

Combining the results of the above studies of fossil track record in Kapoeta with those in grains in the lunar regolith (Arrhenius et al. 1971 *b*), it was suggested (Wilkening et al. 1971) that the individual track-rich grains were most probably irradiated in free space rather than on the surface of a parent body albeit the latter possibility could not be ruled out at that time. We will now summarise the results of our recent work of Arrhenius et al. (1971 *a*), which is a fresh attempt to observe irradiation features in meteorites and in samples from the lunar regolith with a view to delineate minimum necessary physical conditions and situations which must have existed during irradiation and subsequently when the meteorite parent bodies formed. The experimental results (see Section 2) concern the study of fossil track records in olivines, pyroxenes and other silicate crystals in:

(i) grains from the lunar soil (regolith) and meteorites
(ii) sections of lunar regolith grains of mm size
(iii) sections from surface and interior regions of lunar breccia rocks of cm size
(iv) sections of track-rich meteorites.

Studies of sections (ii) through (iv) aimed at *in-situ* observations of the shapes of the irradiated crystal and the extent of preservation of track-rich borders in the compaction, as in previous work (Lal and Rajan 1969; Wilkening et al. 1971). The total surface area of the sections examined carefully is about 10 cm^2.

Table 1. *Characteristic irradiation features in meteoritic and lunar samples*

Feature or Detail	In Meteorites[a]	In lunar grains[b]
Abundance of Irradiated grains (of size $>100\ \mu m$)	Variable (of the order of (10 % in gas-rich meteorites)	Variable (0–80 %)
Whether irradiated on all sides	Yes, as a rule	2π to 4π irradiation
Isotropy of irradiation (with respect to relative time of irradiation of different sides)	Yes, a a rule	Rarely
Estimated time of irradiation	$(10^3–10^4)$ yrs.	0 to $\sim 10^6$ yrs.

[a] The comments in this column refer chiefly to irradiated grains seen in the meteorites Kapoeta and Weston. As discussed in the text, irradiated grain aggregates have now been found (Arrhenius et al. 1971 a) in Kapoeta and Fayetteville. The experimental data on grain aggregates is too limited as yet to allow generalization.

[b] In the case of lunar samples, the information is based primarily on observations of grains from Apollo 11 and 12 "fines". As discussed in the text, usually the irradiation record in lunar breccia is not well preserved.

Results of the present study to be discussed in detail elsewhere (Arrhenius et al. 1971 *a*), are summarised in Table 1. A new observation (Arrhenius et al. 1971 *a*) made in the case of meteorites, Kapoeta and Fayetteville, is that they contain irradiated rocklets (=grain aggregates). The sizes of the track-rich rocklets range between 200–1 000 μm. The so far limited observations on the irradiated *rocklets* are consistent with their irradiation contemporaneously with the *grains* in these meteorites because the deduced energy spectrum and dose based on track density distributions is closely similar. Further work is in progress to check on this point. (It may be pointed out here that the study of the characteristics of the corpuscular radiation to energies of $\gtrsim 100$ MeV/n can be carried out, if rocklets of few mm size can be found). Mention must be made here of the interesting deductions by Pellas (1971) of a complex irradiation history of xenolith fragments in several meteorites. Pellas and his collaborators report that some of the xenolith[1] fragments have had an independent thermal, shock and irradiation history, different from that of the material now surrounding them. Irradiation due to both low energy and high energy particles is reported in these objects. Thus, the observations of Pellas (1971) and Arrhenius et al. (1971*a*) essentially show that the low energy particle irradiation (before accretion), was confined not only to single grains but included also grain-aggregates and fragments of different sizes.

We will first intercompare low-energy irradiation features observed in lunar grains and meteorites. The most important differences is the extent of variability in the case of lunar grains with respect to the time of irradiation

[1] Signifying material of different composition, embedded in the meteorite.

of faces of a grain. The irradiated meteoritic grains (and grain aggregates), on the other hand, all seem to have had a very similar radiation history. Furthermore, it should be remembered here that this comparison is restricted to *free lunar grains*—the irradiation record in grains within the lunar breccias, (results refer to rock 10 018 and a dozen breccia rock fragments from the lunar regolith) is very clearly dominated by a complex shock history. The grains within a lunar breccia strongly enriched in solar wind gases, (cf. results of Marti (1971) and Funkhouser et al. 1970 for rock 10 018) do not generally show a well preserved fossil track record. Based on extensive petrographic studies of lunar breccias (McKay et al. 1971; Mason and Melson 1971), it has been concluded that lunar breccia presumably lithified by sintering during base surge rather than by shock lithification. The "shocking" in breccia grains as evidenced by the lack of properly formed tracks, presumably occurred in most cases before brecciation. Thus it is difficult to draw any meaningful conclusions about the nature of preservation of record of irradiation from the studies of lunar breccias; the individual grains in the lunar regolith are, however, ideal for studying the nature of irradiation on the surface of a large body. And, in the light of above discussions (*viz.* observations in Table 1), it seems quite clear that at least bodies of the size of the moon can be ruled out for consideration as hosts for irradiation of grains in a manner similar to those found in gas-rich meteorites. Hence it is strongly indicated that the irradiation took place in an environment where the turnover of grains was not precluded in the manner observed on the moon. This is in principle possible in any low-gravitational or non-gravitational assemblage of particles such as a small asteroidal or cometary object or their precursors, loose grain aggregates and particle swarms.

The possibility of an isotropic irradiation due to electrostatic turnover of grains (Gehrels et al. 1964; Rhee 1971) has been considered to be attractive (Mazor et al. 1970). A merit of this hypothesis lies in the fact that the grains will receive an irradiation while essentially floating in space, just above the body, and can be irradiated isotropically (Table 1). However, the exchange of grains between those kept aloft and those in the top layer will definitely result in a spectrum of irradiation times. Furthermore, in an asteroidal irradiation model, the integrated irradiation time would be expected to be critical to the grain size. For grains (or aggregates) of 100–1 000 microns size, our observations however, do not show any differences in dose values, within the present uncertainties of estimation (within a factor of two), based on observations of tracks at depths exceeding one micron from the surface.

We have noted above (cf. Table 1, Lal and Rajan 1969; Wilkening et al. 1971) that the borders of track-rich crystals are found to be well preserved as a rule. We have so far seen some 50 examples of irradiated grains in examination of sections of Kapoeta. In two of these cases, all the grain borders are not track-rich. However, in view of our observation of unambiguous cases of irradiated rocklets (grain-aggregates) in Kapoeta and Fayetteville, we cannot rule out that these grains in fact formed part of a rocklet during irradiation. Thus, all evidence suggests that the lithification of meteorites was a very gentle process (cf. Wilkening et al. 1971). We have discussed earlier that moonsized objects are too large to be considered as parent bodies for the accretion of track-rich meteorites. In the case of an asteroidal model, accretion at low velocities could possibly occur only in the case of collisions within members of the asteroidal streams. However, the observed similarity in isotropic and nearly irradiation dosage observed in track-rich grains of different sizes poses problems, as discussed earlier.

The above mentioned difficulties in assuming electrostatic overturn mechanisms operating on asteroids or large size objects e.g. the Moon, make it attractive to investigate the possibilities for the characteristic irradiation to stem from the stage preceding accretion of the grains into large bodies. Information about particle behavior in this situation can only be obtained indirectly from observation of present particle streams (jet streams) in space, and from theoretical considerations of the particle interaction and focussing in such streams (Alfvén 1969, 1970, 1971; Trulsen 1971; Danielsson 1971). It seems that the grain irradiation features observed in meteorites can be satisfactorily explained on most accounts by grain irradiation and focussing in jet streams: the isotropic irradiation of grains and grain-aggregates would occur during their exposure in free space before focussing and (gentle) compaction during accretion. The latter is an intrinsic merit of this mechanism because (relative) accretional velocities are low (Alfvén 1970; Alfvén and Arrhenius 1970, 1971). In the absence of other physically acceptable models, we are tempted to consider the presently available particle stream model as the most satisfactory one to provide sufficient amounts of extra-terrestrial materials having well preserved particle irradiation records of the type observed.

At the present moment, none of the experimental evidence in gas- or track-rich meteorites bears to the question "When did the irradiation occur?" In terms of present considerations both irradiation and accretion of meteorites would be expected to be ongoing processes in the solar system, although proceeding at a smaller rate now than in the early history of the solar system.

In conclusion, we note that although one observes that both solar wind implantation and low energy particle irradiation are a prime ongoing phenomena in lunar surface materials (Anders 1971; Arrhenius et al. 1971 *b*; Crozaz et al. 1971), the physical conditions during irradiation and accretion of the meteorites were very different. In terms of the present state of our knowledge of particle interaction processes, it is tempting to consider particle streams in space, in the past and today, as the irradiation and accretion environment for certain types of meteorites.

Summary

The interaction of corpuscular radiation with matter leads to a characteristic and a rather permanent alteration (not easily obliterated by mild shocks or thermal events). Some of these alterations are very sensitive to the amount of shielding from the corpuscular radiation in space. It seems quite likely that the corpuscular radiation was always present in the solar system. These facts lead one to the conclusion that it should be possible to reconstruct certain facets of the accretional and post-accretion history of parent bodies of meteorites. It should, however, be borne in mind that such deductions may not apply to all types of meteorites because of a possible bias introduced by basing conclusions on the particular meteorites where records of interaction of corpuscular radiation are observable.

In this paper we consider specifically the information based on observations of the fossil record of low-energy particle irradiation of grains and rocklets (=grain aggregates) in meteorites and in lunar materials. This record relates to (i) times scales involved in the accumulation of matter till shielding reached values of 10^{-3}–10^{-2} $g \cdot cm^{-2}$ and (ii) the nature of processes which occurred subsequent to this accretion terminating finally in the formation of a compact body, a fragment of which we examine now in the form of a meteorite.

Observations in several meteorites and in lunar grains and rocklets lead one to the conclusion that accretion followed irradiation of both grains and rocklets in space over periods of the order of 10^3–10^4 years; more if erosion of grain surfaces was important during irradiation. It can also be concluded that compaction during and after accretion occurred at low relative velocities (<0.2 $km \cdot sec^{-1}$) so that the irradiated grains did not suffer any appreciable damage (shock fragmentation or melting).

In terms of our present day knowledge of dynamics of bodies in the solar system, conditions for this type of accretion can be found in a range of sizes of objects/systems not exceeding, however, the order of magnitude of 100 km. Such objects could be similar to but not necessarily identical with asteroids or comets. It seems attractive to consider the latter class of objects as a favorable site for accretion.

There exist arguments today for a cometary origin of meteorites; the low relative velocity accretion required can perhaps be understood on the basis of the astronomically observed behavior of cometary meteoroid jet streams and the theoretical considerations of focussing in such streams. It seems noteworthy that the formation of particular meteorites where fossil records of low energy particle irradiation are found may be going on today but of course at a much smaller rate than in the early history of the solar system.

I would like to thank Dr G. Arrhenius, Mr D. Macdougall and Miss B. Martinek for collaboration and helpful criticisms. Helpful suggestions made by Dr H. Alfvén are gratefully acknowledged. Drs K. Marti and Charles Meyer Jr. made important contributions in selection of lunar samples. We are thankful to NASA for lunar samples; help given by Drs. M. B. Duke, R. B. Laughon, D. S. McKay and C. Meyer in the preparation of samples is gratefully acknowledged.

Samples of Fayetteville were made available for the present study by Drs. P. K. Kuroda, O. K. Manuel and H. E. Suess. To them and several others, we remain beholden for their trust in giving us valuable meteorite specimens for an analysis.

References

Alfvén, H., 1969, Astrophys. Space Sci. *4*, 84.

Alfvén, H., 1970, Astrophys. Space Sci. *6*, 161.

Alfvén, H., 1971, Science *173*, 522.

Alfvén, H. and Arrhenius, G., 1970, 1971, Structure and evolution history of solar system, I and II. Astrophys. Space Sci. *8*, 338 and *9*, 3.

Amin, B. S., Lal, D., Lorin, J. C., Pellas, P., Rajan, R. S., Tamhane, A. S. and Venkatavaradan, V.S., 1968, Meteorite research (ed. P. M. Millman). D. Reidel Publishing Co. Dordrecht, Holland, paper no. 26, 317.

Anders, E., 1971, A. Rev. Astr. Astrophys. *9*, 1.

Arrhenius, G. and Alfvén, H. 1971, Earth Planet. Sci. Lett. *10*, 253.

Arrhenius, G., Lal, D., Macdougall, J. D. and Martinek, B., 1971 *a*, "Studies of low-energy irradiation of grains and grain aggregates in meteorites and lunar samples". In preparation.

Arrhenius, G., Liang, S., Macdougall, D., Wilkening, L. and Bhandari, N., Bhat,S., Lal, D., Rajagopalan, G., Tamhane, A. S. and Venkatavaradan, V. S., 1971*b*, Proceedings of the Second Lunar Science Conf., Geochim. Cosmochim. Acta *3*, 2583.

Barber, D. J., Hutcheon I. and Price P. B., 1971, Science *171*, 372.

Barber, D. J., Cowsik, R., Hutcheon, I. D., Price, P. B. and Rajan, R. S., 1971, preprint "Solar flares, the lunar surface and gas-rich meteorite".

Bhandari, N., Bhat, S., Lal, D., Rajagopalan, G., Tamhane, A. S. and Venkatavaradan, V. S., 1971, Proceedings of the Second Lunar Science Conf., Geochim. Cosmochim. Acta *3*, 2599.

Borg, J., Dran, J. C., Durrieu, L., Jouret, C., and Maurette, M., 1970, Earth Planet sci. Letters *8*, 379.

Comstock, G. M., Evwaraye, A. O., Fleischer, R. L. and Hart, R. H. Jr., 1971, Proceedings of the Second Lunar Science Conf., Geochim. Cosmochim. Acta *3* 2569.

Crozaz, G., Walker, R. and Woolum, D., 1971, Proceedings of the Second Lunar Science Conf., Geochim. Cosmochim. Acta *3*, 2543.

Crozaz G., Haack U., Hair M., Maurette M., Walker R. and Woolum D., 1970, Proc. Apollo 11 Lunar Sci. Conf., Geochim. Cosmochim. Acta Suppl. 1, *3*, 2051.

Danielsson, L., 1969, Astr. Space Sci. *5*, 53.

Danielsson, L., 1971, In Physical studies of Minor Planets (ed. T. Gehrels). Proc.

12th Coll. Int. Astron. Un., held 8–10 March 1971. NASA SP-267. Also see Proceedings (this conference).

Dran, J. C., Durrieu L., Jouret C. and Maurette M., 1970, Earth Planet. Sci. Letters *9*, 391.

Eberhardt, P., Geiss, J. and Grogler, N., 1965, J. Geophys. Res. *70*, 4375.

Fan C. Y., Gloeckler, G., McKibben, B. M., Pyle, K. R. and Simpson J. A., 1968, Can. J. Phys. *46*, 498.

Fleischer, R. L., Maurette, M., Price, P. B. and Walker, R. M., 1967, J. Geophys. Res., *72*, 331.

Fleischer, R. L., Haines E. L., Hart, H. R., Woods, R. T. and Comstock, G. M., 1970, Proc. Apollo 11 Lunar Sci. Conf., Geochim. Cosmochim. Acta Suppl. 1, *3*, 2103.

Fleischer, R. L. and Hart, H. R. Jr., 1971, Fission track dating: Techniques and problems, Proc. Burg Wartenstein Conference on Calibration of Hominoid Evolution, held in July 1971.

Fleischer, R. L., Hart, H. R. Jr., Comstock, G. M., and Evwaraye, A. O., 1971, Proceeding of the Second Lunar Science Conf., Geochim. Cosmochim. Acta 3, 2559.

Funkhouser, J. G., Schaeffer, O. A., Bogard, D. D. and Zahringer, J., 1970, Proceedings Apollo 11 Science Conference. Geochim. Cosmochim. Acta Suppl. 1, *2*, 1111.

Gehrels, T., Coffeen, T. and Owings, D., 1964, Astron. J. *69*, 826.

Henke, R. P. and Benton, E. V., 1966, U.S. Naval Radiological Defense Lab. Rept. No. TR-1102.

Hsieh, K. C. and Simpson J. A., 1970, Ap. J., *162*, 197.

Krishnaswami, S., Lal, D., Prabhu, N. and Tamhane A. S., 1971, Science *174*, 287.

Lal, D., 1969, Space Sci. Rev. *9*, 623.

Lal D. and Rajan R. S., 1969, Nature *223*, 269.

Lal, D., Lorin, J. C., Pellas, P., Rajan, R. S. and Tamhane, A. S., 1969, Meteorite Research (ed. P. Millman) (A Int. Atom. Energy Ag. publication), paper no. *249*, 275.

Lal, D., Macdougall D., Wilkening L. and Arrhenius G., 1970, Proc. Apollo 11 Lunar Sci. Conf., Geochim. Cosmochim. Acta Suppl. 1, *3*, 2295.

Macdougall, D., Lal D., Wilkening, L., Bhat, S., Arrhenius, G. and Tamhane A.S., 1971, Geochem. J. (Japan), in press.

Marti, K., 1971, Personal Communication.

Mason, B. and Malson, W. G., 1970, Proc. Apollo 11 Lunar Sci. Conf., Geochim. Cosmochim. Acta Suppl. 1, *1*, 661.

McKay, D. S., Greenwood, W. R. and Morrison, D. A., 1970, Proc. Apollo 11 Sci. Conf., Geochim. Cosmochim. Acta Suppl. 1, *1*, 673.

Mazor, E., Heymann, D. and Anders, E., 1970, Geochim. Cosmochim. Acta *34*, 781.

Pellas, P., Poupeau, G., Lorin, J. C., Reeves, H., and Audouze, J., 1969, Nature, *223*, 272.

Pellas, P., 1971, Irradiation history of grain aggregates in ordinary meteorites: possible clues to the advanced stages of accretion. Proceedings (this conference).

Price, P. B. and O'Sullivan, D., 1970, Paleontology. Proc. Apollo 11 Lunar Sci. Conf., Geochim. Cosmochim. Acta Suppl. 1, *3*, 2351.

Price, P. B. and Fleischer, R. L., 1971, Identification of Energetic Heavy nuclei with solid dielectric track detectors: applications to astrophysical and planetary studies. To appear in Ann. Rev. Nucl. Sci.

Price, P. B., Rajan, R. S. and Shirk, E. K., 1971, Proceedings of the Second Lunar Science Conf., Geochim. Cosmochim. Acta *3*, 2621.

Rhee, J. W., 1971, Space Res. *XI*, 275.

Suess, H. E., Wanke, H. and Wlotzka, F., 1964, Geochim. Cosmochim. Acta., *28*, 595.

Trulsen, J., 1971, In Physical Studies of Minor Planets (ed. T. Gehrels). Proc. 12th Coll. Int. Astron. Un., held 8–10 March 1971. NASA SP-267. Also see Proceedings (this conference).

Wilkening, L., Lal, D. and Reid, A. M., 1971, Earth and Planet. Sci. Letters *10*, 334.

Discussion

P. Pellas

The crystal detectors are not perfect and it seems to me very dangerous to investigate the isotropy or non-isotropy of the recorded particle irradiation without checking the efficiency of the detectors. We irradiated olivine and used your etching method. We found variations by a factor of 5 to 6 in track-registration efficiency for different crystallographic directions.

D. Lal

I would agree with you that one has to be careful and not work blindfold with crystal detectors. The particular etching recipe used by us leads to an unbiased track revelation in olivines; we have carefully checked on this point.

F. L. Whipple

Where tracks are dense, what is the total energy contributed to the crystal by the high-energy particles?

D. Lal

The highest track density in irradiated grains is of the order of 10^{10} tracks/cm^2. This corresponds to a value of $\lesssim 10^3$ BeV/cm^2 for the total dose due to iron nuclei.

E. Anders

I may have missed something, but I think none of the differences between lunar and meteoritic grains require an irradiation in space; all can be explained by irradiation on the surfaces of asteroids. Gold has shown experimentally that grains exposed to a proton flux move and jump around owing to electrostatic repulsion. This effect will be more important on asteroids than on the Moon because of the reduced gravity. Thus the grains will be more uniformly irradiated. The shorter irradiation times reflect the higher rate of meteorite bombardment in the asteroid belt, which leads to shorter residence times at the surface. As I noted in my talk at the Jet Propulsion Lab conference in March, 1970, this same factor is also responsible for the lower gas content of meteorites relative to lunar soil.

We know from the Apollo lunar samples that gas implantation and track formation are ongoing processes in the solar system. Before we invent an altogether different process for the gas-rich and track-rich grains in meteorites, let us examine carefully how the symptoms of a surface irradiation vary with the size of the body.

H. Alfvén
Dr Lal also said that there are small grains which are isotropically irradiated and then form big aggregates. Does this mean that the whole aggregate is isotropically irradiated after the accretion?

D. Lal
One observes cases of uniform irradiation for both single grains as well as clumps (a sort of fruit cake) which may be called rock fragments or grain aggregates.

E. Anders
There is a semantic problem here. Should we not refer to these as rock fragments rather than aggregates? Aggregates are random collections of unrelated grains which consolidated somehow, while a rock is an organically grown unit. Now what is the petrography of this so-called aggregate or rock fragment?

G. Arrhenius
We should use as neutral a word as possible. Rock certainly does not necessarily mean igneous rock that you speak about now. A sedimentary rock for example is just particles tossed together. The word fragment is a loaded word that means something that is broken apart. That is why I think it is a good idea to use the words "aggregate" or "rock", which are absolutely neutral by specifying neither the mechanism by which the individual crystallites were brought in place nor the mode by which the aggregate reached its present size.

Irradiation History of Grain Aggregates in Ordinary Chondrites. Possible Clues to the Advanced Stages of Accretion

By P. Pellas

Equipe de recherce du C.N.R.S., Laboratoire de Minéralogie du Muséum, Paris

And yet to me, what is this quintessence of dust?

Hamlet, II, 2, 299

I. *Introduction*

The ordinary chondrites are interesting objects not only because they are the most abundant meteorites (78 % of all known falls) but because they also are a "primitive" material, even if carbonaceous chondrites are more "primitive" in composition. On the basis of their bulk chemical composition, their redox state and free metal (Fe–Ni) content (Fig. 1), they are divided into three groups (Urey and Craig 1953; Craig 1964): 1) the *bronzite* chondrites or H-group (17–20 mole % fayalite in olivine; $Fe^{\circ}=15$–17 weight %; total Fe = 25–28 weight %); 2) the *hypersthene* chondrites or L-group (22–25 mole % fayalite in olivine; $Fe^{\circ}=5$–7.5 weight %; total Fe = 21–23 weight %); 3) the *amphoterites* or LL-group (27–32 mole % fayalite in olivine; $Fe^{\circ}=0.3$–3 weight %; total Fe = 20 weight %). The Si/Mg atomic ratio for these common chondrites is of the order of 1.06 ± 0.03 which differs markedly from the same ratio for enstatite chondrites (1.27 ± 0.10) and carbonaceous chondrites (0.95 ± 0.03). As the Earth is mainly collecting for the most ordinary chondrites of H, L (the most abundant falls) and LL-groups, many meteoriticists have pointed out the strong probability that there is a biasing in the recovered chondritic material. This biasing appears to be real even in the case of "ordinary" chondrites because there exists at least 4 objects having the Si/Mg (atomic) ratio of ordinary chondrites, but drastically differing in redox state. Fig. 1 also shows the location where these atypical materials (two of them being included in Woodbine iron meteorite and Cumberland Falls enstatite achondrite) are plotted in the fayalite versus ferrosilite diagram.

It is not the purpose of this paper to give a comprehensive account of the mineralogy and petrology of ordinary chondrites. Those topics have been extensively covered by Van Schmus (1969) and Dodd (1969). It seems however of interest, in order to facilitate the understanding of the results presented in this paper, to summarize some observations or results previously obtained by students of ordinary chondrites.

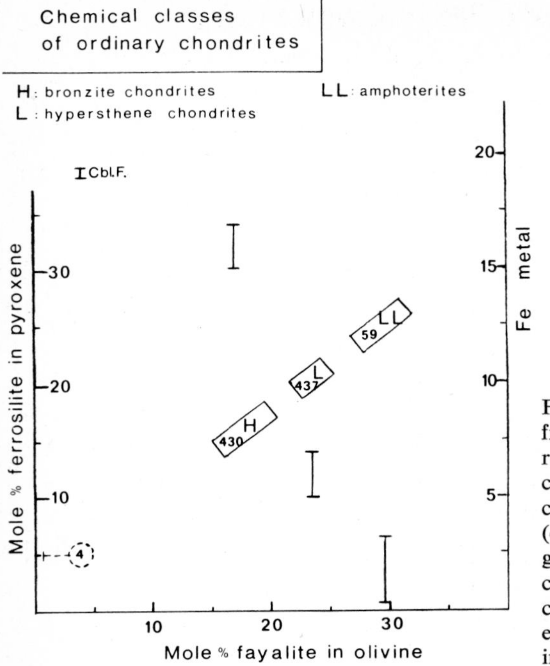

Fig. 1. Diagram showing chemical fractionation (Metal-Silicates) and redox state for the different chemical groups (H, L, LL) of ordinary chondrites. Four chondritic objects (dotted circle) are outside the groupings: *Winona* and *Mt Morris* chondrites; a black chondritic inclusion in *Cumberland Falls* (CblF.) enstatite achondrite; a chondritic inclusion in *Woodbine* iron.

1) Van Schmus and Wood (1967) have subdivided the three discrete chemical groups of ordinary chondrites into four petrographic types, so that type 3, the least crystallized, shows a texture highly chondritic with no apparent feldspar, whereas type 6, well crystallized, contains distinct feldspar crystals. Types 4 and 5 are intermediate types. Marti (1967*a*) has shown that there is a correlation between the abundance patterns of the fractionated heavy rare gases (^{36}Ar, ^{84}K, ^{132}Xe) trapped in objects of different petrographical type, so that the least crystallized type 3, chondrites are enriched and the most crystallized ones (type 6), are depleted over 2 to 3 orders of magnitude in their trapped gases content. Although Marti's results do not allow distinction between the H, L and LL chemical groups, the large variation in trapped heavy rare gases abundances has been found to be correlated semi-quantitatively with equally large variations in concentrations of the volatile trace elements such as In, Bi, Tl, Cs, Br, Pb and C (see Keays et al. 1971, and other references therein). Furthermore clear-cut differences in isotopic ratios (e.g. $^{18}O/^{16}O$) exist between the three chemical groups as those observed by Taylor et al. (1965), Reuter et al. (1965) and more recently by Onuma et al. (1971).

2) It appears from different types of experimental results that ordinary chondrites have had an almost identical protohistory, a fact which in some

ways facilitates comparisons among them. That seems to be the case for the high temperature (1 200–1 600°C) ^{129}I–^{129}Xe retention which was effective in a time interval of roughly 15 million years (perhaps much less: 2–5 million years), as shown by Hohenberg et al. (1967) and Podosek (1970). Similar types of experiments have also demonstrated that the radiogenic ^{40}Ar has been retained at an early stage in the time interval 4.6–4.3 Gy (gigayears) ago, by using the $^{39}Ar/^{40}Ar$ method (Turner 1969; Podosek 1971), if no later events have perturbed the clock. Furthermore precise Rb–Sr "internal" isochrons (Wasserburg et al. 1969 and references therein) as well as isochrons obtained on different objects of a given chemical group (Gopalan and Wetherill 1969 and references therein) show that the last strontium homogenization for the ordinary chondrites occurred approximately 4.63 ± 0.08 Gy ago.

3) Other evidences such as exposure ages or concordant-discordant gas-retention ages strongly suggest that ordinary chondrites come from *at least* three asteroidal "parent-bodies" whose debris cross the orbit of the Earth. Two of those appear to have been affected by shock events of different intensities which occurred at different times in the past and which have not affected the third one (LL parent-body). For instance, a large number of L-chondrites (up to $\frac{2}{3}$ of falls and finds) have been involved in an outgassing event around 500 My ago, as first suggested by Anders (1964) and documented by Heymann (1967) and Turner (1969). Shock features and reheating (up to 1 200°C) symptoms of some of these objects, sometimes blackened, have been verified by mineralogical studies (Heymann 1967; Taylor and Heymann 1969) and shock pressures estimated between 0.5 to >0.8 Mb (megabar) (Taylor and Heymann 1971). On the other hand, one half of the H-chondrites cluster at an exposure age of 3 ± 1 My which indicates that about one third of these objects belong to the peak, if destruction lifetimes against collisional fragmentation for stones less than a meter in radius (Dohnanyi 1969) are taken into account. With respect to the exposure ages, those of L-chondrites seem predominantly to be governed by their lifetimes against collisional fragmentation and planetary capture (Herzog and Anders 1971). The differences between L- and H-groups are even more accentuated by considering the ^{4}He and ^{40}Ar retention ages which are high and concordant for the majority (75%) of the H-group objects (Anders 1964; Müller and Zähringer 1969). Very distinctively, the LL-chondrites differ from the other two groups in that the majority of them ($\simeq 60\%$) are "brecciated" (Mason and Wiik 1964). In spite of this last characteristic they show high and concordant gas-retention ages for most (85%) of them (Heymann 1965; Müller and Zähringer 1969) i.e. an even higher percentage than for the H-group objects. In contrast with the H-group they are not at all affected by the peak exposure age at 3 My. Instead, as far as a small statistic is significant, it seems that

there is a cluster at 9 My for the LL-objects (15 involved objects on 24 cases, which corresponds to $\simeq 7$ objects clearly outside the trend observed for lifetimes against collisional fragmentation).

4) Many ordinary chondrites are *xenolithic breccias*, i.e. they are not simply aggregates of chondrites and matrix but also contain angular to rounded lithic fragments ("xenoliths") of pre-existing chondritic material. According to Binns (1967) the British Museum Collection contains 74 xenolithic chondrites whose distribution among the three chemical groups is the following:

H-group (148 objects) ... xenolithic chondrites: 37 (25 %)
L-group (184 objects) ... xenolithic chondrites: 18 (10 %)
LL-group (29 objects) ... xenolithic chondrites: 18 (65 %)

Binns also found that in all cases the xenoliths were chondritic and that they belong to the same chemical group as the host chondrite. In few cases, however, xenoliths having an "exotic" origin have been found, mostly fragments of carbonaceous chondrites (Van Schmus 1967; Fredriksson 1969). Very recently when sawing a piece of the *St. Mesmin* LL-object an exceptional, presumably bronzite (?) chondrite, xenolith was discovered by our group. The ferrosilite content mole % of pyroxene is 13–17, but the object appears olivine-poor. The study of this peculiar material (shown in Fig. 2 and 3) is in progress.

Fig. 2. Section of *St Mesmin* (LL) chondrite. The brecciated texture of the stone is apparent. The lighter fragments (xenoliths) are imbedded in a greyish host (gas-rich). A few xenoliths (in the mm size) appear black as a result of shock. In the upper right corner, the "exotic" H (?) chondrite xenolith is visible and shows a high metal content (see Fig. 3).

Fig. 3. At this higher magnification, the preferential orientation of the metal patches is clearly apparent, due probably to a violent shock. Note, however, the lack of visible deformation, strain or reaction border at the boundary with the host material.

It is in one H-object belonging to this category of brecciated xenolithic chondrites that Konig et al. (1961) found a high amount of unfractionated (solar type) rare gas, the *Pantar* chondrite. Five years before, Gerling and Levski (1956) had discovered the first "gas-rich" meteorite, the enstatite achondrite Staroe-Pesyanoe.

II. *Gas-Rich Ordinary Chondritic Breccias*

In the years 1961–1971, 29 xenolithic chondrites were found to contain solar-type rare gases (23 H-, 4 L- and 2 LL-objects) i.e. a percentage of 12, 2 and 8 with respect to the number of ordinary chondrites analysed in each chemical group.

The amount of ^{4}He (excluding the radiogenic and cosmogenic components) can vary in these stones from $\simeq 1 \times 10^{-5}$ to 2.2×10^{-2} ccSTP$\cdot g^{-1}$, whereas the $^{4}He/^{20}Ne$ ratio can differ (due probably to diffusion losses) from <70 for *Dimitt* (Eberhardt et al., 1966) up to 540 for Tabor (König et al. 1962). A general and comprehensive account has been given by Pepin and Signer (1965) on the elemental and isotopic ratios of rare gases obtained from these objects. It is worthwhile to note that Suess, Wanke and Wlotzka (1964) were the first to suggest that the implantation of the solar-type rare gases could best be explained by particle radiation such as solar wind. The discovery of very large densities of nuclear tracks ($\simeq 10^9$ cm^{-2}) on the surface of some silicate crystals in gas-rich chondrites (Pellas, Poupeau and Lorin 1968) and gas-rich achondrites (Pellas et al. 1969; Lal and Rajan 1969) gave a very strong support to the solar wind hypothesis and also indicated that a more energetic component emitted during solar flares was present, as suggested by Eberhardt et al. (1965). After the discovery of very high amount of solar-type gas in lunar soil (LSPET, 1969) along with the observation of extremely high track-densities ($\simeq 10^{11}$ cm^{-2}) in the smallest grain-sizes (Borg et al. 1970; Barber et al. 1971*a*) there is rather a general agreement that the rare gases were chiefly implanted by means of solar wind and that the high track-densities were produced by the more energetic solar flare component. Furthermore, the recent finding that extremely high track-densities ($\simeq 10^{11}$ cm^{-2}) are also present in silicate crystals of *Fayetteville* H-chondrite (Barber *et al.* 1971*b*) gave an even stronger support to the reality of the heavy solar flare particle implantation. Finally, the track-density data observed in a glass filter from Surveyor 3 spacecraft (Crozaz and Walker 1971; Fleischer et al. 1971; Barber et al. 1971*b*) has proved definitely and definitively the solar flare origin of the high track-densities such as those measured in lunar soil *and* gas-rich meteorites whatever the latter are, chondrites or achondrites.

One important problem remains to be solved; the definition of the time interval when the solar energetic particles irradiated the gas-rich chondrites. It seems at this moment quite feasible to bracket approximately this time interval. Mazor et al. (1970) and Anders (1970) have postulated that the solar gases could have been acquired at any time during the last 4.5 Gy. This type of truism is rather unsatisfactory to the mind. It is a purpose of this paper to point out that there are strong evidences that the gas-rich xenolithic chondrites were subjected to an irradiation stage in the ancient past, i.e. most probably 4.0–4.3$\pm$0.1 Gy ago, almost at the end of the accretional processes for asteroidal-sized bodies.

Another face of the problem is the duration of the irradiation stage. In fact this "duration" depends on the "chronometers" used and other parameters such as mixing rates or burial depths of materials considered, and thus can vary greatly for different objects, and even for different materials (xenoliths, host matrix) belonging to the same object. The chronometers themselves are not at all faithful, because it is rather probable that the properties of the solar and galactic fluxes for long periods of time in the remote past were not the same as to-day. Thus a precise answer appears difficult to attain at this time, unless all known methods to study a given object would be applied, which up to now has unfortunately not been the case. In a general way, the problem consists to disentangle different types of irradiation which are superposed in gas-rich chondritic material. It will be shown (section III) that in some favourable cases unequivocal solutions have been found which permit to give a *qualitative* description of the multiple irradiation history. To reach these solutions the problem must be approached by using different ways to circumvent it. One of these ways implies to understand fully what type of material the gas-rich xenolithic chondrites are.

On mineralogical and petrographical grounds the gas-rich xenolithic chondrites do not differ from the "gas-poor" ones. They consist in a host material frequently darker in colour and showing a pronounced chondritic texture compared to the generally lighter more crystallized xenoliths imbedded in it (Fig. 2 and 3). The host appears as an aggregate of crystals, crystal fragments and chondrules, within a matrix of smaller grain-size. As the solar-type rare gases were found in the host—and not in the xenoliths—a well-known type of generalization and oversimplification-inducing error was accepted as the truth, i.e. that the solar particles irradiated only the dark host. Thus was born the visual notion of "dark-light" structure which turned out to be rather uninformative. At the deeper level of information which characterizes track research, the following observations should be mentioned: 1) the host material appears *always* to be formed by a much less crystallized chondritic material (types 3 and 4) and it contains a smaller amount of admixed 5 or 5–6 types materials, as revealed sometimes by the presence of feldspar crystals. The amounts of trapped ^{84}Kr and ^{132}Xe are compatible with the "classification" of the host as a mixture of types 3 up to 5–6. This observation is consistent with the enrichment in volatile elements (Rieder and Wanke 1969); 2) the fact that an important component of solar gas is present in the host is only due to the smaller grain-sizes of crystals in this fraction which increases the surface to volume ratio for the solar wind trapping efficiency; 3) only a *fraction* of the crystals from the host appears irradiated by the solar flare iron-group particles (Pellas et al. 1969; Lal and Rajan 1969) in the kinetic energy interval $\simeq 1$–10 MeV/nucleon, the energy range being *only* a function of the crystal sizes; 4) the generally high $^{4}He/^{20}Ne$ ratios indicate that the solar wind implantation occurred when the material was cold ($T < 350K$). Since the rare gas implantation no significant heating has taken place, except for those chondrites which have suffered later shock events. The same conclusion is suggested by the preservation of solar flare tracks, although the temperatures required to anneal nuclear tracks are higher than those necessary to remove ^{4}He (Fleischer et al. 1967).

In contradistinction to the host material, the xenoliths (often whitish) range in size from a few centimeters to a few millimeters (Fig. 2). In fact one could consider that these xenoliths form a *continuum* from the largest fragments to individual grains. Some of the centimeter-sized xenoliths characteristics are the following: A) their chondritic texture is much less apparent or even absent compared to the host, and they belong for the most part to the well crystallized petrographic types 5 and 6 with some exceptions (see f.i. *Weston*, section III) although their bulk chemical composition is *the same* as that of the host material. Their well crystallized state is corroborated by their low content of fractionated trapped gases and volatile elements. This clearly indicates that before their incorporation in the chondritic breccias the xenoliths have had a thermal history which produced the chemical "equilibration" of their olivine and pyroxene crystals; in this connection, a rather good indicator of the thermal history of a chondritic material is given by the relative amount of trapped heavy rare gas released at high temperature: in such a case as Pantar xenoliths (petrographic type 5), although some of them have apparently been partially outgassed before being incorporated into the breccia, Turner (1965) has shown that the greatest fraction of trapped ^{132}Xe is released in the temperature range 1 000°–1 400°C ($\simeq$ 68 %), whereas for the dark host $\simeq$ 48 % only of the gas is released in the same temperature interval; B) a *fraction* (variable for different gas-rich stones, and even variable within the same stone) of the xenoliths appears to have been irradiated by solar flare particles up to 150 MeV/nucleon (the upper limit of the kinetic energy depending only on the largest xenolith size studied up to now) *and/or* in a higher energy range (GeV/nucleon) by a galactic radiation component *not* attributable to the galactic irradiation of the meteoroids. These data will be presented and discussed in section III. However it is worth pointing out the erroneous interpretation of Suess and Wanke (1967) who denied that petrographic type 6 material could be irradiated; C) some of the xenoliths show clear evidence of shock-induced effects, such as metallic or glassy veinlets terminating at the xenolith boundaries. Moreover, it happens that blackened and indurated xenoliths are found, identical in texture to the heavily shocked L-chondrites *Tadjera*, *Orvinio* and *McKinney* studied by Heymann (1967) and Taylor and Heymann (1969, 1971), or *Farmington* (Buseck *et al.* 1966). The significance of these shocked xenoliths, rather frequent in *St. Mesmin* gas-rich amphoterite will be discussed in section IV.

In short, gas-rich xenolithic chondrites differ only from the other xenolithic chondritic breccias by having a *fraction* of their material which was at one time directly exposed to the solar and galactic cosmic radiations. Credit should be given to Wlotzka (1963) for reaching many of these conclusions through petrographical observations only.

III. *Experimental Results on Weston and Djermaia Xenoliths*

Six xenoliths in *Weston*, sampled on a single piece (from Arizona State University, Tempe), and ten xenoliths in *Djermaia* coming from three stones of this fall, were studied by coupling nuclear track-method and rare gas analysis within a collaboration between E.T.H. Zurich group (L. Schultz and P. Signer) and Paris Museum and C.N.R.S. group (J. C. Lorin and P. Pellas). The work is in its final stage and will be published soon (Lorin et al. 1971). At this point it is worthy of note that the interpretation of the results presented below does not involve the scientific responsibility of the Zurich group but only that of the Paris group, and especially of myself.

Weston and *Djermaia* are classical gas-rich H-chondrites, i.e. their host portion is enriched in solar wind ^{4}He with the respective following values: 24.5–171 (Weston) and $2.4–13.8 \times 10^{-4}$ cc STP·g^{-1} (Djermaia). The $^{4}He/^{20}Ne$ ratios for the host materials, sampled far away from the fusion crust, are 280 for *Djermaia* and 400–450 for *Weston*. There seems to exist at least a strong qualitative correlation between the degree of friability of the host material and the $^{4}He/^{20}Ne$ ratios in gas-rich xenolithic chondrites, the more friable hosts showing the highest values of the latter elemental ratio (with the important exception of *Fayetteville*). In this connection, *Djermaia* is a very hardened stone and *Weston* a very friable one. This observation suggests that compaction processes might well have partially outgassed the solar wind implanted helium in the host, and perhaps also perturbed the U, Th–He clock of the xenoliths. The $^{20}Ne/^{22}Ne$ ratios of *Weston* and *Djermaia* gas-rich hosts are respectively 11.7 and 13.0.

1. **Weston Xenoliths**

Five out of six are centimeter-sized white xenoliths of petrographic type 5. The sixth, not yet studied by track-method, is a hardened black material showing a highly chondritic texture and strongly enriched in fractionated ("planetary-type") ^{36}Ar ($\simeq 17 \times 10^{-8}$ cc STP · g^{-1}), presumably H-material of petrographic type 3 (Fa: 17–20, Professor J. Fabriès, personal communication). In Table 1 are shown the results of nuclear track-densities measured in each xenolith (the xenoliths are placed in the first column in the order of their geometric sequence in the meteorite piece), the track-density scatter (third column), the spallagenic ^{3}He and ^{21}Ne content, and their bulk K–Ar age (K content of 815×10^{-6} g/g). Assuming that the lowest ^{3}He and ^{21}Ne contents might correspond to the meteoroid "exposure age" (this seems approximately acceptable—though not at all certain—for three other xenoliths coming from two other *Weston* pieces show the same spallogenic

Table 1. *Track-density variations and spallogenic rare gases content in Weston xenoliths*

Xenolith No	Track-density observed (10^6 tracks. cm^{-2})	Variation factor[a] in track-density and distance from xenolith No 5 (in cm)	^3He	^3He (excess)	^{21}Ne	^{21}Ne (excess)	K-Ar age (Gy)[b]
			(All in units 10^{-8} cc STP·g^{-1})				
Light no 5	0.13–0.17	1.3 (—)	—	—	—	—	—
Light no 3	*0.17–0.31*	1.8 (1.3 cm)	*63*	*7*	*15.2*	*2.5*	4.3
Light no 4	0.11–0.17	1.7 (2.5)	56		12.9		4.0
X–D (black)	—	(3.7)	*−72*	*16*	*17*	*4.2*	>3.5[c]
Light no 1	0.33–0.52	1.7 (6.0)	55		12.9		4.2[d]
Light no 2	0.18–0.38	2.1 (7–8 cm)	56		12.5		4.2

[a] Variations due to the large sampling size (0.5–1 cm) and to the geometric β angle condition (Fleischer et al., 1967).
[b] Assumed K content: 815×10^{-6} g/g, averaged value of seventeen H-chondrites (Kaushal and Wetherill, 1969).
[c] For H-type 3 material, the preceding K content is an upper limit.
[d] U, Th–He age: 4.2 Gy.

^3He and ^{21}Ne contents, i.e. 58–60 and $12.5–13.0 \times 10^{-8}$ cc STP·g^{-1} respectively), then this "age" is 25 ± 3 My. It is apparent from Table 1 that two xenoliths are clearly outside the background values. Those Light no 3 and X–D (black) show a much higher spallogenic rare gas content. It appears very difficult to explain these differences by target effects or a difference in the energy spectrum hardness during the meteoroid exposure age because: 1) the composition of H-5 material in the different white xenoliths is indistinguishable on a mineralogical basis; 2) type 3 H-material (X-D black) does not show such a large difference in target chemistry compared to other H-types; 3) all the xenoliths belong to the same meteorite piece and are contained within a distance of 8 cm with respect to xenolith no. 5 (see Table 1, third column). From the track data on xenoliths, as well as taking into account the *minimum* track-densities measured in the "unirradiated" (i.e. in crystals showing no track-density gradient of solar flare type) fraction of the gas-rich host, an exposure age of $\gtrsim 22$ My may be computed for the meteoroid. Track data allow also one to locate the piece at a depth of 12 cm within a meteoroid having a radius larger than 30 cm. At this point it is worth noting that the track-densities of xenolith no. 3 are clearly higher, with no overlap, compared to those obtained from the *nearby* xenoliths nos. 5 and 4. In other words, xenolith no. 3 shows an *excess* of nuclear tracks, outside the background tracks attributable to the "exposure age" of the meteoroid. That this track-excess ($\approx 10^5$ cm^{-2}) correlates with the spallogenic ^3He and ^{21}Ne

excesses (7 and 2.5×10^{-8} cc STP·g^{-1}) points out a fundamental contradiction with respect to the "exposure age" meaning assumed thus far. Still, the case of the black xenolith (X–D) is even more obvious, although track measurements have not been carried out yet. For it, the excess spallogenic rare gases are so abundant that there must be something wrong in the "exposure age" idea commonly assumed.

The simplest way to explain these anomalies—which in fact are not "anomalies" at all—appears as the most realistic one also. It is to accept that these xenoliths have been subjected to an irradiation stage before their incorporation in the *Weston* breccia. Since no track-density gradient has been observed inside xenolith no. 5, the track-excess can be explained by a galactic cosmic-ray type of energy spectrum. This last statement, *based on observations*, implies clearly that this irradiation stage occurred when xenolith no. 3 was shielded, i.e. with a 2π irradiation geometry. In such a case, the track-excess could be produced by an irradiation stage of 4×10^9 y. at a depth *no more* than 40 cm (Fleischer et al. 1967). At this depth however, the interaction mean free path of galactic protons and the secondaries cascade would produce a greater amount of $^3He_{sp}$, and especially of $^{21}Ne_{sp}$ and $^{22}Ne_{sp}$ by means of the large integrated neutron flux (Eugster et al. 1966), compared to the spallogenic excess measured. Thus, a so long duration for this exposure stage—based only on nuclear track data—should be discarded if rare gas excesses are taken into account. In fact, a rather acceptable agreement between track and rare gas data—which by no means constitutes a certainty—would be to consider that the irradiation occurred at a depth of $\simeq 12$ cm during 10 My roughly. Yet, the high content of neutron-produced ^{80}Kr and ^{82}Kr from bromine[1] observed by Eugster et al. (1969) indicates presumably that the mean depth of irradiation for xenolith no. 5 could well be of the order of 20 cm. From track data this latter depth implies an irradiation stage of 60 My which conflicts with the relatively small excess of $^3He_{sp}$.

Whatever is the true duration of the anterior exposure stage, it is worthwhile noticing at this point that the chronometer used in the former computations is the galactic flux, in the energy range 1.5–2 GeV/nucleon, whose constancy in the past it is acceptable to assume, even if the prior irradiation occurred such a long time as 4 Gy ago, at the *ancient* orbit of the *Weston* "parent-body".

The case for xenolith X–D (black) remains unclear, due to the lack of track data. This fragment contains a great amount of solar gases (8×10^{-4} cc STP·g^{-1}) which can hardly be explained by a host contamination (in view of the care taken to avoid it). On the other hand its largest dimen-

[1] It is worthwhile to note that the Eugster et al. (1969) data have been obtained on 0.1- to 1-g-samples of mixed *Weston* material. This fact is especially important when results have to be interpreted, especially in case of xenolithic chondrites (see Section V).

sions ($1.8 \times 1.2 \times 0.6$ cm) cannot eliminate the possibility of a preponderant solar flare implantation for the rare gases. At this stage of study, it is only possible to say—on a qualitative basis—that the fragment was subjected to a much longer first irradiation stage than xenolith no. 3 (taking into account all the possible depths as defined before, plus a non-negligible surface irradiation stage).

2. Djermaia Xenoliths

In some of the xenoliths the records of two, or even three, different irradiation stages have been found. A) One irradiation, the most recent, corresponds to that received by *all* xenoliths during the exposure age of *Djermaia* meteoroid. It lasted 16 My (computed from $^{21}Ne_{sp}$ content in some xenoliths), or 17 ± 3 My by galactic tracks data. The $^{3}He_{sp}$ values cannot be considered; the sampling was made too close to the fusion crust. B) Inside some xenoliths there are track-density variations in orthopyroxene crystals up to a factor of 30 over a few millimeters distance (Table 2 and Fig. 4). These track-density variations cannot be attributed either to the galactic iron-group nuclei from the meteoroid exposure age or to a fission component. Thus another irradiation event has to be postulated. By subtracting the track-density profile of the

Table 2. *Track-density variations and spallogenic* ^{21}*Ne content in Djermaia "light" xenoliths*

Sample No	Xenolith No	Track-density variation[a] (10^6 tracks cm^{-2}) over the distance (cm)	Variation[a] factor	^{21}Ne (10^{-8} ccSTP. g^{-1})	K-Ar[c] ages (Gy)
2422	L-1	0.21–1.6/0.3 cm	7	9.2	4.0
	{L-2 al	0.24–1.3/0.25	5.4	10.0	{4.3
	{L-2 total	0.24–6.7/0.4	28	10.0	
	L-3	0.74–4.3/0.5	6	9.5	4.2
2423	L-1	0.7–1.7/1.0	2.5	7.8	4.4
	L-2	2.5–13.0/1.0	5.2	—	
	{L-3	1.4–2.6/0.6	1.8	—	
	{L–3 (chondrule)	1.8–2.5/0.08	1.4[b]		
2424	L-1	2.7–4.7/1.0	1.7	7.8	4.3
	L-2	6.0–8.6/0.6	1.4	—	
	L-3	1.7–2.8/0.7	1.6	—	
	L-4	8.7–11.4/0.5	1.3	—	

[a] not corrected for meteoroid "exposure age", i.e. for the recent galactic component.
[b] variation explained by the geometric β-angle condition (Fleischer et al., 1967).
[c] Assumed K content: 815×10^{-6} g/g, averaged values of seventeen H-chondrites (Kaushal and Wetherill, 1969). The K–Ar ages of three samples taken from the gas-rich dark host are: 4.0, 4.1, 4.3 Gy.

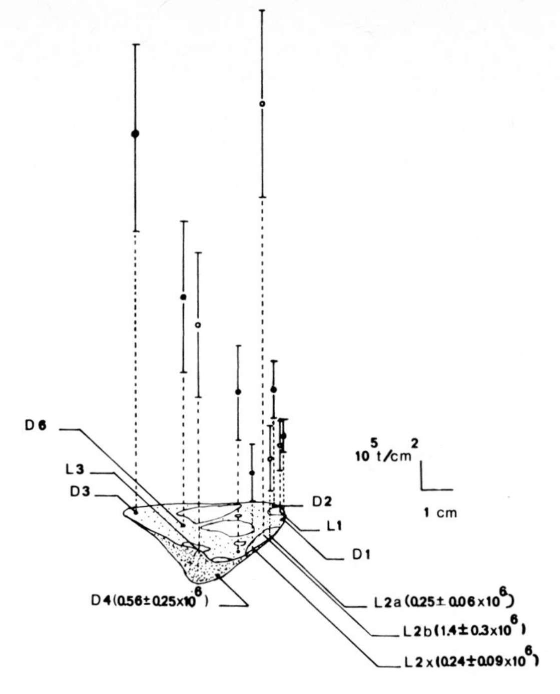

Fig. 4. *Minimum* track-density profile in *Djermaia* xenolithic H-chondrite (piece no. 2422). The *galactic* track-density gradient is visible over a distance of about 5 cm. Note the sudden increase in track-density (*minimum values*) at the xenolith boundary (0.05 cm *inside*). From the xenolith boundary towards the center, the track-densities decrease by a factor of 28 over a distance of 0.4 cm (track-density data *not* corrected for the galactic component). (L-2 xenolith.)

meteoroid exposure age, the L-2 xenolith (Fig. 4) shows track-density variations by more than a factor of 50 over a distance of 4 mm. These track data are interpreted in terms of an irradiation by particles with a power-law kinetic energy spectrum with $\gamma > 2.2$ in the energy interval 40–150 MeV/nucleon for iron nuclei, quite similar to the *present time* solar flare radiation as determined by track studies on the glass filter from *Surveyor* 3 spacecraft (Crozaz and Walker 1971; Fleischer et al. 1971; Barber et al. 1971*b*) and also to the solar flares ^{4}He ($\gamma = 3.0$) measured by Hsieh and Simpson (1970) on the satellite IMP-IV during the year 1967. The significance of this result is that at the time of the irradiation, the long time average of the solar flare radiation had approximately the same E^{-3} functional relation as to-day, i.e. that solar flare radiation was predominant over the entire kinetic energy range from 1 MeV/nucleon (case of crystals irradiated in the gas-rich host material) up to 150 MeV/nucleon (case of xenolith L-2, Fig. 4) *at the orbit of Djermaia parent-body*. This result shows the importance of determining at what time this solar flare irradiation took place (Section IV). C) In this exceptional *Djermaia* xenolith, as well as in some others, there is a clear $^{21}Ne_{sp}$ excess (Table 2) which appears difficult to explain (at least in the present state of our knowledge) by the solar flare radiation described in B) or by any target-effect (*Djermaia* xenoliths belong to petrographical types 5 and 6 which have the same major element chemistry). Therefore that would suggest that these xenoliths experienced also a *shielded* irradiation, of the same type as for *Weston* xenolith no. 3, during which $^{21}Ne_{sp}$ was accumulated but essentially no tracks were registered ($\lesssim 10^4$ cm^{-2}).

In order to obtain information on the durations of the two earlier irradiation stages, some assumptions have to be made *which are by no means certain* but which enable us to calculate these first "exposure ages". For instance, a normalization to the *Surveyor* glass data implies that a maximum of 4×10^5 y are needed to explain the track-density variations resulting from the *surface* exposure B). As far as the duration of the exposure which produced the excess $^{21}Ne_{sp}$ is concerned, a period of about 10 My with a shielding of the order of 30 cm is evaluated assuming 2π irradiation geometry and using "conventional" production rates and fluxes (Herzog and Anders 1971). The latter calculation is subject to the same uncertainties as for *Weston* xenolith no. 3, since the same amount of excess $^{21}Ne_{sp}$ ($\simeq 2.2 \times 10^{-8}$ cc STP$\cdot$g^{-1}) could be produced with more shielding during a much longer period of time.

Perhaps, in the near future, it will be possible to check more precisely the first stages irradiation duration by means of methods which give a measure of the integrated neutron fluxes for different energy range (thermal, epithermal and high energy neutrons) such as Gd isotopic ratios (Eugster et al. 1970), ^{128}Xe excess over the trapped and spallation components, ^{80}Kr and ^{82}Kr from neutron capture on bromine, and ^{40}K from (n, p) reaction on ^{40}Ca. The (apparent) paradox is that large integrated neutron fluxes have been found in small recovered chondrites which are mostly xenolithic chondritic breccias or the so-called "unequilibrated ordinary chondrites", including carbonaceous chondrites and one "apparently" homogeneous chondrite (Marti et al. 1966; Marti 1967; Heymann and Mazor 1968; Eugster et al. 1969; Reed and Jovanovic 1969; Mazor et al. 1970). All these authors, except Marti (1967) and Reed and Jovanovic (1969), assumed explicitly or implicitly that the neutron fluxes they measured were produced during the meteoroid "exposure age". Not only is this assumption strongly uncertain, it is probably wrong in the majority of the cases. It requires indeed very large pre-atmospheric masses (of the order of tons) which in all cases do not correspond to the recovered masses, (except for *Abee* for which Begemann and Vilcsek (1965) have found an ablation loss during atmospheric entry of the order of 30 %).

A similar amount of ablation loss has been found by Cantelaube et al. (1969) for St. Severin chondrite (LL-group) by means of cosmic-ray track-density data obtained over all the surfaces of the reconstructed object. The total ablation loss (30 %) corresponds, in fact, to ablation thickness varying for the different surfaces from less than 1 cm to less than 5 cm. It would be rather surprising that the only chondrite (St. Severin) whose abalation loss has been accurately measured should reveal itself to be an exceptional low-ablated object. That meteoriticists have to do in most of the cases with small pre-atmospheric sized-objects is substantiated indeed by the very frequent low-level of cosmogenic neutron-produced radionuclide ^{60}Co. Further aspects of this subject and their important inferences will be discussed in Section V.

From the experimental data presented some conclusions can already be stated. There exist in gas-rich ordinary chondrites, xenoliths having the same chemistry as the host, which *must* have had independent thermal histories before being incorporated into these brecciated chondrites. Some of the xenoliths have experienced direct and/or shielded irradiation before being compacted into the material which, in a later stage, will form the meteoroid. In very many cases the compaction processes have not perturbed either the very sensitive $^{4}He/^{20}Ne$ ratio or the radiogenic argon clock since the argon retention began. These observations, put together, seem to preclude the possibility that this fragmental material was irradiated *in free space* at the time the pre-compaction irradiation occurred. That does not rule out, but in fact enhances, the very true possibility that "exotic" materials coming from other asteroidal parent-bodies (having presumably very different orbits) have been subjected to an irradiation *in space* as single objects, before their incorporation into chondritic breccias. If the previous observations are well interpreted, the fact that the xenoliths belong to the same chemical group as the dark host seems to preclude an irradiation in space. In this connection, the presence of *unpreirradiated* xenoliths appears even more significant: it points out that—as for the unpreirradiated fractions of the dark host—turnover and stirring mechanisms were actively operating at the asteroidal parent-bodies' surfaces, thus producing a *mixing* of all the fractions. These observations seem to rule out the interpretation put forward by Pellas et al. (1969) and by Lal and Rajan (1969), who attributed the isotropy frequently observed in irradiation geometry on single crystals from gas-rich chondrites and achondrites to and irradiation in space before their accretion and imbedding in the relatively large bodies (meteorites) in which they are found. The large neutron integrated fluxes measured in typical xenolithic chondrites by other groups appear to be a very strong proof for shielded irradiation conditions, along with mixing, and such a clear evidence has to be accepted. On the other hand, it is the author's opinion to-day that neither the isotropy nor the anisotropy in irradiation geometry constitutes a proof of any kind. Crystals, chondrules or centimeter-sized xenoliths gravitationally bound at the surface of small ($<10^7$ cm dia.) asteroidal parent-bodies may show either isotropy or anisotropy in irradiation geometry. Moreover, these materials having suffered extensive brecciation processes, the cases could not be significant if anisotropy is observed. Exemplary in this respect are the lunar fines. All track groups have observed that the irradiation characteristics of lunar grains, although similar to those found in gas-rich meteorites, differ in some respects. Due to the high track density observed in lunar grain-centers, the track density gradient in individual crystals is rather difficult to estimate unless by using up-to-date techniques such as high voltage electron microscopy (Borg et al. 1970; Barber et al. 1971). However, as some authors have contended (Arrhenius et al. 1971) that the crystals from the lunar regolith are in most of the cases not irradiated uniformly on all their surfaces, it is worthwhile recalling that the lunar material in its largest grain-size, e.g. lunar rocks, show cases of both uniform and nonuniform irradiation.

IV. *Preliminary Results on Saint-Mesmin Xenoliths*

Except for the presence of a large "exotic" xenolith (Figs. 2 and 3) presumably a type 6 (feldspar apparent) bronzite chondrite, *St. Mesmin* is a typical LL-chondrite which was found to be gas-rich by Heymann and Mazor (1966). Samples of ten xenoliths were taken from a single *St. Mesmin* stone (no. 368). Some of the rare gas results, obtained by Schultz and Signer, are reported below. Three grey host locations show variations in the solar wind ^{4}He content from 11.8 to 53×10^{-4} cc STP$\cdot$g^{-1}. The ^{4}He/^{20}Ne ratio for the host material varies from 250 to 285, whereas Heymann and Mazor (1966) measured in their piece a value of 480. Further in the Paris sample the ^{20}Ne/^{22}Ne ratio is 12.5 whereas the ^{20}Ne/^{36}Ar ratio, all corrected for spallation and "fractionated" components, differs from 21 to 32, pointing out the inhomogeneity of the material clearly visible in Figs. 2 and 3. The ^{3}He "exposure age" of the meteoroid is $\leqslant 10.5$ My. The U, Th–He versus K–Ar ages of nine xenoliths are shown in Fig. 5, along with the bulk radiogenic ages measured by different groups on *St. Séverin* (LL-type 6) chondrite. There exists a very apparent scatter for the *St. Mesmin* xenoliths data, compared to *St. Séverin* data, in spite of the fact that eight out of nine xenoliths were measured by the Zurich group. Two xenoliths were discarded and the results not reported in Fig. 5 because one shows an important host contamination and the other gives a meaninglessly high ^{4}He age of 4.9 Gy. The scatter is due mainly to the ^{4}He ages of the xenoliths, while their argon ages do not differ markedly from the *St. Séverin* argon age scatter, for which Podosek (1971) has shown by the $^{39}Ar/^{40}Ar$ method that this last object was subjected to a simple cooling history since the ^{40}Ar retention. With the petrographical knowledge of the uneven distribution of K-rich phases (feldspar) and U-rich phases (whitlockite) in ordinary chondrites, especially true for LL-material, it is better to compare the averaged values of *St. Mesmin* xenoliths and *St. Séverin* samples, corresponding in fact to a better sample homogenization. Within the margin of errors, the two stones retained their argon 4.3±0.25 Gy ago, despite the fact that the former is a highly brecciated chondrite and the latter a well crystallized type 6 breccia. On the other hand, the radiogenic helium retention differs between the two objects, being 4.0±0.25 for *St. Mesmin* xenoliths, and 4.2±0.2 for *St. Séverin* (see in Fig. 5 caption the constants used). It is not the author's purpose to dwell on the precision of the mean radiogenic age values but only to consider them as very strong qualitative indications that the last complete outgassing of *all St. Mesmin* xenoliths occurred at least about 4 Gy ago. From this view-point the case of the three blackened xenoliths (Fig. 5) is highly significant. These three fragments

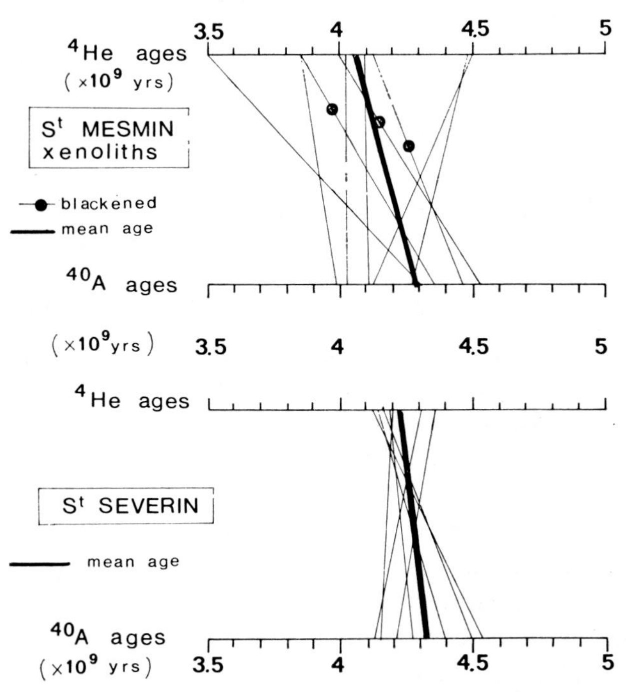

Fig. 5. U, Th-He versus K-Ar ages for 9 *St Mesmin* xenoliths and 7 *St Séverin samples.* Error margins are ±15 % for individual measurements (mean K content: 950×10^{-6} g/g). Constants used: $^{40}K/K = 0.0118$; $\lambda = 5.32 \times 10^{-10} y^{-1}$; $\lambda_K/\lambda_\beta = 0.123$; $\lambda^{238}U = 1.537 \times 10^{-10} y^{-1}$; $\lambda^{235}U = 9.77 \times 10^{-10} y^{-1}$; $^{232}Th = 0.499 \times 10^{-10} y^{-1}$; $U = 11 \times 10^{-9}$ g/g; Th/U = 3.6. (*St Mesmin data*: L. Schultz and P. Signer 1970–1971; Heymann and Mazor 1966. *St Séverin data*: Funkhouser et al. 1967; Eberhardt, Geiss and Graf 1966; Zähringer 1968; Marti et al. 1969; Marti 1969).

have been heavily shocked, heated or even melted (as shown by petrographical study) and consequently they were totally outgassed of their helium and lost, partially at least, their argon. Nevertheless they show high helium and argon ages.

These observations point out the fact that about 4.3 Gy ago the *St. Mesmin* material as a whole became a closed system for the K–Ar clock, whereas the system was effectively closed for U, Th–He about 4.0 Gy ago. Then, if the facts are logically interpreted, the time interval 4.3–4.0$\pm$0.1 Gy has to correspond to the brecciation processes which occurred on the asteroidal parent-body of LL-chondrites. At this point it is worthwhile recalling that the majority of LL-chondrites (60–65 %) show a high degree of brecciation and, in spite of this characteristic, 85 % of them present high and concordant gas-retention ages (Heymann 1965; Müller and Zähringer 1969). In such a perspective the *St. Mesmin* case is not uncommon.

The upper limit (4.3 Gy) of the time interval is set by using *St. Séverin* as a reference material, and applying the conclusions attained to the well crystallized *St. Mesmin* xenoliths (of petrographical type 5 and 6). The reasoning is the following: there exists a strong qualitative correlation between fission tracks from ^{244}Pu and fission Xe in *St. Séverin* whitlockite (Wasserburg et al. 1969; Cantelaube et al. 1967). Wasserburg et al. (1969) have shown that the expected amount of ^{136}Xe calculated from fission tracks data is lowered by a factor of about 25 from what is observed. This difference appears well explained by a later retention time for tracks in whitlockite than for fission Xe. In fact, the "fission tracks–fission Xe" interval is of the order of 400 My. As said, this "time interval" appears fully explained by the low track-retention temperature in whitlockite (a temperature of 250°C during 1 My would anneal all the tracks, as shown by Cantelaube et al. 1967). Thus, assuming that fission Xe began to be retained 4.70–4.65 Gy ago in *St. Séverin* whitlockite, the track retention became effective roughly 4.30–4.25 Gy ago. In the conditions of a slow cooling, a relatively small fraction of radiogenic helium is expected to be retained at 250°C for 1 My period in olivine and pyroxene, as shown by Hunecke et al. (1969) for $^{3}He_{sp}$. However, at a temperature of 100°C during 1 My olivine and pyroxene retain 99 % and 82 % of their ^{3}He. As a great fraction of radiogenic helium produced in whitlockite was trapped in olivine and pyroxene, due to the great abundance of these mineral phases (85–90 %) in LL-type 6 material, and taking into account the large mean range of alpha particles in whitlockite ($\simeq$40 μ) (Cantelaube and Pellas, 1971), it seems very likely that when *St. Séverin* cooled down to 100°C all the helium was essentially retained in the meteorite as a whole. So the bulk He age of 4.2 Gy for *St. Séverin* is well as expected.

If the preceding reasoning is valid, then it can be extended to the *St. Mesmin* xenoliths which, belonging to the LL-group, should have had a similar (though not necessarily identical) thermal history. Furthermore the two stones belong to the same 9 My "exposure age" peak of the LL-chondrites. The only variation between the two materials is based on their average helium retention age (4.2 Gy for *St. Séverin*; 4.0 Gy for *St. Mesmin* xenoliths). The margin of errors do not erase the strong qualitative picture: *in the time interval* $4.3-4.0\pm0.1$ *Gy the brecciation processes due to a continuum of impacts of different intensities occurred on the LL-asteroidal parent-body in order to produce the St. Mesmin LL-mixture observed to-day.*

This picture does not signify at all that before 4.3 Gy there were neither impacts nor brecciation processes on the LL-asteroid. In this matter *St. Séverin* itself is a brecciated material, like *Ensisheim* or *Jelica.* In fact, strong qualitative differences can be expected in the resulting breccia if a projectile collides with a hot material (*St. Séverin* was at a high temperature before the ^{40}Ar retention began, during the time interval 4.6–4.4 Gy ago) or with a cold material. In the former case, the chemical equilibration between the silicate phases will not be perturbed; only relics (faults, slippages, metal veins) will subsist occasionally and the macroscopic appearance will show large or small clasts immerged in a coherent and fine-grained well-crystallized matrix, i.e. the type of breccia named "*brèches de friction*" by Cristophe–Michel–Lévy (1971). During this time interval (4.6–4.4 Gy) *St. Séverin* material was an open system for radiogenic argon (same as for *St. Mesmin* xenoliths). Since ^{40}Ar began to be retained in the feldspar (at about 4.3 Gy), *St. Séverin* was never more subjected to strong shock events till its capture by the Earth. On the other hand, impacts on a cold material will produce the so-called "polymict brecciated structure" characteristic of xenolithic ordinary chondrites. This peculiar structure, not obliterated even by the subsequent compaction processes, has to be formed very gently in order not to perturb either the radiogenic helium clock or the $^{4}He/^{20}Ne$ solar ratio of the gas-rich host material (see f.i. Wilkening et al. 1971).

Two words on the projectiles. Those must be materials moving on same original Keplerian orbits during the accreting stage(s) in order to maintain the record of chemical fractionations (established earlier) observed in the different chemical groups of ordinary chondrites. The accreting stage(s) for asteroids must be considered as the result(s) of concurrent accretion and breakup processes, as proposed by Kuiper et al. (1958), Hartmann (1968), Alfvén (1964), Anders (1965), and Alfvén and Arrhenius (1970, 1971).

The accretion processes for asteroids have had to begin very soon ($\simeq 4.7$ Gy ago) after the earlier condensation of iron and Mg–Fe silicates (whatever the condensation processes assumed) in order to explain the sharp isochronism in ^{129}I–^{129}Xe high temperature correlation observed by Hohenberg et al.

(1967), Hohenberg (1967) and Podosek (1970). Furthermore, it seems possible to consider that the *earlier* accretion processes occurred all along the cooling history of materials moving in their own original Keplerian orbits (short time scale 0.1–10 My?). This qualitative picture, or cartoon, has at least the advantage of explaining the preservation of the original chemical fractionation which *is observed* in an object like *Hedjaz*, formed by a *mixture* of chondrules belonging *all* to the L-group, showing however petrographic types 3, 4, 5 and 6 (Kraut and Fredriksson 1971).

During this earlier time, the gases were presumably still present in the protosolar medium in order not to allow the low-energy (1–10 MeV/nucleon) solar flare iron nuclei registration in single crystals or chondrules, with the steep differential energy spectrum observed by Pellas et al. (1969) and Lal and Rajan (1969) (see also the case of *Djermaia* xenolith L-2 in Section III) in the host crystals of gas-rich chondrites and achondrites. This picture would also explain the smooth-running accretion (gases still are present in the protosolar medium) and compaction processes because of the low approach velocities of materials (planetesimals, embryos) accreting on the same Keplerian orbits. In proportion as time is going on, and as a consequence of Jupiter's catastrophic accretion which induces great orbital perturbation on small bodies originally formed in Keplerian orbits, it might be expected that many asteroidal orbits will become eccentric, and *not yet* accreted materials coming from *different* primeval orbits will be trapped in between Kirkwood gaps (or even *in* the actual "Kirkwood gaps" which, of course, at this earlier time may well have not yet existed). At that time (or shortly before or after?) the gases should be removed as a consequence of the formation of the sun and the beginning of the T-Tauri phase (solar wind and solar flares). This possible mechanism would produce objects like *Chainpur*, *Sharps* (impossible to "classify") or even some (perhaps all?) carbonaceous chondrites, characterized by a *mixing* of different chemical materials (showing the entire spectrum of different thermal histories) on the chondrules and matrix basis. In this respect, Osborn and Schmitt (1971) working on single chondrules from *Allende*, *Al Rais* and *Karoonda* have found "exotic" materials *never* observed before. In *Allende* also, Fireman et al. (1970) reported recently very large *correlated* variations between $^{128}Xe/^{132}Xe$ and $^{129}Xe/^{132}Xe$ ratios in different types of chondrules.

On the basis of these facts it is possible to expect that chondrules, as single objects in the size range 0.1–1 cm dia., would have been subjected to different relative irradiation prehistories (solar flare irradiation) and that they should show different "exposure ages". This latter effect is observed *indeed* not only in chondrules from carbonaceous chondrites (see Allende, f.i.) but also in such an object like *Chainpur* which shows *three* different ^{21}Ne "exposure ages" in different samples: 10 My (Zähringer 1968), 18 and 23 My (Heymann and Mazor 1968), whereas the ^{3}He "exposure ages" in the same samples are respectively 7, 17 and 19 My. Same types of differences are observed for objects such as *Krymka* (L-3) and even *Baratta* (L-3) and *Goodland* (L-4) (pyroxene and olivine are retentive mineral phases for $^{21}Ne_{sp}$) though the latter two have been partially outgassed as a consequence of later shock events (Heymann and Mazor 1968).

This last type of "heterogeneous chemical material" resulting presumably from mixing processes would also explain the case of the large and black (shocked) metal-rich xenolith found in *St. Mesmin* (Figs. 2 and 3). The fact that this "exotic" xenolith contains apparent plagioclase feldspar (petrographic type 6) indicates it has been subjected to a preceding thermal history before it came to rest on the LL-asteroid. Thus it can be expected—when rare gas data will be available—that the trapping should have occurred at a relatively late stage in the interval 4.3–4.0 Gy, in order to fit the requisite time interval to produce the well crystallized petrographic type 6 material inside its original asteroidal parent-body (one among other H-group asteroids?). If the cartoon is correct, this peculiar xenolith must also show "excess" spallation effects. There exists an other case—the black chondrite inclusion in *Cumberland Falls* aubrite—for which excess spallation effects must be found compared to the aubritic material which trapped it. If these excess spallation effects were found, then they could give us information on the travel time in interplanetary space from one asteroidal orbit to another one in which the "exotic" projectile is captured and incorporated in the target-asteroid. Thus far, the "exposure radiation age" idea, restricted to day by the inability of the human mind to imagine all the possibilities which occurred in such a complex system as our solar system, will in the near future—it is the author's opinion—enlarge our understanding of all the possible orbit-transfers which took place $\simeq 4.0$ Gy ago, and even perhaps in older times. A careful search for these "exotic" fragments trapped in many meteorites of all types, not only will bring interesting information on the transfer-times from orbits to orbits in the asteroidal belt, but they also offer us the possibility to study very atypical materials whose asteroidal parent-bodies are not in the present time in Earthcrossing orbits.

Examplary in this respect are the silicate inclusions in *El Taco* iron. They show $^{20}Ne/^{22}Ne$ ratio of solar type, and $^{4}He/^{20}Ne$ ratio much lower than 40 (Hintenberger et al., 1969). This observation indicates clearly that the silicates were irradiated by energetic solar particles before being trapped by the iron. Furthermore, Podosek (1971) was able to show that the ratio $(^{40}Ar/^{36}Ar)_{trap.}$ during the *high temperature release* (1 500–1 600°C) was roughly 2. This observation put at least some doubts on the complex mechanisms proposed in order to explain the so-called "^{40}Ar excess" observed in lunar fines (Eberhardt et al. 1970; Manka and Michel 1970; Heymann and Yaniv 1970). A better explanation of the "^{40}Ar excess" would perhaps be given by considering the production rates from solar flare nucleons on targets like Ca, Sc, Ti, Cr, Mn and Fe.

To return to gas-rich chondrites, one fact is clear: they show high argon retention ages in the range 3.8–4.7, except for 4 objects out of 21, which belong to the 3 My exposure age peak of the H-chondrites. This fact is also verified for the L and LL-gas-rich objects. It is always possible to explain these high ages by compaction processes which occurred a few My ago on the parent-object. However, this "theoretically" possible assumption requires a so great number of favourable circumstances that it appears exceedingly improbable. In particular, one should always produce gas-rich xenolithic chondrites showing the same petrographic texture (Section II). This prospect seems so remote that it has to be discarded. One fact points to this remote prospect: the very low abundance of gas-rich L-chondrites (4 objects out of 176 studied for rare gas). If this low abundance has something to do with the 500 My violent outgassing event, there is then no reason why we observe such a paucity of gas-rich objects in the L-group. For, 500 My appears indeed a sufficient long time to produce gas-rich L material in a steady-state rate on the surface of the L parent-body.

Before concluding, it is worthwhile noticing the relatively long time required to produce the xenolithic chondritic breccias (4.3–4.0 Gy ago). The fact that the chondritic asteroids began their accretion $\simeq$ 4.7 Gy and ended it at $\gtrsim$3.9 Gy may perhaps be connected—as suggested by Alfvén and Arrhenius (1970)—to the low density of matter in the asteroidal region (four or five orders of magnitude smaller than that in the regions of the giant or terrestrial planets). Hence, the preceding authors expect the time scale of the asteroidal accretion to be four or five orders of magnitude larger than the accretion time scale of the planets. If the accretion time of the planets is of the order of 0.1 My, then the 800 My value obtained from the observations reported in this paper, appears to be in rather good agreement with the estimate of Alfvén and Arrhenius (1970).

V. *Concluding Remarks—New and Old Ideas*

In summary, clear evidences for superimposed irradiations have been found in several gas-rich xenolithic ordinary chondrites, both by track-method and rare gases analysis. These facts should influence our understanding of the so-called "exposure ages", at least for certain meteoritic objects.

In recent years, very accurate experiments by prominent teams of scientists have essentially obliterated the hypothesis of an early solar irradiation of the small planetesimals. However, the severe limitations on the irradiation fluxes imposed by these experiments should perhaps be subjected to a critical analysis. On the first hand, the meteoritic objects which were used to obtain these results may not represent the best choices in order to establish the "severe limitations". In particular the selected samples were often *mixtures* of materials with possibly different "pedigree" resulting from the complex mixing processes which accompanied the birth of the asteroids. Furthermore, the experimental procedures which were instrumental in establishing these limitations, may not have been totally appropriate.

Meteoritic fragments exist which show such wild irradiation evidences [a new life for the FGH model ? ! ... (1962)] that they are often discarded by the dumfounded observer (see f.i. the study of *Allegan* chondrite by Podosek 1970; see however Reed and Jovanovic 1969).

Fortunately, the well understandable anthropomorphic instinct has been conducive to the creation of such sterilizing concepts as the notion of "AVCC" for fractionated heavy rare gases, the main result of which being the erasing of all interesting features by a general averaging effect leading to a comfortable peace of mind.

Further progresses can only come from an alert hunting of the *anomalies* which will be brought to attention either by chance or by educated guessing.

Looking before and after, gave us not
That capability and god-like reason
To fust in us unused.
Hamlet, IV, 4, 37

This work made in collaboration with J. C. Lorin and G. Poupeau has profited from harsh discussions with Drs. E. Anders, G. Arrhenius, Y. Cantelaube, M. Cristophe Michel-Lévy, H. Reeves, L. Schultz, P. Signer and especially L. Wilkening.

Special thanks are due to the E.T.H. Zurich team (L. Schultz and P. Signer) which gave us permission to use their data, and J. L. Birck for isotopic dilution analysis. Finally, gratitude is due to Prof. J. Fabriès for his constant help and cooperation.

References

Alfvén, H., 1964, Icarus, *3*, 52.
Alfvén, H. and Arrhenius, G., 1970, Astrophys. Space Sci., p. 186 and 282.
Alfvén, H. and Arrhenius, G., 1971, Preprints "Arguments for a Mission to an Asteroid" and "Asteroidal Theories and Experiments".
Anders, E., 1964, Space Sci. Rev., *3*, 583.
Anders, E., 1965, Icarus, *4*, 399.
Anders, E., 1971, Ann. Rev. Astron. Astrophys, *9*, 1.
Arrhenius, G., Liang, S., Mc Dougall, D., Wilkening, L., Bhandari, N., Bhat, S., Lal, D., Rajagopalan, G., Tamhane, A. S. and Venkatavaradan, V. S., 1971, Apollo 12 Lunar Sci. Confer., unpublished report.

Barber, D. J., Hutcheon, I., and Price, P. B., 1971*a*, Science, *171*, 372.

Barber, D. J., Cowsik, R., Hutcheon, I. D., Price, P. B. and Rajan, R. S., 1971*b*, Proceedings Second Lunar Science Conference, Vol. *3*, 2705. MIT Press.

Begemann, F. and Vilcsek, E., 1965, Z. Naturforsch., *20a*, 533.

Binns, R. A., 1967, Earth and Planet. Sci. Letters, *2*, 23.

Borg, J., Dran, J. C., Durrien, L., Jouret, C. and Maurette, M., 1970, Earth and Planet. Sci. Letters, *8*, 379.

Buseck, P. R., Mason, B. and Wiik, H. B., 1966, J. Geophys. Res., *30*, 1.

Cantelaube, Y., Maurette, M. and Pellas, P., 1967, Radioactive Dating and Methods of Low Level Counting, (I.A.E.A., Vienna) p. 213.

Cantelaube, Y., Pellas, P., Nordemann, D. and Tobailem, J., 1969, Meteorite Research (ed. P. M. Millman) Dordrecht–Holland: D. Reidel Co., p. 705.

Cantelaube, Y. and Pellas, P., 1971, in preparation.

Craig, H., 1964, in Isotopic and Cosmic Chemistry (eds: H. Craig, S. L. Miller and G. J. Wasserburg). North-Holland, p. 401.

Cristophe Michel-Lévy, C., 1971, Bull. Soc. Fr. Minéral. Cristallogr., *94*, 89.

Crozaz, G. and Walker, R. M., 1971, Science, *171*, 1237.

Dodd, R. T., 1969, Geochim. Cosmochim. Acta, *33*, 161.

Dohnanyi, J. S., 1969, J. Geophys. Res., *74*, 2531.

Eberhardt, F., Eugster, O. and Geiss, J., 1965, J. Geophys. Res., *70*, 4427.

Eberhardt, P., Geiss J. and Graf, H. P., 1966, unpublished results.

Eberhardt, P., Geiss, J. and Grögler, N., 1965, J. Geophys. Res., *70*, 4375.

Eberhardt, P., Eugster, O., Geiss, J., and Marti, K., 1966, Z. Naturforsch. *21a*, 414.

Eberhardt, P., Geiss, J., Graf, H., Grögler, N., Krähenbühl, U., Schwaller, H., Schwarzmüller, J. and Stettler, A., 1970, Proceedings of Apollo 11 Lunar Science Conference, Vol. *2*, 1037.

Eugster, O., Eberhardt, P. and Geiss, J., 1969, J. Geophys. Res., *74*, 3874.

Fireman, E. L., De Felice, J. and Norton, E., 1970, Geochim. Cosmochim. Acta, *34*, 873.

Fleischer, R. L., Maurette, M., Price, P. B. and Walker, R. M., 1967, J. Geophys. Res., *72*, 331.

Fleischer, R. L., Hart, H. R., Comstock, G. M., 1971, Science, *171*, 1240.

FGH: Fowler, W. A., Greenstein, J. L. and Hoyle, F., 1962, Geophys. J., 6, 148.

Funkhouser, J., Kirsten, T., Schaeffer, O. A., 1967, Earth Planet Sci. Letters, *2*, 185.

Fredriksson, K., 1969, Meteorite Research (ed. P. M. Millman) Dordrecht-Holland: D. Reidel Co., p. 155.

Gerling, E. K. and Levskii, L. K., 1956, Dokl. Akad. Nauk. S.S.S.R., *110*, 750.

Gopalan, K. and Wetherill, G. W., 1969, J. Geophys. Res., *74*, 4349.

Hartmann, W. K., 1968, Ap. J., *152*, 337.

Heymann, D., 1965, J. Geophys. Res., *70*, 3735.

Heymann, D., 1967, Icarus, *6*, 189.

Heymann, D. and Mazor, E., 1966, J. Geophys. Res., *71*, 4695.

Heymann, D. and Mazor, E, 1968, Geochim. Cosmochim. Acta, *32*, 1.

Heymann, D. and Yaniv, A., 1970, Proceedings of Apollo 11 Lunar Science Conference, Vol. *2*, 1247.

Herzog, G. F. and Anders, E., 1971, Geochim. Cosmochim. Acta, *35*, 239.

Herzog, G. F. and Anders, E., 1971, Geochim. Cosmochim. Acta, *35*, 605.

Hintenberger, H., Schultz, L. and Weber, H., 1969, Meteorite Research (ed. P. M. Millman). Dordrecht-Holland: D, Reidel Co., p. 895.

Hohenberg, C. M., 1967, Earth and Planet. Sci. Letters, *3*, 357.

Hohenberg, C. M., Podosek, F. A. and Reynolds, J. H., 1967, Science, *156*, 202.

Hsieh, K. C. and Simpson, J A., 1970, Ap. J. (Letters), *162*, L 191.
Hunecke, J. C., Nyquist, L. E., Funk, H., Köppel, V., and Signer, P., 1969, Meteorite Research (ed. P. M. Millmann) Dordrecht-Holland: D. Reidel Co., p. 901.
Kaushal, S. K. and Wetherill, G. W., 1969, J. Geophys. Res., 74, 2717.
Keays, R. R., Ganapathy, R. and Anders, E., 1971, Geochim. Cosmochim. Acta, *35*, 337.
König, H., Keill K., Hintenberger, H., Wlozka, F. and Begemann, F., 1961, Z. Naturforsch., *16a*, 1124.
König, H., Keil, K. and Hintenberger, H., 1962, Z. Naturforsch., *17a*, 357.
Kraut, F. and Fredriksson, K., 1971, Meteoritics, *6*, 284.
Kuiper, G. P., Fujita, Y., Gehrels, T., Groeneveld, I., Kent, J., Van Biesbroech, G. and Van Houten, 1958, Astrophys. J. Suppl. Ser., 3, 289.
Lal, D. and Rajan, R. S., 1969, Nature, *223*, 269.
Lorin, J. C., Pellas, P., Schultz, L. and Signer, P., 1971, Preprint, to be submitted Earth and Planet. Sci. Letters. See also EOS 1970, *50*, (abstract).
L. S. P. E. T., 1969, Science, *165*, 1211.
Manka, R. H. and Michel, F. C., 1970, Science, *169*, 278.
Marti, K., Eberhardt, P. and Geiss, J., 1966, Z. Naturforsch., *21a*, 398.
Marti, K., 1967*a*, Earth and Planet. Sci. Letters, *2*, 193.
Marti, K., 1967*b*, Earth and Planet. Sci. Letters, *3*, 243.
Marti, et al., 1969, Meteorite Research (ed. P. M. Millman) Dordrecht-Holland: D. Reidel Co., p. 246.
Marti, K., 1969, unpublished results.
Mason, B. and Wiik, H. B., 1964, Geochim. Cosmochim. Acta, *34*, 781.
Müller, H. W. and Zöhringer, J., 1969, Meteorite Research (ed. P. M. Millman). Dordrecht-Holland: D. Reidel Co., p. 845.
Onuma, N., Clayton, R. N. and Mayeda, T. K., 1971, EOS, *52*, 270.
Osborn, T. W. and Schmitt, R. A., 1971, 34th Annual Meeting, Meteoritical Society, Tübingen, Germany, August 20–28, abstract; see also Osborn, T. W., Ph.D. Thesis, Oregon State University.
Pepin, R. O. and Signer, P., 1965, Science, *149*, 253.
Pellas, P., Poupeau, G. and Lorin, J. C., 1968, Bull. Soc. Fr. Miner. Cristal., *91*, L (abstract).
Pellas, P., Poupeau, G., Lorin, J. C., Reeves, H. and Audouze, J., 1969, Nature, *223*, 272.
Podosek, F. A., 1970, Geochim. Cosmochim. Acta, *34*, 341.
Podosek, F. A., 1971, Geochim. Cosmochim. Acta, *35*, 157.
Reed, G. W. and Jovanovic, S., 1969, J. Inorg. Nucl. Chem., *31*, 3783.
Reuter, J. H., Epstein, S. and Taylor, H. P., 1965, Geochim. Cosmochim. Acta. *29*, 481.
Rieder, R. and Wanke, H., 1969, Meteorite Research (ed. P. M. Millman). Dordrecht-Holland: D. Reidel Co., p. 75.
Sanz, H. G. and Wasserburg, G. J., 1969, Earth and Planet. Sci. Letters, *6*, 335.
Schultz, L. and Signer, P., 1970–1971, unpublished results.
Suess, H. E., Wanke, H. and Wlotzka, F., 1964, Geochim. Cosmochim. Acta, *28*, 595.
Suess, H. E. and Wanke, H., 1967, J. Geophys. Res., *72*, 3609.
Taylor, H. P., Duke, M. B., Silver, L. T. and Epstein, S., 1965, Geochim. Cosmochim. Acta, *29*, 489.
Taylor, G. J. and Heymann, D., 1969, Earth Planet. Sci. Letters, *7*, 151.
Taylor, G. J. and Heymann, D., 1971, J. Geophys. Res., *76*, 1879.
Turner, G., 1965, J. Geophys. Res., *70*, 5433.

Turner, G., 1969, Meteorite Research (ed. P. M. Millman). Dordrecht-Holland: D. Reidel Co., p. 407.
Urey, H. C. and Craig, H., 1953, Geochim. Cosmochim. Acta, *4*, 36.
Van Schmus, W. R., 1967, Geochim. Cosmochim. Acta, *31*, 2027.
Van Schmus, W. R. and Wood, J. A., 1967, Geochim. Cosmochim. Acta, *31*, 747.
Van Schmus, W. R., 1969, Earth-Sci. Rev., *5*, 145.
Wasserburg, G. J., Hunecke, J. C. and Burnett, D. S., 1969, J. Geophys. Res., *74*, 4221.
Wasserburg, G. J., Papanastassiou, D. A. and Sanz, H. G., 1969, Earth Planet. Sci. Letters, *7*, 33.
Wilkening, L., Lal, D. and Reid, A. M., 1971, Earth and Planet. Sci. Letters, *10*, 334.
Wlotzka, F., 1963, Geochim. Cosmochim. Acta, *27*, 419.
Zähringer, J., 1968, Geochim. Cosmochim. Acta, *32*, 209.

Discussion

F. L. Whipple
Do the low cosmic-ray ages of some meteorites limit their premature exposure to galactic cosmic rays by time, intensity or shielding?

P. Pellas
As discussed in detail in my paper this problem is in the first approximation undetermined; however, there are in principle possibilities to improve the situation by using additional methods.

F. L. Whipple
Can you set a limit either on the time which these particles were in space or upon the density of the galactic cosmic ray flux during the formation of the system? I guess your answer is "no"!

P. Pellas
It depends! At least for the gas-rich chondrites I am 99 per cent sure that they were irradiated on the surface of an asteroid because we have always to maintain the chemical record, the same internal differentiation, the same petrographical texture. But in other material we have exotic fragments which are interesting, because they will give us the possibility in the near future to obtain information on the transfer time from a primeval orbit to another primeval orbit in the asteroid belt, for asteroids which are not today in Earth-crossing orbits, or which might even have been destroyed in the past.

E. Anders
I agree that the formation rate of gas-rich meteorites was higher 4–4.5 Gy ago, but I do not think it has declined as steeply as you suggest You are basing your argument on the fact that many gas-rich meteorites have U–He or K–Ar ages of 4–4.5 Gy. But this does not imply that the brecciation

event which consolidated these meteorites from an asteroidal regolith happened 4–4.5 Gy ago. In general, brecciation does not outgas the meteorite; the heat release in the impact is too slight and too rapid. The elemental ratios of noble gases show that not even the solar He, Ne, Ar are lost during brecciation, and since radiogenic He^{4}, Ar^{40} are more firmly bound than the solar gases, the U–He, K–Ar clocks are not reset. A gas-rich breccia with an age of 4–4.5 Gy thus may have formed quite recently.

The only cases where brecciation seems to have been accompanied by gas loss are the L-chondrites and the shergottites. But here one finds evidence for sustained heating in contrast to the gas-rich meteorites. A likely explanation is that the parent bodies of these meteorites were disrupted by collision with another body of comparable size, so that the energy release per unit mass was unusually high. In a more normal case the kinetic energy of the projectile is spread over a 100-fold greater mass of ejecta, so that little heating results. Furthermore I think the chemical composition through the asteroid belt probably changes only gradually with semi-major axis. A more important point is that lunar breccias have formed in recent times and probably are still forming today. And most lunar breccias have not lost significant amounts of solar-wind gases, so it is possible, even on a body with a fairly high escape velocity where impact velocities are at least 2.4 km/sec, to cement grains into a breccia without outgassing.

P. Pellas

For a detailed discussion of your first question, see particularly the second paragraph from the end of Section IV in my paper. Now, concerning the lunar breccias you have to be aware of the differences between them and the ordinary chondritic breccias. In the former case you have mixture of materials of different chemical composition. Clearly that is not the case of the ordinary chondritic breccias. Moreover the He^{4}/Ne^{20} ratio is much lower in the lunar breccias.

E. Anders

In most of them it is not much different from the ratio in soil. Only in some few breccias which apparently were thermally welded together are the ratios lower.

P. Pellas

Yes, but the lunar soil itself is a mixture of multiple components. Furthermore the moon is a big object and the asteroids are small objects; so when you have an impact on the moon you observe very strong impact metamorphism features which exist also in ordinary chondritic breccias, except that they are very rare.

D. Lal

I would like to remark first than in the case of several xenolith "fragments" which you said to be irradiated, the extreme variations in track densities are of the same order as the errors in individual measurements.

Also may I ask how many of the "fragments" have high track density gradients similar to those seen in irradiated grains in Kapoeta, for example? And finally, does your work provide answers to questions such as fragmentation of borders of "xenolith fragments" in compaction in the matrix, extent of isotropy of irradiation for different faces of fragments, differences in the time of irradiation, energy spectra, etc. for different meteoric xenoliths.

P. Pellas

You will find the answer to all of these questions in my paper, particularly in the last paragraph of Section III where the discussion is more detailed than in this oral presentation. In addition, in answer to your question of ten xenoliths studied in Djermaia, four show solar flare signature. In Weston eight xenoliths were studied. One has been pre-compaction irradiated by galactic cosmic rays, and a second probably contains solar flare rare gases.

As for border effects in the grains, these are not important for our xenoliths which are as large as 0.3 to 1 cm.

G. Arrhenius

The commonly used term "asteroidal parent body" makes most people think of a rather massive, cohesive, and well defined object. Or to say that "irradiation took place at the surface" infers that there was a well defined boundary between this object and space. I think that all of the apparent discrepancies between the types of information we have heard originate from these unfortunate semantic associations.

The key to a more realistic concept may be found in Professor Kopal's comment that if you make the objects smaller and smaller the gravitation plays less and less of a role and one would have to assume a gradual transition from cohesive bodies to loose aggregates, clouds and swarms of particles smeared out over an entire orbit. The interesting experimental problem then lies in establishing the stage in this sequence where each individual irradiation took place.

A Gasdynamical View on the Motion, Heating and Accretion of Solid Bodies in the Solar System

By Hiroshi Sato

Institute of Space and Aeronautical Science, University of Tokyo, Japan

1. *Introduction*

There are a large number of stories and theories about the evolution of the solar system. Some stories are very interesting but not realistic from scientific view points. Some theories are successful in explaining some aspects of the solar system but not all of them.

After all we may not be able to find only one perfect theory for explaining the whole history of the solar system. We should be satisfied with assigning a probability of occurrence to each theory or process. This is always true when we trace the past from the present.

Here we present a gasdynamical view on the formation of planets and satellites. We try to find relations between facts in modern gasdynamics and possible evolutionary processes of the solar system.

2. *Primeval Atmosphere*

We start from the hot, gaseous Galaxy, which is composed of electrons, ions and neutral atoms of various light and heavy elements. The Galaxy is turbulent, namely, the velocity, density and temperature fluctuate randomly in space and time. Also the electric field and magnetic field fluctuate. The Galactic turbulence may be similar to the turbulence we encounter on the Earth in many respects except the effect of the gravity. Since the scale of the Galaxy is very large, the gravitational force is greater than any other forces such as inertia, pressure and electromagnetic forces. Due to the gravitational force the fluctuation level of the density is very large. In other words, many isolated high-density spots are formed. On the other hand, the temperature fluctuation results in many hot and cold spots. When the temperature of the Galaxy goes down, the condensation of the gas takes place in these cold spots. The vorticity defined by rot $\mathbf{v}$ ($\mathbf{v}$: velocity vector) is not zero in the Galaxy and each spot has angular momentum. The solar system might be formed from one of these high-density, cold spots.

The primeval solar atmosphere is also turbulent. The primeval Sun exists at the center. There are fluctuations of velocity, density, magnetic field, etc. The structure of the turbulence is determined by the gravitational, electromagnetic and gasdynamical forces. Although the relative magnitude of the gravitational force is not so great as in the Galactic turbulence, a large-amplitude density fluctuation may still exist. Because of the temperature fluctuation, cold and hot spots are also formed. The angular momentum of the whole atmosphere is not zero. This however, does not necessarily mean that the atmosphere rotates as a whole. Besides random turbulent fluctuations, there are periodical waves such as electromagnetic waves, sound and various plasmadynamic and gasdynamic waves. They propagate in various directions and interact with random fluctuations.

One of the significant contributions of the turbulent fluctuations in the atmosphere is the enhancement of the spatial transport of various quantities such as mass, momentum and energy. For instance, the forced convection of heat due to the velocity fluctuation is much larger than the molecular conduction. This is also true for the viscosity. The "turbulent viscosity" may be several orders of magnitude larger than the molecular viscosity. The conversion of energy is also enhanced by turbulence. One example is the high-rate dissipation of the kinetic energy of the gas into heat (Hinze 1959). Another example is the production of the fluctuation of the magnetic field due to the velocity fluctuation (Betchov 1963).

As the total mass of the primeval atmosphere we assume 2×10^{31} kg which is ten times the mass of the present solar system. About 10 % of the total mass belongs to the Sun and the rest is used for constructing planets and satellites. In the evolutionary process a large amount of gas escapes. This is the reason why we assume the large initial mass. As the scale of the atmosphere we take 30 AU radius and 0.1 AU thickness. These figures are quite arbitrary and we use them only for an order-of-magnitude estimate. The mean density of the atmosphere is 2×10^{-5} kg/m^3. If the average molecular weight of netural particles and ions is assumed to be 10, this density corresponds to the number density of 1×10^{21} particles/m^3. This high number density gives a basis of treating the gas as continuum. Because of the stratification due to the gravity of the Sun and the confinement by the magnetic field, the density near the center of the atmosphere is expected to be much higher than the average value. Moreover, heavy elements concentrate near the center and light elements such as helium and hydrogen are rich at the outer part.

The velocity of the gas in the atmosphere is decomposed into two parts, the time-mean velocity in a certain averaging time and the instantaneous fluctuation. The mean velocity is a function of time in a long scale. Among three spatial components of the mean velocity the radial component, U and the circumferential component, V are significant. We assume that the mean

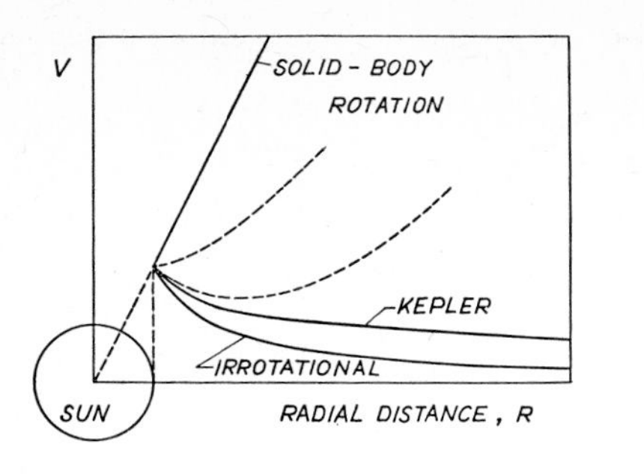

Fig. 1. Radial distributions of circumferential velocity.

motion is almost axisymmetric. Then, V is a function of the radial distance from the center, R. If the motion is irrotational (rot $\mathbf{v}=0$, everywhere), V is proportional to R^{-1}. For the solid body rotation, $V \propto R$ and for the Kepler motion, $V \propto R^{-\frac{1}{2}}$. Those relations are illustrated in Fig. 1 and the actual distribution may be expressed by broken lines. The high value of V at large R is resulted from the high turbulent viscosity. The radial component U may be small except some special periods, in which a strong blowing from the Sun is present. The amplitude of the velocity fluctuation is determined by the balance of rates of the production of turbulence energy and the dissipation into heat.

The temperature of the gas near the center is higher than that around the edge of the atmosphere. The gasdynamical heating of the atmosphere is mainly by shock waves and turbulence. The radiation loss and the local adiabatic expansion are significant cooling processes. Hot spots and cold spots are formed and they are mixed by large-scale turbulent fluctuations. The condensation of the gas takes place in cold spots and small-size grains are formed there. Those grains may be the basic material for the construction of the solar system. We assume the coexistence of gas and grains and place an emphasis on the gas-grain interaction rather than on grain–grain collisions.

3. *Force and Moment*

When the condensation takes place and a small solid body is formed, various kinds of force and moment act on the body. Among those the gravitational force and the gasdynamical force predominate. There is a force due to the electric and magnetic fields. Although this force is significant in the large-scale motion in the atmosphere, the contribution to the local flow around the body may be small. The reason is that in cold spots the recombination of ions and electrons takes place and the gas around the body may be very weakly ionized. The centrifugal force appears as a consequence of a curved orbit of the body. In the final stage of the evolution the centrifugal force is in balance with the gravity. Thus, we consider three forces, which are gravitational, gasdynamical and centrifugal forces.

In order to express the relative significance of these forces we introduce nondimensional numbers. One is the "gravity number, *Gn*", which is the ratio of the gravitational force, F_G and the gasdynamical force F_D. The former, F_G is expressed as

$$F_G = GMm/R^2,$$

in which G: the gravitational constant, M and m: masses of two bodies and R: the distance between the two bodies. The body with mass M can be a large solid body or a lump of gas. For planets, M is the mass of the Sun and R is the distance from the Sun, and for satellites both refer to a planet. The gasdynamical force is expressed as

$$F_D = C\tfrac{1}{2}\varrho U^2 A,$$

in which ϱ: the gas density, U: the relative speed of the solid body to the gas, A: the cross-sectional area of the body and C: the force coefficient. The gravity number is defined by

$$Gn = \frac{F_G}{F_D} = \frac{\theta}{\varrho}\frac{a}{U^2}\frac{GM}{R^2},$$

in which θ and a are the density and the radius of the solid body, respectively and the coefficient, C and a constant of order of one is omitted. At the early stage of accretion, a is very small, so Gn is small compared to one. For instance, with $R = 10$ AU, $\theta/\varrho = 10^8$, $a = 10^{-3}$ m, $U = 100$ m/s, Gn is 6×10^{-4}. At the final stage, Gn is much larger. Under the same conditions except the radius, $a = 10^5$ m, Gn is 6×10^4. Thus, Gn is an important parameter in the accretion process.

Concerning the gasdynamical force the Knudsen number, Kn is significant. It is the ratio of the mean-free-path of gas particles, λ, and the size of the solid body, that is, $Kn = \lambda/a$. If Kn is small, the gas is considered to be continuum. On the other hand, if $Kn \gg 1$, the gas should be treated as discrete particles. It must be emphasized here that the continuity of the gas is distinguished by the ratio, Kn rather than by the mean-free-path itself. At the early stage of the accretion a is small and Kn is large. As a increases, Kn becomes smaller. At the final stage, the gas density is very low and λ is large. Then again Kn becomes large.

The whole accretion process can be illustrated in Gn–Kn diagram as shown in Fig. 2. The process starts with $Gn \ll 1$, $Kn \gg 1$, (Region I) and ends with $Gn \gg 1$, $Kn \gg 1$ (Region IV). We notice other two regions, Region II in

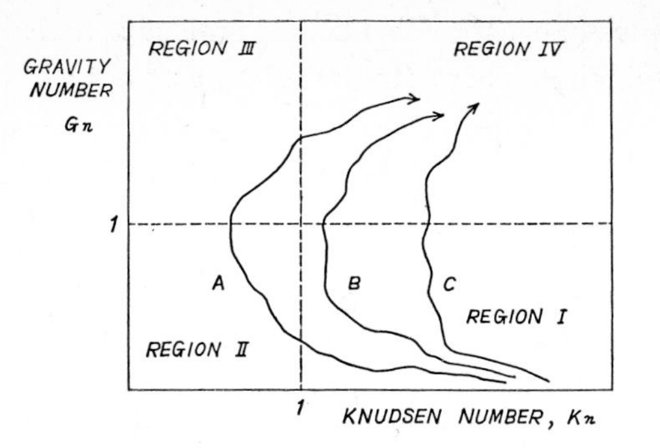

Fig. 2. Accretion process expressed by Gravity Number and Knudsen Number.

which $Gn \ll 1$, $Kn \ll 1$, and Region III in which $Gn \gg 1$, $Kn \ll 1$. In Region I we use the rarefied gas dynamics for estimating forces on the body. Conventional continuum fluiddynamics is applicable in Region II. In Region III the gravity term must be added into the equation of motion. We may call it the gravitational fluiddynamics (GFD). In Region IV the orbit mechanics is applicable. Possible accretion processes are expressed by solid lines, A, B and C. It is not yet clear which path is most realistic. Path A is characterized by the dense atmosphere for a long period of time. Path C denotes that the whole process occurred in teneous gas. These paths can be drawn not only for planets but also for satellites. We may be able to select the most probable path by detailed investigations of dynamical and chemical processes. We consider all four regions and point out some features relevant to each region in Section 5.

The basic scheme of the flow around a spinning solid sphere is shown in Fig. 3. The force acting on the sphere is separated into the drag force parallel to the flow, the side force normal to the flow and the moment around the center of gravity. When $Gn \ll 1$, the relative speed between the body and the gas, U, originates from the turbulence, and the drag force always reduces the relative speed. Therefore, U may not be too large and the Mach number —the ratio of the relative velocity and the sound velocity—will not exceed

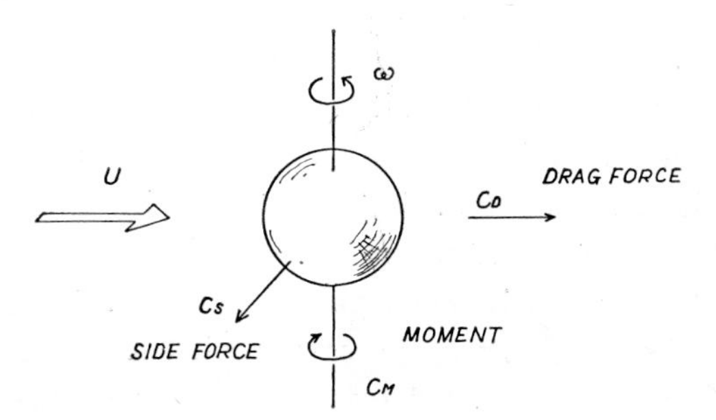

Fig. 3. Forces acting on a spinning sphere.

one. Thus we consider an incompressible subsonic flow around a sphere. The drag and side forces on the spinning sphere are expressed as

$$\text{drag force} = C_D \tfrac{1}{2}\varrho U^2 \pi a^2,$$
$$\text{side force} = C_S \tfrac{1}{2}\varrho U^2 \pi a^2,$$

and the moment around the center of gravity is $C_M \tfrac{1}{2}\varrho a^5 \omega^2$

in which ω is the angular velocity of the body. The drag coefficient of a non-spinning sphere is a function of Reynolds number Re $(=2aU\varrho/\mu$, μ: viscosity) and values are found in text-books. The side force on a spinning sphere (Magnus effect) is either in the direction, $\mathbf{v}\times\boldsymbol{\omega}$ ($\mathbf{v}$, $\boldsymbol{\omega}$ are vectors of flow velocity and spin, respectively) or reverse depending on Re and $a\omega/U$. An example of coefficients for a sphere with rough surface is $C_D=0.3$, $C_S=0.25$ at $Re=10^5$ and $a\omega/U=0.24$ (Maeda, Okazaki, Kondo, Kuzuoka and Imahori 1969) On the value of C_M we do not have enough information.

The drag force in the circumferential flow of gas around the Sun contributes to the orbiting motion of planets. If there is a radial flow, the acceleration due to the side force is circumferential. Therefore, the side force serves to the establishment of the orbiting motion. The same force acts on satellites around planets in a smaller scale. Establishments of spin and orbiting motion of planets and satellites are closely related.

4. *Heat Exchange*

In the atmosphere there are various forms of energy, such as kinetic energy, electromagnetic energy and heat. There are various kinds of energy conversion processes such as particle acceleration and thermalization. If the Sun exists, the solar radiation is a significant heat input to the inner part of the atmosphere. On the other hand, at the boundary of the atmosphere the radiation loss may exceed the energy supply from inside and the temperature goes down. Due to the turbulence, the cold part is mixed with the hot part and cold and hot spots are formed.

If we record the temperature of a small lump of gas, it may change as shown in Fig. 4. The temperature fluctuates and the average value gradually decreases as shown by a broken line. At point A the temperature is equal to the melting point. But at point B the temperature exceeds the melting point again and the evaporation takes place between points B and C. The temperature goes down again and the condensation process continues.

Solid bodies formed in the atmosphere are heated by the solar radiation,

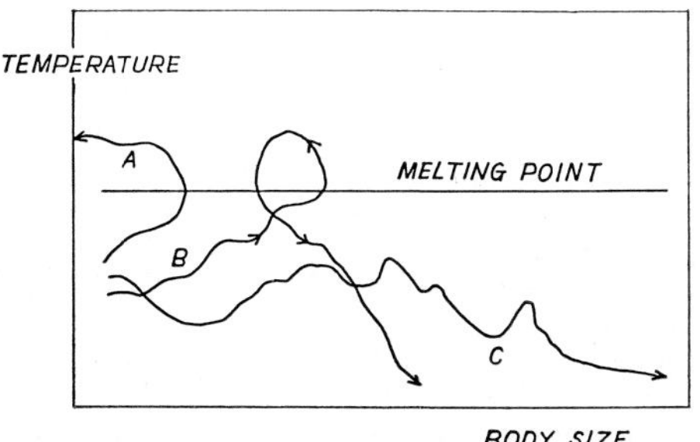

Fig. 4. Temperature variation of a small lump of gas.

by the latent heat of condensation and by the kinetic energy of colliding small grains. The heat is lost by the radiation from the surface and by the forced convection due to the gas flow around the body. The temperature of the body is determined by the balance of those inputs and outputs of heat. Only bodies which experience more cooling than heating can grow. Since the turbulence is a random process, the accretion of the body takes place in a statistical manner. Fig. 5 illustrates three typical temperature histories in the accretion process. The abscissa is the size of a solid body and the ordinate is its temperature. The path A shows a case in which temperature goes up with time. When the temperature exceeds the melting point, the evaporation starts and the body size is reduced. If the temperature stays above the melting point, the body finally disappears. If the temperature—size history is like path B, the body grows inspite of a short evaporation period. The path C indicates a favorite temperature history for the formation of a large planet.

5. *Accretion and Spin*

As an example of the gasdynamical treatment of the accretion process the increase of the angular momentum due to the condensation is considered. The fundamental equation for the change of angular momentum of a solid body in the gas is

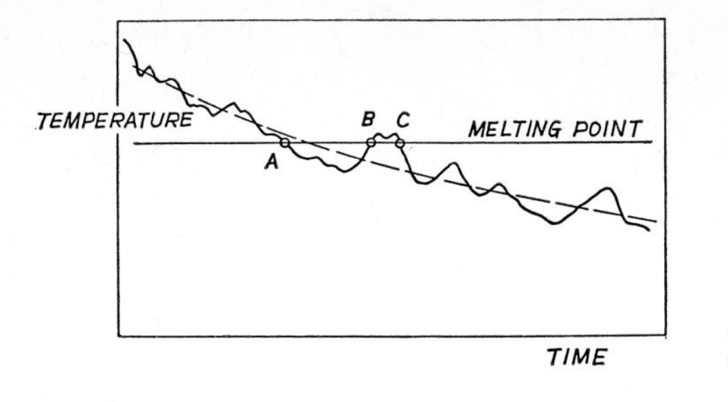

Fig. 5. Temperature history of a solid body in solar atmosphere.

$$\frac{dJ_s}{dt}=\frac{dJ_a}{dt}+M,$$

in which J_s is the angular momentum, dJ_a/dt is the increase of the angular momentum due to the condensation and M is the gasdynamical moment acting on the body. The angular momentum of a solid sphere with homogeneous density is

$$J_s=\frac{2}{5}ma^2\omega=\frac{8}{15}\pi\theta a^5\omega,$$

in which θ is the density, and a and ω are the radius and the angular velocity, respectively and both are functions of time. For the non-gravitational, collision-free accretion ($Gn \ll 1$, $Kn \gg 1$, Region I in Fig. 2) atoms hit the sphere with the thermal velocity, v_{th}. If all incident particles are spherically symmetrical and condense on the surface, $M=0$. Denoting the angular velocity of the gas as ω_g, changes of mass and angular momentum are

$$\frac{dm}{dt}=\frac{d}{dt}\left(\frac{4}{3}\pi a^3\theta\right)=\pi a^2\varrho v_{th},$$

$$\frac{dJ_s}{dt}=\frac{dJ_a}{dt}=\frac{\pi}{3}\omega_g\varrho v_{th}a^4.$$

From these equations we have

$$a=\frac{v_{th}}{4}\frac{\varrho}{\theta}t+a_0,$$

$$a\frac{d\omega}{dt}+5\omega\frac{da}{dt}=\frac{5}{8}\frac{\varrho}{\theta}\omega_g v_{th},$$

in which a_0 is the radius at $t=0$.
Substituting a the change of ω is expresssed as

$$\frac{d\omega}{dt}+\frac{5}{4}\frac{\varrho}{\theta}\omega v_{th}=\frac{5}{8}\frac{\varrho}{\theta}\omega_g v_{th}.$$

The solution is

$$\omega=\frac{1}{2}\omega_g+\left(\omega_0-\frac{1}{2}\omega_g\right)\exp\left(-\frac{5}{4}\frac{v_{th}}{a_0}\frac{\varrho}{\theta}t\right),$$

in which ω_0 is the initial value of ω. This solution indicates that ω starts from ω_0 and approaches to $\frac{1}{2}\omega_g$ at large t.

In the case of non-gravitational, collision-dominant accretion ($Gn \ll 1$, $Kn \ll 1$, Region II), we consider a sphere placed in the incompressible shear

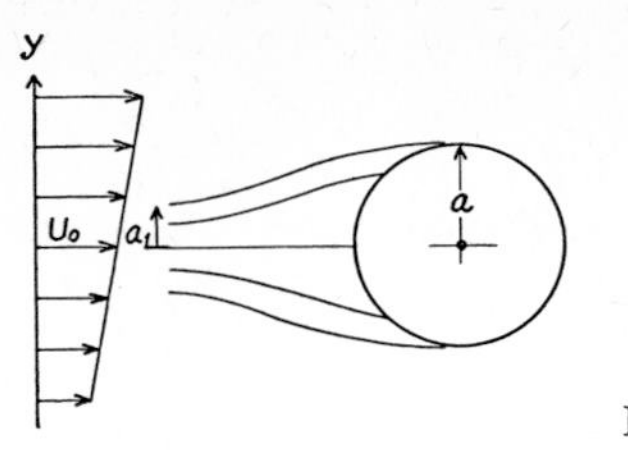

Fig. 6. A sphere in shear flow.

flow as shown in Fig. 6. The velocity distribution of the approaching flow is expressed as

$$U = U_0 + \omega_g y.$$

Concerning the condensation we assume that the gas inside radius a_1 condenses on the surface of a sphere and we define the accretion coefficient $\eta_a = a_1^2/a^2$. Then the mass flux to the sphere is $\varrho U_0 \pi a^2 \eta_a$, U_0 being the flow velocity on the center-line of the sphere. Obviously, η_a is determined by the condensation rate on the surface which depends on the surface conditions such as temperature, roughness and surface chemical reactions. Streamlines in front of the sphere with various values of η_a are sketched in Fig. 7. These are for a stationary sphere. If the sphere is spinning, streamlines are modified. Due to the condensation, streamlines inside radius a_1 end on the surface. Streamlines behind the sphere can not be drawn precisely because the flow is turbulent there. The value of η_a may exceed unity if the condensation capability is very high.

The rate of increase of mass of the sphere, dm/dt is given by

$$\frac{dm}{dt} = 4\pi a^2 \theta \frac{da}{dt} = \varrho U_0 \pi a^2 \eta_a,$$

from which

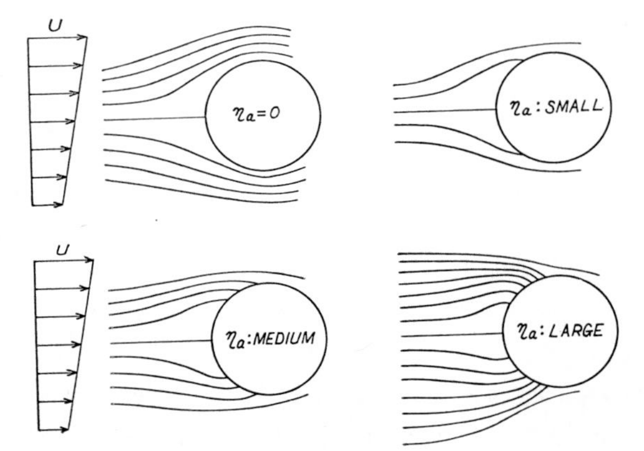

Fig. 7. Streamlines around a stationary sphere in shear flow with various values of accretion efficiency, η_a. $Gn \ll 1$, $Kn \gg 1$.

$$a = a_0 + \frac{U_0 \varrho \eta_a}{4\theta} t.$$

This means that the radius increases linearly with time. The increase of the angular momentum due to accretion is

$$\frac{dJ_a}{dt} = 2 \int_{-a_1}^{a_1} \varrho (U_0 + \omega_g r)^2 \sqrt{a_1^2 - r^2}\, r dr$$

$$= \frac{\pi}{2} \varrho U_0 \omega_g a^4 \eta_a^2.$$

The moment due to the shear flow is

$$M = C_M \frac{\varrho}{2} a^5 (\omega_g - \omega)^2.$$

The balance of the angular momentum is given by

$$\frac{8}{15} \pi \theta a^4 \left(a \frac{d\omega}{dt} + 5\omega \frac{da}{dt} \right)$$

$$= \frac{\pi}{2} \varrho U_0 \omega_g a^4 \eta_a^2 + C_M \frac{\varrho}{2} a^5 (\omega_g - \omega)^2.$$

We substitute a to this equation and non-dimensionalize ω and t by the relation

$$\omega = \omega_g \omega^*, \quad t = t^*/\omega_g,$$

then,

$$\frac{d\omega^*}{dt^*} = A\omega^{*2} - B(t^*)\omega^* + C(t^*),$$

in which

$$A = \frac{15}{16} \frac{C_M}{\pi} \frac{\varrho}{\theta},$$

$$B = 2A + \frac{5}{s + t^*},$$

$$C = A + \frac{15}{4} \eta_a \frac{1}{s + t^*},$$

$$s = \frac{4 a_0 \omega_g \theta}{U_0} \frac{1}{\varrho \eta_a}.$$

No analytical solution for this differential equation can be obtained but we may consider the nature of the solution. If $d\omega^*/dt^*$ is plotted against ω^*, we

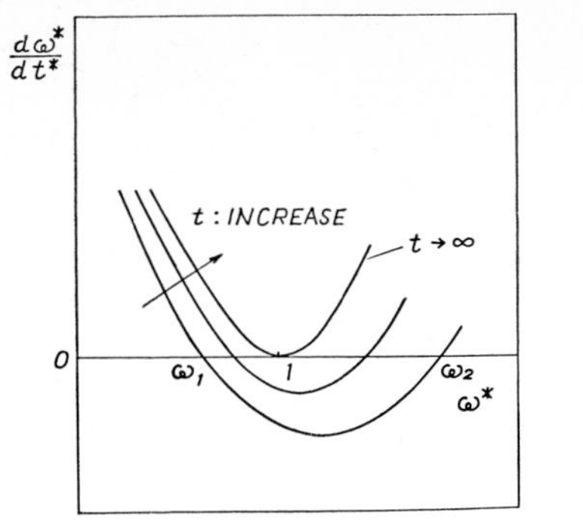

Fig. 8. Angular velocity, ω^* and angular acceleration, $d\omega^*/dt$. Gn$\ll$1, Kn$\ll$1.

have curves as shown in Fig. 8. The curve shifts as indicated by an arrow with time. The increase of ω is possible for $\omega^* < \omega_1$, or $\omega^* > \omega_2$. Since the accretion process starts from small ω, the flow accelerates the spin if $\omega^* < \omega_1$, but decelerates between ω_1 and ω_2. The value of ω_1 is obtained as

$$\omega_1 = 1 + \frac{5}{2} C_1 \left[1 - \sqrt{1 + \frac{4}{25 C_1}(5 - C_2)} \right],$$

in which

$$C_1 = \frac{4}{15} \frac{\pi}{C_M} \eta_a \frac{U_0}{a_0 \omega_g} \frac{1}{1 + t^*/s},$$

$$C_2 = \frac{15}{4\pi} \eta_a.$$

Small t corresponds to large C_1 and at $t \to \infty$, ω_1 becomes unity. Since the acceleration of spin is possible only when $\omega^* < \omega_1$, ω_1 gives an upper bound of ω^*. A numerical example is, by taking $a = 0.5$, $C_M = 0.01$, $U_0/a_0 \omega_g = 10$ and $t = 0$, $\omega_1 = 0.12$. This means that the acceleration takes place for the initial nondimensional angular velocity $\omega_0^* < 0.12$. If ω_0^* is larger than 0.12, ω^* decreases at first and then, due to the increase of ω_1, ω^* increases up to unity. If ω^* is larger than 1, it decreases and approaches unity. In other words, the final angular velocity is $\partial U/\partial y$ and the further increase of spin angular momentum due to the accretion in Region II is accomplished only by the increase of mass rather than the increase of the angular velocity.

As the solid body grows by the accretion the gravitational force on the gas around the body becomes large. The flow field is modified by the gravity. Because of the large size of the body, *Kn* may still remain small. Thus the condition, $Gn \gg 1$, $Kn \ll 1$ (Region III in Fig. 2) is realized. At the same time the body has attained the orbital speed so the relative speed with the gas may be high subsonic or even supersonic. The gravitational force will result

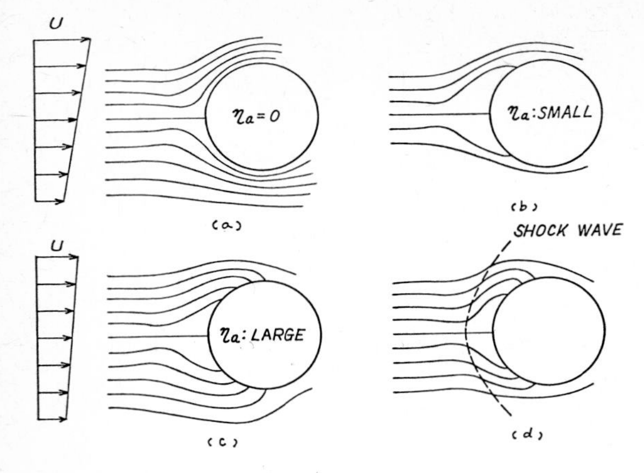

Fig. 9. Streamlines around a stationary sphere in shear flow. $Gn \gg 1$. $Kn \ll 1$.

in an increase in the accretion efficiency η_a but the overall picture of the accretion process in this region may not be too much different from that in Region II. Fig. 9 illustrates possible streamlines around a stationary sphere placed in a shear flow for values of η_a, zero (*a*), small (*b*) and large (*c*). The difference from those in Fig. 7 is the acceleration toward the center by the gravity. The Bernouilli equation for the compressible, adiabatic flow in the gravitational field is written as

$$\frac{1}{2}v^2 + C_p T + \Phi = \frac{1}{2}v_0^2 + C_p T_0 + \Phi_0,$$

$$\Phi = -\frac{Gm}{a},$$

in which Φ is the gravitational potential and the subscript $_0$ denotes the condition in the gas close to the surface of the body. The continuity equation is

$$\varrho v A = \varrho_0 v_0 A_0$$

in which A is the area of a small stream tube. Denoting values at a position far enough from the body by subscript 1, the condition of being supersonic near the body is

$$v_0^2 = v_1^2 + v_{es}^2 + 2C_p(T - T_0) > C_0^2 = \gamma RT,$$

in which v_{es} is the escape velocity defined by $\sqrt{2Gm/a}$. Due to the v_{es}-term, a subsonic flow can become supersonic. Around a sphere in a supersonic flow, we expect a shock wave as shown in Fig. 9 (*d*). Streamlines change the direction on the shock wave and the heat transfer from the gas to the body is enhanced. As a result the condensation rate decreases and for high supersonic flow, the ablation (evaporation) will result. As we have stated earlier the accretion is a statistical process. Condensation and evaporation might be repeated and only bodies which happen to be in a favorite condition can grow into planets or satellites.

In Region IV ($Gn \gg 1$, $Kn \gg 1$), the gas density is very low and motions of planets or satellites are almost established. Therefore, the gasdynamical force and moment are only small perturbations. The relative flow speed might be supersonic but no shock waves due to neutral particles will be formed because the mean-free-path is large. Only plasmadynamic shock wave may be possible as observed at present around the Earth.

6. *Concluding Remarks*

We started from the gaseous primeval solar atmosphere and considered the evolutionary process with an emphasis on the gas—grain interaction. The importance of turbulence was pointed out and the gasdynamical force and the accretion rate were estimated. Some of the gasdynamical processes might really have taken place in the history of the solar system.

References

Betchov, R., 1963, J. Fluid Mech. Vol. 17, Part 1, 33.
Hinze, J. O., 1959, Turbulence, McGraw-Hill, New York.
Maeda, H., Okazaki, S., Kondo, Y., Kuzuoka, T. and Imahori, J., 1969, Proc. 19th Japan NCTAM, 167.

Discussion

H. Alfvén

As a general remark about the strategy of approach to the problem of the origin and evolution of the solar system, the uncertainty increases rapidly when we go backward in time.

We can approach the problem in fundamentally two different ways—one, which is the classical way of Laplace and essentially the approach of Dr Sato. It starts from an assumption of an early "primeval" state. The assumption is necessarily extremely uncertain, because what we know about the primeval dynamic state of the Galaxy, and our region in the Galaxy, 4 or 5 billion years ago is next to nothing. Hence we start from uncertainty, and at the development of the theory the uncertainty grows still more.

The other approach is to proceed backwards in time. We start from what we know about the solar system of to-day and try to extrapolate further and further backwards. Space research nowadays decreases the uncertainty in our knowledge of the present state and our strategy should be to decrease it further and further backwards.

Another point is that we should not aim only at a theory of the formation of planets—as is usually done—but of both planets and satellites. The planetary system and the satellite systems are so similar that it is difficult to believe that they have not originated by essentially the same processes. We have only one planetary system but at least three well developed satellite systems: the Jovian, Saturnian and Uranian systems. This means that we can check a theory on all these four cases, and by numerical majority the emphasies lies on the satellite systems.

Hence for the evolution of the solar system it is essential that we concentrate our attention to the satellite systems. For example—from our point of view—the main interest of the Grand Tour is if we could get a thorough investigation of the satellite systems.

Especially the last five years of magnetospheric research have given us so much information about how the plasmas behave in space. It is generally recognized that these properties have very little to do with all the earlier speculations.

As Sir Harrie has stated, man is not ingenious enough to derive theoretically how nature works. We must have access to experiments. A revolution in cosmical plasma physics is coming now because of the information we get now, how a plasma really behaves.

Concerning Dr Sato's speculation about the rotation of the celestial bodies he ought to study the detailed quantitative theory by Tom Giuli (1968) about the rotation of planets which is a good approach to the understanding why planets rotate.

Giuli, R. T., 1968, Icarus *8*, 301.

H. Sato

If we can proceed backwards in time, it may be the best way but the present state of the solar system is entirely different from the gaseous state which I assume to be original. Because of the big gap between the past and the present, I think it can be justified to assume the original state and investigate what happens later. The work by Giuli is concerned with the process in Region IV, which I have not covered in my paper.

P. Pellas

In a general way I agree with Professor Alfvén's statement, however, with some important restrictions. In fact there is a lot of information on the period preceding the formation of the solar system. The information comes from monitors of nuclear processes like elemental and isotopic ratios of lithium, beryllium and boron, as well as from radioactive nuclides having mean-lives in the range of 2.5×10^7 years (^{129}I) up to 2×10^{10} years (^{232}Th).

Z. Kopal

I should like to make a comment on some of the conclusions which Professor Alfvén and Professor Pellas have drawn. I agree with them in general, but should draw the "curve of uncertainty" even more steeply in the past. We simply do not know what the solar system looked like at the time of its origin. The present properties of the system are exptrapolable by the laws of celestial mechanics to time-intervals of the order of 400–500 million years in the past, but not for a time ten times as long which separates us from its origin. In particular, we cannot extract from the present data on planetary orbits anything valid about the scale of the solar system at the time of its origin. We do not know how far the Earth (or Jupiter) may have been from the Sun at the time of its formation.

The approximate constancy of the terrestrial paleo-temperatures for the past 3 000 million years cannot give us any assurance that our distance from the Sun has not changed appreciably in the course of time. The global paleo-temperatures are the resultants of at least three independent factors (greenhouse effect of the terrestrial atmosphere, distance from the Sun, and the solar energy output) which are all apt to vary with time; and since the last one has indubitably changed in the course of the solar evolution on the Main Sequence (the zero-age Sun was probably by some 40 % less luminous than it is at the present time), either one (or both) of the two others must have changed as well in order to keep the global paleo-temperature approximately constant.

My second comment concerns the influence of the total mass of the system. This morning Professor Pellas pointed out that what happens in ε Aurigae may not give much of a clue to what happened to our Sun, because of a large disparity in mass of the two bodies: the combined mass of ε Aurigae being about 50–60 times as large as that of the Sun. In passing over from the solar planetary system to that of the Jovian satellites, however, we cross a mass-ratio of close to 1 000. This may not matter critically for the outcome, but we also possess no assurance that it does not.

B. Lehnert

I agree with Alfvén that plasma physics is likely to enter the problems discussed in this paper. You have only treated the state of neutral gas and not that of a plasma. Still I think that many of your general conclusions and points can be "translated" into plasma physics and have important applications there. This applies in particular to the influence of turbulence (and instabilities) on the state of the fluid, and to the dynamics of the grains which then has to include electric phenomena and should be based on magneto-fluid and plasma dynamics.

F. L. Whipple

I do not know whether the coordinate in Alfvén's diagram is logarithmic or linear, so I cannot comment on it. While we are discussing the philosophy of our problem I must point out that we do not know that our physical laws or their constants have changed in 4.6×10^9 years. We do know that pleochroic halos of α-particles in mica made today are the same as they were $2\text{–}3 \times 10^9$ years ago. The constant of gravity may have changed significantly. The spectral lines have maintained their relative positions for longer than 4.6×10^9 years, but perhaps they have shifted proportionally to wavelength. All in all we must look for possible changes in physical laws.

S. K. Runcorn

The philosophical method which will prove to be useful in this field is not, I think, as clear, as Professor Alfvén argues. Professor Alfvén seems to be sceptical about the value of starting with assumptions about the initial state of matter from which the development of the solar system took place and by contrast emphasizes the importance of drawing inferences about the past by studies of the fully developed bodies of the solar system. However, although the latter may seem both more accessible than the primitive material 5×10^9 years ago, it may be possible that we can use physical laws with more confidence in the early history than in the present. For the physics of the solid state of matter is not yet well developed: the physics of gases, or even dust clouds, is much simpler. Indeed this is the reason why understanding of the internal constitution of stars developed so successfully when we are still only beginning to understand the physical processes in the earth's interior and our knowledge of the internal constitution of the planets is quite negligible. It does not seem to me unlikely that we can understand the physics of the processes which formed the solar system at such a remote time, just as we can understand in general terms the dynamics of remote galaxies. Once matter condenses into large cold objects short range forces become the controlling influence in their physical properties and the "predictive" power of general physical theories sharply decreases. The principles on which we infer the original conditions from the present state of the solar system are by no means secure.

H. Alfvén

The discussion reminds me of an old story from a symposium, when the speaker talked about the interior structure of stars and said: "In principle the structure of a star is extremely simple". A voice from the audience: "If looked upon from a distance of a few light-years even you, Mr Speaker, would look very simple."

E. Anders

In response to professor Kopal's remark, I think where celestial mechanics fails, paleo-climatology helps us out. For if the Earth's semi-major axis had changed appreciably in the past, this would have had some very drastic effects on the Earth's climate. If it had been a little bit larger than at present, there would have been very massive, major glaciations, which in turn increase the albedo and bring the temperature down even further. The fossil record does not show such glaciations. If the semi-major axis had become smaller, by as little as 6–10 million kilometers, the Earth would have become a runaway greenhouse like Venus, and again this would have had very drastic effects on the fossil record, let alone the survival of life (Rasool and de Bergh 1970).

Rasool, S. I. and de Bergh, C., 1970, Nature, *226*, 1037.

Z. Kopal

This remark cannot pass without an answer. There are several other independent variables which bear on the matters which you mention. The climate of the Earth can change with changing semi-major axis of the orbit around the Sun, but we also know for sure that the energy output of the Sun is changing secularly. The Sun could not have been shining with constant luminosity in the past few thousand million years. In fact, astronomers are certain that four thousand million years ago—when the Sun was a nearly zero-age Main Sequence star—its luminosity was about 40–45 % smaller than it is now. At our present distance from the Sun, and with the present composition of our atmosphere, this would expose the Earth to a global glaciation of which there is no evidence in the geological strate going back 3.5 billion years or even more. So the only way to reconcile an absence of such a glaciation with the relatively low luminosity of the zero-age Sun is to assume that either the youthful Earth possessed a more efficient way of storing heat through its atmospheric greenhouse effect (greater contents of CO_2?); or that we were, at that time, appreciably nearer to the Sun intercepting a greater amount of its feebler radiation; or a combination of both sufficient to keep the mean temperature of the terrestrial surface above the freezing point of water—at least in the past 3.5 billion years—when the preserved sedimentary strata attest to the existence of liquid water on the Earth.

On the Conditions for Cosmic Grain Formation

By B. Lehnert

Royal Institute of Technology, Stockholm, Sweden

Abstract[1]

The balance of matter and heat is investigated for cosmic grains being exposed to radiation and being surrounded by a plasma or a neutral gas. The grains are able to increase their mass only within certain ranges of density and temperature of the surrounding gas, of the radiation intensity, and of the physical properties of matter. The results are summarized as follows:

(I) Grains consisting of different elements or chemical compounds differ strongly in their behaviour.

(II) Grains of certain elements or compounds should be absent in interplanetary matter within certain regions of the solar system.

(III) The conditions obtained put certain limits on the density, temperature, and the degrees of dissociation and ionization of the gas out of which grains can condense to form the precursors of planets during the formation of the solar system. Corresponding limits are also placed on the intensity of solar radiation. This provides a method of estimating the solar radiation and the gas densities during the development of the solar system.

(IV) The conditions deduced for grain balance should in principle apply to any gas composition. However, in detailed calculations of grain condensation of chemical compounds the equivalent evaporation rates, sticking probabilities and other pertinent physical data have to be determined. This requires further investigations beyond the scope of this paper which only gives a few simplified examples to illustrate the mechanism of grain balance.

Discussion

P. Pellas

In your abstract you write: corresponding limits are also placed on the intensity of solar radiation. Can you give us some information on this assumption, because it is very important, if actually checked by results.

[1] This paper has been published as B. Lehnert, 1970, Cosmic Electrodynamics *1*, 218.

B. Lehnert
What I mean by this statement is that, if you vary the intensity of the solar radiation, the critical boundaries will move in n–T space as well. Thus, there are some constraints put on the condensation process by the radiation intensity which could, in their turn, be used to estimate the solar radiation intensity in some cases.

P. Pellas
I understand. Now, let us suppose that the yong Sun was a very active star which passed at a given time through a T-Tauri phase. There are indications that some fractions of "primitive" meteoritic material have been irradiated by strong neutron fluxes. For instance, recently Fireman et al. (1970, see the reference in my paper) have found in the Allende carbonaceous chondrite that different types of chondrules present large *correlated* variations of $^{128}Xe/^{132}Xe$ and $^{129}Xe/^{132}Xe$ ratios. A very plausible mechanism to explain this type of correlation is neutron capture on Iodine. To produce ^{128}Xe from ^{127}I by neutron capture is no problem; however, to obtain ^{129}Xe from ^{129}I implies that the latter isotope was not extinct when neutron irradiation occurred. Now, the chondrules in Allende show large grain-sizes from about 0.1 up to 0.8 cm (dia.): in other words, they are not "grains" but already "objects". Thus far we cannot reject the possibility that the chondrules are secondary objects resulting from disruption of pre-existing materials as suggested many years ago by Urey. Conversely, chondrules could well be primitive objects having been subjected to neutron fluxes produced by particle irradiation from a very active Sun. In such a case, could your constraints define *where* (i.e. at what distance from the central body) the chondrules have been formed? It should be of great interest to check that by means of your theoretical constraints since already there are some data, and there will be more within the next few years.

B. Lehnert
I first wish to point out that this theory has been developed as a tool for investigation of the condensation process in general terms. Thus, I cannot tell you for the moment how detailed calculations would come out for the specific cases of the isotopic ratios you mention. At the same time it should be pointed out that such isotopic ratios may be the result of several different processes operating at various periods of the development of cosmic matter. One process is that due to neutron radiation which you have just mentioned. However, there may also exist other mechanisms for the production of anomalies of isotope abundances, in local regions of the gas itself, from which the grains are formed. Thus, there is one possible pro-

cess of isotope separation produced by the centrifugal force in a rotating plasma as suggested by Bonnevier, and another mechanism in the transition region between a magnetized plasma and a neutral gas as suggested by myself. This is just to mention a few additional possibilities.

However there is of course another point too, which I have not mentioned here. You might have one process which produces the grains originally. Then, if the grains are allowed to exist for 5 million years under certain other conditions, they might become modified by radiation. We have to think of the time-scales of the different processes involved. It is certainly very difficult to say how all these come into play.

D. Lal
The xenon isotopic changes probably occurred after the grain formation and accretion processes discussed.

P. Pellas
The processes were probably simultaneous. The chondrules in Allende are closed systems since 4.4–4.5 Gy (bulk K-Ar ages of Fireman et al. 1970, whereas Rb-Sr data converge to 4.6 Gy). In fact we do not know the processes which took place during a period of roughly 10^8 years at the beginning, except that this period may well be of the order of 10^7 years or even less.

S. K. Runcorn
How about the time-scale?

B. Lehnert
That is a good point. I have not figured it out explicitely. But I think it could be done, and Professor Arrhenius has made some estimates of such a transient process. You could extend this analysis also to a time-dependent case. I have so far only looked into marginal steady states.

A Computer-Generated Motion Picture Showing a Rendezvous Mission with Comet Encke[1]

By H. E. Newell

U.S. National Aeronautics and Space Administration, Washington D.C.

Perhaps before we show the motion picture I should say a few words about it. Roger Bourke and his colleagues at the Jet Propulsion Laboratory have been working intensively on the calculation of possible trajectories for a number of missions and have succeeded in producing some computer-generated movies showing some of the characteristics of the more interesting missions. On Thursday morning I intend to show in connection with my paper some of the trajectory calculations corresponding to the Grand Tour missions. It had occurred to me that Dr Whipple or perhaps Tom Gehrels might have wanted to show this film in connection with the discussions tomorrow, but more time appears available today.

The film shows a rendezvous mission with comet Encke, where by rendezvous we mean a mission in which the spacecraft in arriving at the comet in effect becomes a companion with the comet and moves along with it. There are several types of missions that one can talk about; fast fly-bys in which the spacecraft moves at many km/sec by the object and slow fly-bys in which the spacecraft moves at many tens or hundreds of meters per second by the object, and then rendezvous in which the spacecraft moves at perhaps a few meters per second by the object. This one is a rendezvous mission and corresponds to the launching of the spacecraft in March of 1978 to arrive in the vicinity of Encke in 1980 some 20 days before perihelion of Encke. The mission can be carried out by using the Titan 3D Centaur, — which is a combination of the Titan vehicle with the Centaur, — both of which exist and are being put together right now for the Grand Tour mission, and using solar electric power for propulsion after the initial launching from earth.

Solar electric propulsion does not exist at present and will have to be developed in order to make this kind of mission possible. But once that capability is developed, not only the rendezvous with Encke but also with other comets or with asteroids will be possible. The specific impulse assumed for the propulsion system is 3 000 seconds; and the power level for propulsion, 16 KW corresponding to one astronomical unit from the Sun, the power varying inversely with the square of the distance from the Sun.

[1] Introduction to the NASA film on comet Encke.

As we begin the film the viewer is assumed to be 12 AU above the ecliptic plane and viewing the 4 innermost planets.

Discussion

B. A. Lindblad
What is the nearest approach distance that is planned? I presume that you do not dare go too near the comet.

H. E. Newell
I do not have the data here.

H. Alfvén
Comets are usually assoicated with meteor streams. How risky then is a rendezvous with a comet?

F. L. Whipple
I have calculated the meteor hazard for comet d'Arrest, somewhat like comet Encke. In a fly-by at 10 km/sec the hazard to the spacecraft becomes serious between 10 000 and 1 000 km miss distance. The ejection velocity is only meters per second for meteoroids so that a slow enough approach would not be dangerous. I note, as a point of interest, that the mass cannot be measured gravitationally because ejected gases acting on the spacecraft will exert more force than the gravity of the nucleus. A gravity-gradient device will be necessary.

T. Gehrels
Would it be safer to go before the comet rather than behind it?

F. L. Whipple
Oddly enough, as Brian Marsden has shown, comet Encke is unique in that it shows no net repulsion from the sun because of non-gravitational forces caused by jet action. Thus I cannot predict which direction of approach is safest. Perhaps more detailed studies can answer the question but I doubt it. For almost all comets the sunward side would be most dangerous at close quarters.

P. M. Millman
Perturbations of the meteoroids over a long time scale are also important. Often these cannot be predicted without detailed computer calculations. They may place meteoroids either ahead of or behind the comet.

Chemical Effects in Plasma Condensation

By Gustaf Arrhenius

Scripps Institution of Oceanography and Institute for Pure and Applied Physical Sciences
University of California, San Diego, La Jolla, California

Limiting Conditions

Few facts are known about the formation of the solar system. However, it is generally conceded that the material, now gathered into planets and satellites, originally was emplaced in the form of a gas from which the solids condensed.

This seems to be not only a possible, but a necessary assumption, since we know no other path which by established physical processes leads to the present dynamic state of the solar system.

The reconstruction of the conditions under which condensation took place must take into account not only the chemical properties of the limited samples that we have of the condensed material. It must also explain, or at least not conflict with the fundamental dynamic features of the solar system in order to have a claim to being realistic.

Among the physical facts which severely limit the possible conditions for emplacement and condensation (Alfvén and Arrhenius 1970*a*) are foremost:

Analogous Structure of Planetary and Satellite Systems

We have not only one but four systems where a central body is surrounded by companion bodies with similar geometric and dynamic distributions. The regularities common to the planetary and satellite systems would seem to exclude any hypothesis that distributes the planets by some random process such as turbulence in the circumsolar plasma. And, most important in the present context, condensation conditions assumed on the basis of such hypotheses must then also by necessity be unrealistic.

The Isochronism of Spin

All known bodies in the solar system which have not been tidally braked, from the asteroids to the giant planets, have spin periods which are very similar. In the majority of cases they vary only by a factor of two from an average of about 9 hours.

This appears to exclude theories which attempt to generate individual planets and satellites from collapsing gas clouds. Such theories fail for small planets and satellites also because of insufficient gravitation (Alfvén and Arrhenius 1970*a*).

Angular Momentum Distribution

Less than one per cent of the angular momentum of the solar system resides in the Sun. Any hypothesis that explains the Sun by a contractive process consequently must include an explanation of the transfer of angular momentum from the Sun to the circumsolar gas in whose progeny the momentum now resides. Similarly the orbital momenta of satellites must derive from the spins of the central bodies. At present, the only known possible mechanism for such a transfer, is magnetohydrodynamic coupling (Alfvén 1954, 1967).

This dependence, however, also introduces limits on the density of the circumsolar gas; the mechanism is considered to fail above average number densities of the order of 10^9 cm^{-3}. Experience of plasma behavior in space suggests however, that the actual distribution cannot have been homogeneous, but was most likely filamentary (Alfvén 1971). The densities in such filaments can exceed the average by orders of magnitude in proportion to the magnetic field strength. Assuming that this strength did not exceed an upper limit of the order of 1 gauss at 1.5 AU (cf. Brecher 1971), filamentary densities of the order 10^{13} would appear to be a safe upper limit (further discussion will appear in Alfvén and Arrhenius, in preparation).

Thermal Balance Between Gas and Grains

Many hetegonic discussions address themselves primarily to chemical questions. Foremost among these is the problem of explaining the observed chemical heterogeneity of the inner solar system. Some of these discussions ignore crucial physical consequences of the solutions proposed. One such discrepancy arises from the attempted use of the Sun to heat the condensing grains (at least tripling the present radiation temperature in the asteroidal region) without any effect on the state of excitation of the gas which is assumed to remain at the same temperature as the grains. Remedail action by interposition of sufficient optical thickness from dust or gas leads to equally severe conflict with the physical limitations discussed above and in the following.

Propositions of thermal equilibrium between grains and gas suffer from lack of realization that the normal situation for circumstellar gas is in excited and partially ionized states. This fact is evidently of major importance in considering the physical and chemical behavior of the gas.

Furthermore, solid grains embedded in an optically thin gas medium and in thermal balance with the gas, maintain a temperature which may be an order of magnitude below the electron-ion-atom temperature of the gas (Lindblad 1935; Spitzer 1968; Wickramasinghe and Nandy 1970; Arrhenius and Alfvén 1971). Typically the dust-gas star envelopes, now being studied by infrared astronomical techniques, show grain temperatures of 500–700 K assoicated with partially ionized gas at a few thousand degrees.

Energy and mass balance equations for plasma-grain systems in steady state have been developed by Lehnert (1970, 1971). A numerical solution based on similar considerations is shown in Fig. 1 for the heat balance of a grain in plasma, illustrating the low grain temperature (of the order of hundreds of degrees) at plasma temperatures of the order 10,000°, and in the density range permitted by the momentum transfer requirement.

The heat source for the plasma most commonly considered is electromagnetic

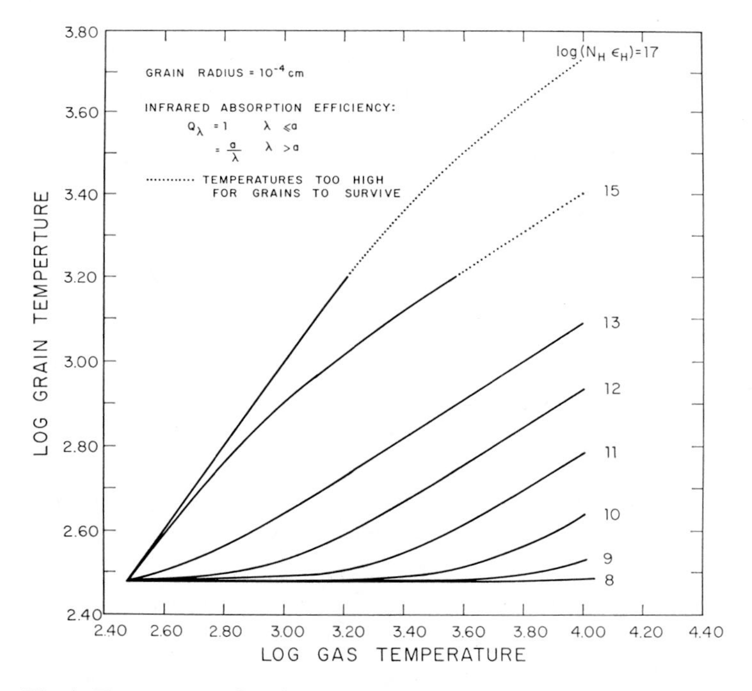

Fig. 1. Temperature of an isolated grain (x-axis) as function of gas temperature (y-axis) and gas pressure (individual curves marked with the exponent of the number density in units of cm^{-3}). The curves represent approximate solutions of the steady state equations of Lehnert (1970). The finite grain temperature, approached at decreasing gas pressure results from the power input by radiative heating, primarily from the Sun; the present day value at 1 AU is arbitrarily used here.

The graph illustrates the relatively low efficiency of gas heating even at the largest local pressures conceivable ($n \lesssim 10^{12}$ cm^{-3}) in a situation where magnetohydrodynamic momentum transfer can take place. At a number density of 10^{12} atoms/cm^3 and a plasma temperature of 10^4, for example, the grain temperature due to the combined effects of solar radiation and gas heating is more than ten times lower (about 830 K). Data from De and Honda (in preparation).

radiation from the Sun; this contributes also directly to the heating of the grains (cf. Lehnert op. cit.). It has furthermore been suggested (Brecher and Arrhenius 1971) that recombination of atomic hydrogen remains of importance up to 500–600 K as a source of heat, particularly on grains consisting of transition metals, primarily nickel-iron and their compounds.

The emplacement and condensation events during the formative stage of

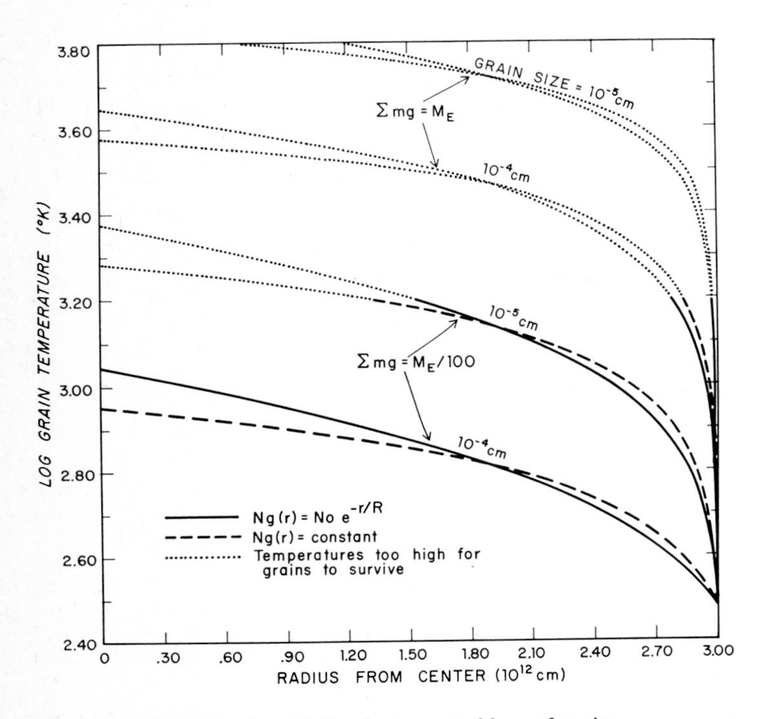

Fig. 2. Heating by back radiation in an assemblage of grains.

An arbitrary set of conditions has been chosen for the sake of illustration of the effect. Matter is assumed to be distributed in a toroid with a large radius of 1 AU and a small radius of 1/4 of the distance between Mars and Venus (representing the domain of the Earth). In this volume is distributed a mass of grains (Σ M_g) which in the upper two curve pairs equals the mass of the Earth, and in the lower two pairs is one percent of this mass. The upper curves in each of these two sets represent a grain size of 0.1 μm, the lower ones 1 μm. Of the two curves in each pair one (dashed) corresponds to a linear distribution of grain concentration (N_g), the other (full line) to concentration falling exponentially to 1/e times its central value at the surface of the toroid. The infrared absorption efficiency (Q_λ) of the grains is assumed to be unity for wavelengths, λ less or equal to the grain radius a, and $Q_\lambda = a/\lambda$ for $\lambda > a$.

Thermal energy is being supplied to the grains throughout the cloud mainly by electrodynamic heating of the gas. If back radiation from the grains were absent, this would result in an equilibrium grain temperature here set at 300 K.

When in addition back radiation from grains is taken into account, the situations arise which are depicted in the graph. From this is seen that under the conditions assumed all grains in the center of the toroid would vaporize, except in the case of the smaller amount of material distributed as comparatively large particles. A situation similar to this case, and with still less pronounced temperature gradient between surface and center of the particle cloud would prevail in the asteroidal region, where the total mass is only 10^{-5} of that in the Earth region. (From De and Honda, in preparation.)

the solar system were presumably locally and temporally associated with high gas pressures ($\lesssim 10^{13}$ cm^{-3}), the upper limit being determined by momentum transfer considerations. Under these conditions the most important heat source in the fields of the Sun and the giant planets was probably ohmic heating of the plasma assoicated with angular momentum transfer (Alfvén 1954, 1967). An important characteristic of this mode of energy dissipation is that it is practically insensitive to optical opacity due to solid grains, and hence penetrating through dust and plasma.

Finally, a potentially important source of heat for an individual grain is thermal emission from other grains in the surroundings. This effect is exemplified in Fig. 2.

These considerations, together with those of Lehnert (op. cit.), outline the more important controlling factors in condensation of simple two component, chemically inert systems under space conditions. They show how important it is to consider the temperature differential between grains and surrounding plasma and emphasize that average total pressures below the range of 10^{-7} to 10^{-8} atmospheres are a necessary requirement to permit the momentum transfer deemed necessary to explain the present momentum distribution.

Condensation in Multicomponent Systems

The composition of our solar system provides about twenty elements abundant enough to structurally determine the solid phases. The rest of the elements are mainly distributed in solid solution. The heterogeneity of physical and chemical properties results in a fractionation proceeding as a function of grain temperature, gas temperature, pressure and degree of ionization. The results of this fractionation can be seen directly in short term changes in the solar wind plasma and, at least qualitatively, in spectral observations of circumstellar grain-gas systems. The results of such fractionation can also be inferred from meteorites, from bulk differences in planetary compositions, and from laboratory experiments.

The effect of grain temperature on fractionation is essentially by control of the vapor pressure of the solid phases, mostly silicates, oxides, metals and sulfides. Here the picture is relatively simple at least in comparison with the situation in the gas phase. Thorough consideration of the vapor pressure relationships of the solids involved has been given (Anders 1971*a*, *b*; Larimer 1967; Larimer and Anders 1967, 1970), hence only minimal discussion is devoted to this parameter in the following. We will instead turn our attention to the variable parameters in the vapor phase, which profoundly affect the nature of the condensing solid, but which are not preserved and therefore more elusive.

When solid-gas reactions applicable to space conditions were first discussed on a quantitative basis (Latimer 1950; Urey 1952), it was practical to consider for guidance idealized cases, assuming thermodynamic equilibria in closed systems between a minimum number of simple molecular species in their ground states. Upper atmosphere chemistry, plasma chemistry and atomic physics have demonstrated the applicable reaction systems in reality typically to be open (in ideal cases perhaps in steady states) and to involve an array of reactive ions, atoms and molecules in various states of excitation. The importance of plasma reactions for synthesis of complex metastable organic molecules under simulated space conditions was demonstrated in the laboratory by Miller already in 1953; since then plasma synthesis of complex compounds has become an important sector of organic chemical technology. Actual production in space of molecules involving so far H, C, N, O, S, etc. as increasingly complex compounds has recently been discovered by radiowave spectrometry

These developments notwithstanding, the main reactions in the condensation process are today still commonly discussed in the same terms as the original approximations. An example is the important reaction system Fe(Ni)-O-H. Of these components hydrogen occurs mainly as neutral atoms, hydrogen ions, diatomic molecules, water molecules, cluster ions with H_2O groups, at least one metastable atomic state with lifetime long enough to make it an important and highly energetic reactant, molecular ions, and an array of molecular radicals formed in combination of hydrogen with carbon, oxygen, nitrogen and sulfur. A similar situation with a multitude of species prevails for oxygen.

Nickel and iron exist in the gas phase, besides as elemental molecules, atoms and ions, also as stable carbonyl- and chloride-complexes. Much evidence points at such volatile complexes as important components in the condensation of metallic nickel, iron and compounds of these elements, now found in meteorites (Bloch and Müller 1971), notably in carbonaceous and ordinary chondrites.

In contrast these relationships are still commonly discussed in terms of the reaction

$$Fe_3O_4 + 4H_2 \rightleftarrows 3Fe + 4H_2O.$$

From this equilibrium relationship and from assumed partial pressures of the gas components is calculated a metal-oxide conversion temperature which has been one of the cornerstones in the interpretation of meteorites. As is readily demonstrated by experiments, however, the balance between oxide and metal condensation is highly sensitive to the state of atomic dissociation, radicalization, and excitation of hydrogen and oxygen.

Although reasonable predictions of trend and order of magnitude of grain temperature can perhaps be made, taking these facts into account, the complexity of the situation calls for extensive experimental measurements before a meaningful quantitative model can be constructed.

Noble Gas Fractionation

Experiments and observation also suggest that the state of ionization of circumsolar atoms and molecules plays an important role in fractionation in the gas phase and at condensation. The large differences in ionization potential among the noble gases lead to preferential retention of ionized species in magnetic fields and differential escape of the neutral species (see references in Arrhenius and Alfvén 1971). This differentiation mechanism is probably responsible for the large variations in hydrogen-noble gas composition of the solar wind, and perhaps for some specific noble gas fractionation features found on Earth (Jokipii 1964) and in meteorites (Arrhenius and Alfvén 1971).

The noble gases have attracted great interest as indicators of the physical conditions of condensation in view of their relative chemical inertness (in their ground states), and their sensitivity to the temperature of the growing grains (Signer and Suess 1963; Suess 1949; Urey 1952).

Due to the their large size and their limited ability to form bonding orbitals with neighboring atoms, the equilibrium solubility of noble gas atoms in atomically close packed minerals is vanishing with increasing perfection of the crystal. (In contrast, open crystal structures with molecular tunnels and cavities of dimensions commensurate with the noble gas atoms have definable noble gas solubilities.) Nonetheless considerable quantities of non-radiogenic, noble gas atoms are found incorporated in atomically dense structures such as spinels in meteorites (Mazor and Anders 1970). Two modes of incorporation may be visualized and are experimentally verified. One is by implantation of accelerated noble gas nuclei into already existing crystals (Eberhardt et al. 1965; Wänke 1965).

The other, less generally recognized, is by occlusion in defects developing at the surfaces of crystals growing from the gas phase. This process depends on the quasi-equilibrium concentration of gas-accommodating defects which is steeply dependent on the substrate temperature, hence the occluded noble gas content is probably the most sensitive indicator of grain temperature in condensation.

This mode of incorporation is likely to be responsible for the noble gas component or components in meteorites referred to by Suess and Signer (op.

Table 1. *Argon occlusion and isotope fractionation obtained by Brecher and Marti (unpublished) at growth of nickel-iron with a solid temperature of about 350 K from sputtered atoms of iron and nickel in an argon plasma at about 10^{-4} atm.*

The argon isotope composition of the gas source for the plasma is indicated for comparison as well as the isotope composition in minerals from two carbonaceous chondrites, measured by Mazor and Anders (1970)

Sample	Ar^{40} concentration (ccSTP/g)	Ar^{38}/Ar^{36}	Ar^{40}/Ar^{36}
$Fe_{.8}Ni_{.2}$	$.28 \pm .05$	.171	250
$Fe_{.9}Ni_{.1}$	$.29 \pm .02$	.176	236
Gas source (atmospheric Ar)	—	.187	295.5
Allende (C3–4)	$.16 \cdot 10^{-4}$	.172	—
Orgueil (C1)	$.3\text{–}.5 \cdot 10^{-5}$	.177–.208	—

cit.) as the "planetary" component. They would presumably represent the range of ambient gas compositions at condensation resulting from plasma fractionation processes such as those reviewed in Arrhenius and Alfvén (1971). The ambient noble gas composition is likely to have been further modified elementally and isotopically by those specific fractionation processes which are known to operate at gas-solid interaction, and which hence are likely to be important at noble gas occlusion in grains growing in space. Among such gas-solid interaction processes are several modes of isotope separation; a well known effect is the one due to multiple desorption. Another one, due to kinetic effects in the gas phase (Brecher and Marti 1971), is illustrated in Table 1, and compared to a numerically similar effect observed in a spinel phase (magnetite) separated from carbonaceous chondrite material.

Ionization Effects in Reactive Elements

Other element distributions in meteorites suggest the possible effect of the degree of ionization in the plasma from which the solid grains presumably grew. Such effects are most pronounced in elements with extreme contrast between neutralization temperature of the ions on one hand and bond strength of the solid compounds on the other. (As neutralization temperature is taken the ion temperature where the number densities of ions and neutrals are equal in a pure elemental plasma.)

One such example is provided by the element pair thallium–mercury. Both of these elements are highly volatile—mercury even more so than thallium. They differ, however, by a factor two in neutralization temperature of the pure elements (Fig. 3); T_c (Hg) $\approx$ 11,500 K compared to 5,900 K for thallium.

The difference may be enhanced considerably by charge exchange in mixed plasmas (Fite 1971). Larimer and Anders (1970) have demonstrated that both elements, due to their volatility, are concentrated in such meteorite material (Cl) which for several reasons appears to constitute comparatively low grain temperature condensates. In comparison to thallium, however, this fractionation is found to be much less pronounced for mercury. It has been postulated by Arrhenius and Alfvén (1971) and by Anders (1971) that this difference is related to the marked differential ionization of the two elements. Presumably the mercury as neutral atoms becomes available for occlusion in condensing grains of sufficiently low temperature already at a high plasma temperature (~10,000 K), while thallium is retained in the gas phase and condenses appreciably only in a plasma temperature range well below 5,000 K.

Another group of elements, whose distribution can be interpreted as the effect of ion fractionation at condensation contains calcium, yttrium, titanium, zirconium and aluminum. Their oxygen compounds in fully oxidized states have extremely low vapor pressures; magnesium as a component of such

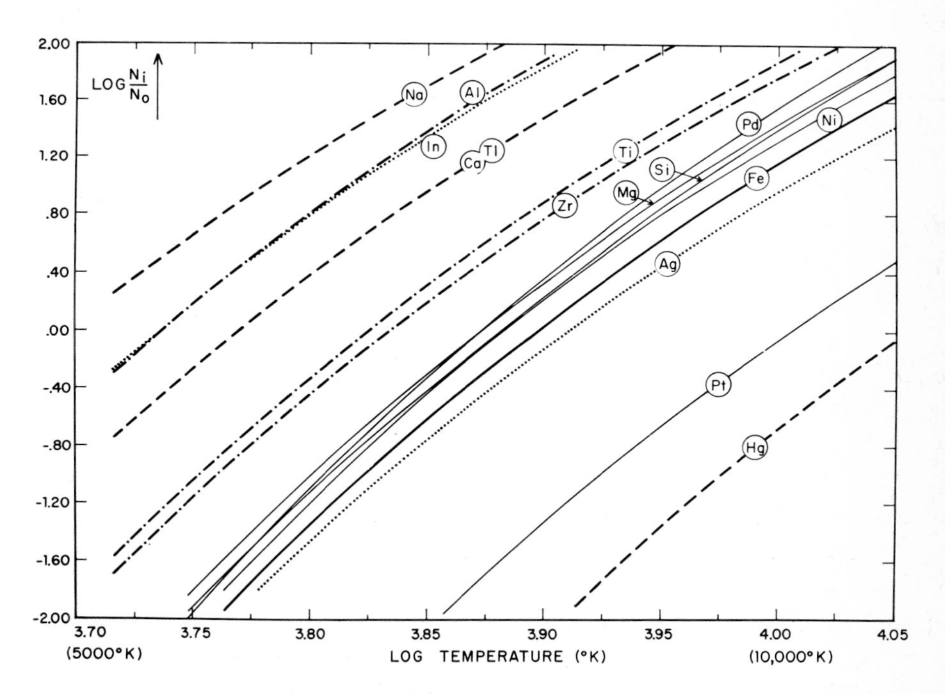

Fig. 3. Degree of ionization (y-axis) as function of temperature (x-axis) in single-element plasma.

Of the two elements Hg and Tl (which have similar vapor pressures) the former is practically neutral at a temperature where the latter remains almost fully ionized.

The place of elements such as Ca, Al, Ti and Zr in the sequence of condensation would be determined partly by the fact that they form stable compounds already at high grain temperature, and partly by effects (working in the opposite direction) due to their high degree of ionization even at low plasma temperatures, as shown by the diagram.

compounds also shares this property. In a cooling neutral gas mixture or in a neutral gas with a thermal gradient one would for this reason expect oxygen compounds of these elements such as aluminum oxide, magnesium aluminum oxide, calcium titanate, and calcium-magnesium aluminosilicates to be the first solid phases to appear (Lord 1967).

In a partially ionized gas, such as in the solar system, this condensation sequence would be expected to be permuted by the degree of ionization of the component elements. Those of the group concerned here all have a comparatively low neutralization temperature (Fig. 3). Hence in a cooling plasma, or in a region with a temperature gradient they would become concentrated into the ionized fraction.

If, as suggested by the pair thallium–mercury, a mechanism is active which leads to preferential re-emission of energetic species resulting from impact of ions in contrast to neutrals, this effect would be expected to lower the rate of condensation of the "refractory" elements in question on grains growing in the plasma. Such an effect is suggested by the actual distribution in meteorites of these elements. They are typically concentrated in discrete bodies (rounded irregular botryoidal aggregates of silicate and oxide crystals) in type 3 carbonaceous chondrites (see f. ex. Clarke et al. 1971; Kurat 1971). Contrary to what would be the case if these aggregates had grown at extremely high grain temperatures, however, the "refractory" crystals are intimately intergrown, often in the form of concentric structures, with components which are unstable already at moderate temperatures. Among these are sodium chloride as inclusion molecule in the aluminosilicate cages of the mineral sodalite (Clarke et al., op. cit.; Fuchs 1969, 1971), unidentified potassium bearing phases, calcium carbonate and uniquely high concentrations of xenon-129, presumably incorporated as iodine-129 (half-life 15×10^6 yrs) shortly after nucleosynthesis (Marti 1971). Attempts to explain these facts by secondary impregnation and transformation processes appear incompatible with the observed concentration gradients, the composition of the surrounding crystal powder, and other features of the volatile containing "refractory" aggregates.

These circumstances instead point at the element- and phase distribution in the calcium-aluminum rich aggregates as primary features, originating at growth of these bodies from a gas phase with a residual ionized fraction enriched in elements with low neutralization temperature, such as Ca, Al, Ti, Y, and Zr and at grain temperatures low enough to permit effective condensation of volatile components such as K, I, Cl and CO_2. In the terminal stage of growth the temperature was low enough and the speciation and concentration of oxygen were such as to permit growth of calcium-aluminum garnet with trivalent iron as a stoichiometric component (Fuchs 1971).

In principle a selective re-emission of atoms, impinging as ions, can be understood by mechanisms such as suggested by Massey (1971); further study is, however, necessary in order to verify the importance of this phenomenon in grain-gas interactions of the type expected in space.

Experimental Approach

In general it is clear that the process of condensation from circumstellar plasma in the range of compositions that we see represented in the solar system must be a complex process. Chemistry is in physical terms, already complex, and complex chemistry is beyond the reach of fundamental physical prediction. A purely theoretical approach, even on satisfactory premises, will consequently never alone give the answer to the problems we are concerned with here. Controlled experiments are of utmost importance. Useful experimental information, pertinent to the present problems, can already be derived from chemical studies of the upper atmosphere and space (Massey, op. cit.). In addition laboratory experiments are potentially elucidating; in our group we have made attempts to study the solid phases resulting from condensation under relevant conditions. Our emphasis has been on crystal structure, composition, isotope and element fractionation (Arrhenius and Alfvén 1971; Meyer 1970, 1971; Brecher and Marti 1971), magnetic properties (Brecher 1971) and occlusion of insoluble gases in the growing grains (Brecher and Marti, op. cit.); in order to relate the results to actual condensation and transformation processes in space, comparative analyses are made of meteorites and lunar materials.

We hope in our continued work to further emphasize spectrometric control of the species distribution in the plasma, including radical and complex formation. These aspects of the work have obvious applications also to the formation and evolution of comets, and in condensates of sufficiently low temperature possibly also to the the longer-term heat budget of bodies accreted from condensate grains (Urey and Donn 1956).

References

Alfvén, H., 1954, On the origin of the solar system. Oxford University Press, London.

Alfvén, H., 1967, Partial corotation of a magnetized plasma. Icarus, *7*, 387.

Alfvén, H., 1971, Discussion in this volume.

Alfvén, H. and Arrhenius, G., 1970*a*, Structure and evolutionary history of the solar system, I. Astrophys. Space Sci., *8*, 338.

Alfvén, H. and Arrhenius, G., 1970*b*, Structure and evolutionary history of the solar system, II. Astrophys. Space Sci., *9*, 3.

Anders, E., 1971*a*, Meteorites and the early solar system. A. Rev. Astr. Astrophys., *9*.

Anders, E., 1971*b*, This volume.

Arrhenius, G. and Alfvén, H., 1971, Fractionation and condensation in space. Earth Planet. Sci. Letters, *10*, 253.

Bloch, M. R. and Müller, O., 1971, An alternative model for the formation of iron meteorites. Paper given at 34th Ann. Meeting of the Meteoritical Society, Tübingen, August 1971.

Brecher, A., 1971, On the primordial condensation and accretion environment and the remanent magnetization of meteorites, Proc. I. A. U. Colloq. 13: "The evolutionary and physical problems of meteoroids", Albany, N.Y.

Brecher, A. and Arrhenius, G., 1971, Hydrogen recombination by nonactivated chemisorption on metallic grains. Nature, *230*, 107.

Clarke, R. S. Jr, Jarosewich, E., Mason, G., Nelen, J., Gómez, M. and Hyde, J. R., 1970, The Allende, Mexico, meteorite shower. Smithsonian Contrib. Earth Sciences 5.

Eberhardt, P., Geiss, J. and Grögler, N., 1965, Über die Verteilung der Uredelgase im Meteoriten Khor Temiki. Mineral. Petrog. Mitt. (Tschremaks) *10*, 535.

Fite, W., 1971, Personal communication.

Fuchs, L. M., 1969, American Mineralogist *34*, 1645.

Fuchs, L. M., 1971, Manuscript in preparation.

Geiss, J., 1971, On elemental and isotopic abundances in the solar wind. Proc. Asilomar Conf. on Solar Wind, March 1971.

Jokipii, J. R., 1964, The distribution of gases in the protoplanetary nebula. Icarus, *3*, 248.

Kurat, G., 1970, Zur Genese der Ca-Al-Reichen Einschlüsse im Chondriten von Lancé. Earth and Planetary Sci. Letters, *9*, 225.

Larimer, J. W., 1967, Chemical fractionation in meteorites—I. Condensation of the elements. Geochim. Cosmochim. Acta, *31*, 1215.

Larimer, J. W. and Anders, E., 1967, Chemical fractionation in meteorites—II. Abundance patterns and their interpretation. Geochim. Cosmochim. Acta, *31*, 1239.

Larimer, J. W. and Anders, E., 1970, Chemical fractionation in meteorites—III. Major element fractionations in chondrites. Geochim. Cosmochim. Acta, *34*, 367.

Latimer, W. M., 1950, Astrochemical problems in the formation of the Earth. Science, *112*, 101.

Lehnert B., 1970, On the conditions for cosmic grain formation. Cosmic. Electrodyn., *1*, 218.

Lehnert, B., 1971, This volume.

Lindblad, B., 1935, A condensation theory of meteoric matter and its cosmological significance. Nature, *135*, 133.

Lord, H. C., 1967, High temperature equilibria and condensation, Part 1. Ph. D. thesis, Univ. of Calif.

Marti, K., 1971, Personal communication.

Massey, H., 1971, This volume.

Mazor, D. H. and Anders, E., 1970, Noble gases in carbonaceous chondrites. Geochim. Cosmochim. Acta, *34*, 781.

Meyer, C. Jr, 1969, Sputter-condensation of silicates. Ph. D. thesis, Scripps Institution of Oceanography, La Jolla, Calif.

Meyer, C. Jr., 1971, An experimental approach to circumstellar condensation. Geochim. Cosmochim. Acta, *35*, 551.

Miller, S. L., 1953, A production of amino acids under possible primitive Earth conditions. Science, *117*, 528.

Signer, P. and Suess, H., 1963, Rare gases in the Sun, in the atmosphere and in meteorites. Earth Science and Meteoritics (ed. J. Geiss). North-Holland Publishing Company, Amsterdam.

Spitzer, L. Jr, 1968, Diffuse matter in space. Interscience, New York.

Suess, H., 1949, Die Häufigkeit der Edelgase auf der Erde und im Kosmos. J. Geol. Res., *57*, 600.
Wickramasinghe, N. C. and Nandy, K., 1970, Interstellar extinction by graphite, iron and silicate grains. Nature, *227*, 51.
Wänke, H., 1965, Der Sonnenwind als Quelle der Uredelgase in Steinmeteoriten. Z. Naturforsch., *20 a*, 946.
Urey, H. C., 1952, The planets—their origin and development. Yale Univ. Press, New Haven.
Urey, H. C. and Donn, B., 1956, Ap. J., *124*, 307.
Urey, H. C., 1952, Chemical fractionation in the meteorites and the abundance of the elements. Geochim. Cosmochim. Acta, *2*, 269.

Discussion

T. Gehrels

What is the temperature range?

G. Arrhenius

So far Aviva Brecher has tried this type of experiment ($\alpha-\gamma$ lamellar crystallization) only at substrate temperatures in the range 300–400 K. But other experiments now under way and theory in general suggest that the phenomenon would be increasingly pronounced at higher temperatures because the diffusion length of the atoms on the surface increases.

L. O. Lodén

Does the mechanism mentioned by you also imply separation of the fractionated elements?

G. Arrhenius

Yes, there are several mechanisms that potentially may contribute to the fractionation effects observed in plasmas in space and in meteorites. In general all of these should become increasingly efficient with increasing extent of cycling.

L. O. Lodén

Thank you. Did you also indicate that the magnetic fields may play an important role in the separation process?

G. Arrhenius

In one of these mechanisms certainly, where you retain the ionized species in the magnetic field, while the neutral gas fraction can diffuse freely.

Other fractionation mechanisms, involving neutral species only, would however be independent of the presence of a magnetic field, or at least there would not be any primary, direct effects.

F. L. Whipple
What happens if you put Ni–Fe in the Mg-silicate?

G. Arrhenius
We would like to investigate this but we have not done so yet.

F. L. Whipple
I assume that you have a solar mix modified by ionization to some extent.

G. Arrhenius
Most of our condensation experiments have for obvious reasons aimed at the phase relationships in simple systems such as Mg–Si–O–H and Fe–Ni–O–H. But there are also some results with complex solar system type mixes. The condensates then consist of multiphase rocks, including magnesium-iron silicates. We have not yet conducted these more complex experiments under such dissociation–association conditions in the hydrogen–oxygen component that iron condenses as metal. The studies by Brecher of the relatively simple system Fe–Ni–O–H have, however, largely been concentrated on the physical and chemical properties of the metal phases.

Y. Öhman
In a brief article in "The Observatory" (*62*, 150, 1939) Professor P. Swings and I drew attention to the possibility that amorphous metals might be formed in interstellar grains. We based our suggestion on earlier laboratory work made by the German physicist J. Kramer. Kramer claimed that the crystallization was accompanied by ultraviolet emission. Have you found any trace of such an ultraviolet emission in your work?

G. Arrhenius
Much work has been done since then on the interesting problem of complete disorder in metals and alloys. The two controlling parameters are found to be temperature of formation, rate of condensation, and complexity of the unit cell. Simple structures such as most metals always seem to crystallize even when quenched at high speed to liquid helium temperatures. Metalloids (such as germanium) and alloys with complex crystal structures can be condensed into complete long range disorder at different critical temperatures. Nickel-iron is presumably the only quantitatively important metal phase in space, and here one would expect ordering even if condensation took place at very low temperature (although with small crystallite size and high degree of imperfection). Short wavelength radiation released at phase transformations would

also suffer terribly from internal extinction in iron or typically iron-rich natural compounds.

In the complex silicate structures, however, the amorphous glass state is easily achieved even at several hundred K, particularly at high silica content and high flux of atoms at condensation (high gas pressure).

Z. Kopal

Can you estimate the efficiency of the process? How long will it take in space?

G. Arrhenius

The rates depend on the mobility and number density of the gas atoms. With the proper values in the kinetic equations you get the order of magnitude growth of 10–100 μm per day at gas temperatures around 10^4 K and densities of $10^{12}-10^{13}$ cm^{-3}, which means that the condensable species are three orders of magnitude less (10^9-10^{10} cm^{-3}).

The approximately $0.5-5$ μm size crystals, which besides chondrules make up the bulk of the mass of C3-type chondrites, would under such conditions form in a small fraction of a day; the crystals and crystal aggregates in achondrites would require days or weeks if formed by vapor growth, and a large iron meteorite would take the order of centuries. Alternatively, and perhaps more likely, large bodies of γ-iron formed by coalescence of small, magnetized grains as proposed by Anders (1971).

Z. Kopal

Do you think this is the mechanism to grow dark nebulae in space?

G. Arrhenius

Since in these cases we do not have the same detailed dynamic information, the pressure constraints are perhaps not as rigorous as in the circumsolar case (although I think that observed line broadening in such nebulae suggests pressures below 10^{13} cm^{-3}). However, in general, since the grains of the dark nebulae are presumably formed by condensation, the considerations presented would apply directly also here.

P. Pellas

1. You obtain in some of your experiments enstatite together with a phase which might be olivine. We know, however, that in enstatite chondrites olivine is mostly absent. This would seem to present a disagreement between your synthesis experiments and what we observe in chondrites.

2. My second question refers to the information one can derive from the

presence of a refractory phase. There are many processes by which such phases, for example zircon, can be formed in a wide range of temperatures and in vapor and liquid media. Hence it would seem that your experimental results do not provide a proof that the meteorites were necessarily formed by the processes you propose, although this remains a possibility.

3. Finally, there remains the major problem, i.e., to explain by your model the exceedingly large variations in elemental and isotopic ratios we observe in fractionated noble gases (Ne, Ar, Kr, Xe) compared to the unfractionated (solar types) rate gases, in different chondritic and achondritic objects and even in silicate inclustions in some irons (f.i. El Taco).

G. Arrhenius

1. The experiments related here are in no way intended to model specifically the formation of enstatite chondrites. The purpose is to clarify the phase relationships in the Mg–Si–O system below the triple point. Depending on the composition of the source gas, different phases are obtained, ranging from pure silica to magnesium oxide. As expected, two phases may coexist. There is no disagreement (or relation) between these results and the fact that enstatite chondrites have little or no olivine.

2. Obviously the temperature where bond breaking occurs in a solid has little or nothing to do with the kinetic energy of the condensing components or the temperature of the growing crystal, except that the latter has to be below the decomposition temperature.

Hence not only zircon, but all solids can be condensed at any temperature below this critical temperature; that is just the point. The mere fact that a solid is found in a natural sample does in itself not give any further information on the mode of formation or on the actual temperature of growth. No such claim is made.

Information on the maximum possible and actual substrate temperatures at the time of growth, and on the mode of formation of the solid comes from quite different considerations discussed in this and in Prof. Anders' paper.

3. On this point I wish to refer back to my paper. The discussion there attempted to demonstrate how it is possible, in principle and experimentally, by fractionation in the plasma, and at condensation of solids to obtain element and isotope separation effects of at least the magnitude observed in meteorites. To disentangle the various possible effects from each other, and to study them quantitatively under a variety of conditions remains an experimental challenge.

Conditions in the Early Solar System, as Inferred from Meteorites

By Edward Anders

Enrico Fermi Institute and Department of Chemistry, University of Chicago, Chicago, Illinois

I. *Introduction*

Traditionally one has assumed that meteorites and planets originated from a solar nebula of about 0.1 to 1 $M_\odot$. Solid grains condensed on cooling and accreted into larger bodies, leaving the gas behind. Condensation proceeded largely under conditions of thermodynamic equilibrium (Urey 1952*a*, *b*, 1954; Lord 1965; Larimer 1967; Larimer and Anders 1967, 1970; Grossman 1972).

This view has been challenged by Arrhenius and Alfvén (1971) in a very stimulating paper. They propose that condensation took place from a plasma, not a neutral gas. A major fractionation mechanism was separation of ions from neutral atoms, preceding the separation of solids from gas. Pressures in the inner solar system were orders of magnitude lower than those assumed in the equilibrium model ($10^{-4\pm2}$ atm), and temperatures were much higher (4 000 to 10 000 K vs. 350 to 2 000 K).

The ultimate test of any such model is its ability to account for the empirical evidence, principally meteorite data. It has become apparent during the last decade that most properties of chondrites were established *prior* to accretion of their (asteroidal?) parent bodies. Thus the chondrites contain a detailed record of the pre-accretionary stage of the solar system. The problem is to decipher this record; to translate structure, composition, and mineralogy into temperature, pressure, time, and chemical environment.

I have discussed the meteoritic evidence in terms of the equilibrium model in several recent reviews (Anders 1968, 1971*a*) and papers (Larimer and Anders 1967, 1970; Keays et al. 1971). I shall therefore limit myself to a fairly brief summary, omitting many qualifications and details, but stressing quantitative predictions and inferences. These inferences, about pressures, temperatures, and time scales in the early solar system, themselves constitute a test of the model. They must be internally and externally consistent, not only with each other but also with independent evidence from other sources.

The paper begins with a brief glossary, followed by a discussion of the condensation sequence of a cosmic gas. § IV to VI review chemical evidence from meteorites, while § VII discusses the chronology of the early solar system. § VIII attempts a critical comparison of the two models in the light of the evidence presented.

II. *Glossary*

Chondrites are stony meteorites containing *chondrules*, mm-sized silicate spherules that appear to be frozen droplets of a melt. They consist largely of *olivine* [(Mg, Fe)$_2$SiO$_4$], *pyroxene* [(Mg, Fe)SiO$_3$], and *plagioclase feldspar* [solid solution of $CaAl_2Si_2O_8$ and $NaAlSi_3O_8$]. In the more primitive chondrites, glass is often found in place of crystalline feldspar.

Chondrules are embedded in a ground-mass or *matrix*. In the less primitive chondrites, the matrix is somewhat more fine-grained than the chondrules, but otherwise has the same mineralogy and composition. Millimeter-sized particles of *metal* (nickel-iron with 5–60% Ni) and *troilite* (FeS) are also present. In the more primitive chondrites, the matrix is very fine grained (to $\sim 10^{-6}$ cm) and richer in Fe^{2+} than the chondrules.

Five chondrite classes are recognized, differing in the proportions of oxidized to reduced iron. *Enstatite* (=E) chondrites are highly reduced, containing iron only as metal and FeS. *Carbonaceous* (=C) chondrites are highly oxidized, containing mainly Fe^{2+}, Fe^{3+}, and little or no Fe^0. The middle ground is occupied by 3 classes of intermediate oxidation state and total iron content: H, L, and LL chondrites. Collectively, they are often called *ordinary* (=O) chondrites.

Each of these classes is further subdivided into "*petrologic types*" numbered from 1 to 6 (Van Schmus and Wood 1967). These types were originally designed to reflect increasing chondrule-to-matrix intergrowth (probably due to metamorphic recrystallization in the meteorite parent bodies), but have turned out to correlate well with compositional trends, e.g. volatile content.

III. *Condensation Sequence of a Cosmic Gas*

Two condensation diagrams based on the work of Larimer (1967 and unpublished) are shown in Fig. 1. They were calculated for a total pressure of 10^{-4} atm and Cameron's (1968) cosmic abundances. This pressure is thought to be representative of the asteroid belt. At higher pressures (or lower H/metal ratios) the curves shift to the right, by varying amounts.

The upper diagram applies to conditions of rapid cooling, where substances condense in successive layers without interdiffusion or alloy formation. The lower diagram, on the other hand, assumes complete diffusional equilibration,

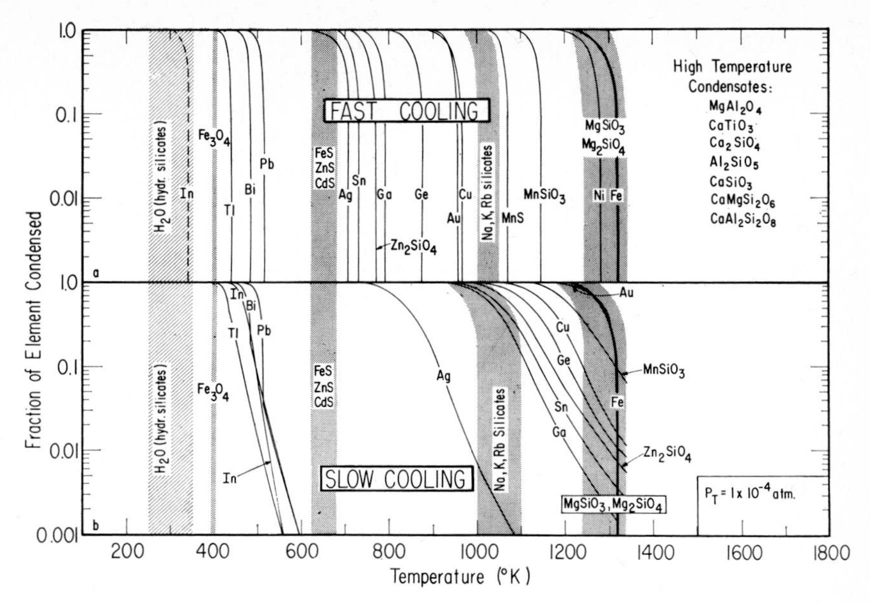

Fig. 1. Condensation sequence of a gas of cosmic composition (Larimer 1967; Anders 1968, with minor revisions). Slow-cooling sequence assumes diffusional equilibrium between grain surfaces and interiors, with formation of solid solutions, while fast-cooling sequence corresponds to deposition of pure elements and compounds, without interdiffusion. Shaded areas represent condensation or chemical transformation of major constituents. The formation range of hydrated silicates is poorly known.

with formation of alloys and solid solutions to the limit of solubility. The effect of such alloy formation is to raise the condensation temperatures of minor elements, and to widen their condensation intervals. Of course, all intermediate situations are possible.

A cooling gas of cosmic composition should thus condense in the following order. Below 2 000 K, some highly refractory compounds of Ca, Al, Mg, and Ti appear, followed by magnesium silicates and nickel-iron at ~1 350 to 1 200 K and alkali silicates at 1 100 to 1 000 K. Up to this point, some 90% of chondritic matter has condensed. Only H, C, N, O, S, and some volatile trace elements still remain in the gas phase. At 680 K sulfur begins to condense on solid Fe grains by the reaction $Fe + H_2S \rightleftarrows FeS + H_2$, followed by Pb, Bi, Tl, and In. Any remaining Fe reacts with H_2O at 400 K to give Fe_3O_4. Finally, water is bound as hydrated silicates at some temperature between 250 and 400 K.

IV. *Chemical Fractionations in Chondrites*

Refractory Elements

The elements heading the condensation sequence (Ca, Ti, Al, Mg) all are fractionated among the major chondrite classes, by small but definite factors.

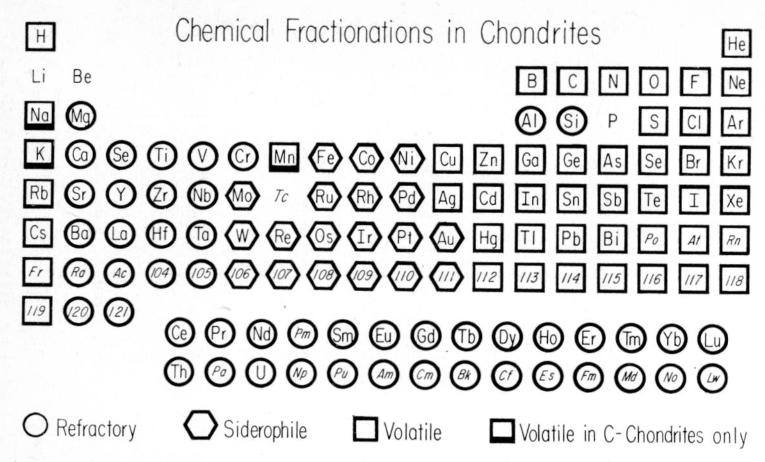

Fig. 2. Elements fall into 3 groups, based on their fractionation behavior in meteorites. In terms of the condensation model, these reflect 4 main cosmochemical fractionation processes: loss of refractories, loss of metal, loss of volatiles during remelting, and loss of volatiles during accretion. Refractories and siderophiles each tend to vary in unison, and by rather small factors (less than 2). Volatiles show greater and more variable depletions (to 10^{-3}) and at least the more strongly depleted elements do not vary in unison.

Analogous fractionations are found for some 20-odd refractory elements (Fig. 2). Their abundances increase in the order $E < O < C$, in the average ratio 0.56 : 0.73 : 1.00. All these elements are quite non-volatile, and would be expected to concentrate in an early condensate or a volatilization residue at $\geq$1 300 K.

Larimer and Anders (1970) have suggested that the trend from C to E chondrites reflects progressively greater loss of an early condensate ($MgAl_2O_4$, $CaTiO_3$, Mg_2SiO_4, etc.), comprising 15–30 % of the total condensable matter. Inclusions of such composition are in fact found in C3 chondrites but not in E or O chondrites. Apparently only the C chondrites retained their early condensate.

Arrhenius and Alfvén state that the refractory inclusions in C3's are "entirely unexpected and the extreme opposite of what would be expected on the basis of condensation in thermal equilibrium". This is not the case; the existence of such compounds was predicted by Lord in 1965 and Larimer in 1967, before they were actually discovered in meteorites. That such minerals occur in carbonaceous chondrites is not at all surprising. These meteorites have nearly cosmic composition (Anders 1971*b*), and must, *ipso facto*, have collected their proportionate share of every successive condensate, high-, medium-, or low-temperature. The mineralogy of the refractory inclusions likwise poses no paradoxes, as shown in detail by Grossman (1972). Some features seemingly contradicting a high-temperature origin (Na, K, Cl, I, Fe^{3+} content) appear to be of late, metamorphic origin.

Herbig (1970*a*) has noted that the interstellar gas is deficient in these same elements (Ca, Ti, Al) and suggested that this might be due to partial condensation of refractories. Such condensation cannot have taken place at the low densities of interstellar space. Herbig therefore suggests that a major part of interstellar matter has once passed through solar nebulae. It would seem that partial condensation of refractories is a fairly universal feature, not limited to our own solar system.

Some of the least volatile transition metals (Ir, Os, Re, Pt) may have been involved in the same fractionation. Their abundances in iron meteorites vary by 3 orders of magnitude and are *anti*-correlated with those of equally reducible but more volatile elements, e.g. Pd and Ni. Similar though much smaller variations have been observed in chondrites. Arrhenius and Alfvén attribute this fractionation to differences in ionization potential, but it can equally well be explained by differences in volatility (Anders 1971*a*).

Metal-Silicate Fractionation

The elements marked siderophile in Fig. 2 correlate with iron in each chondrite class. Total variation is again small (less than a factor of 2) but definite (Urey and Craig 1953). Density differences among the inner planets also imply a difference in metal/silicate ratio (Urey 1952*a*). Apparently a metal-silicate fractionation took place in the early solar system. A likely mechanism is preferential loss of metal grains in the nebula.

Two attempts have been made to estimate the temperature during metal-silicate fractionation (Larimer and Anders 1970). Upper limits of $\leqslant$1 050 K and $\leqslant$985$\pm$50 K for O and E chondrites were based on the observation that elements less volatile than Ge correlated with metal content and hence apparently were already condensed at the time of fractionation, while more volatile elements (Ga, S) were not. Lower limits of $\geqslant$650 K and $\geqslant$680 K were derived from the Fe^{2+} content of silicates and absence of FeS during fractionation, as inferred from regressions of Ni/Mg and Fe/Mg ratios in chondrites. The former value may be too high; new calculations by Grossman (1972) suggest 500–550 K.

Interestingly, the upper limits are close to the ferromagnetic Curie points of the metal in E and O chondrites, 940 and 900 K. Perhaps this is a coincidence, but then it is hard to understand why the temperature had to fall 400° after the refractory-element fractionation, before the metal-silicate fractionation began. More probably, the relationship is causal, the fractionation being triggered by the appearance of ferromagnetism in the metal grains (Wood 1962).

Volatile Elements, Formation of Chondrules and Matrix

Some 37 volatile elements are fractionated in chondrites (Fig. 2). C chondrites show the simplest pattern: abundances decrease uniformly from C1's to C3's, by mean ratios of 1 : 0.55 : 0.33. This has been explained by a two component model (Wood 1963; Anders 1964, 1971*a*; Larimer and Anders 1967). Chondrites are a mixture of a volatile-rich, low-temperature component (=matrix) and volatile-free, high-temperature component (=chondrules+metal grains), both formed in the solar nebula. The decline in volatile content from C1 to C3 is attributed to increasing chondrule content which goes from ~0 in C1's to ~70–80% in C3's.

Matrix is thought to be the original condensate, whose small grain size (~10^{-5} to ~10^{-6} cm) enabled it to equilibrate with the gas down to quite low temperatures. Chondrules probably represent remelted matrix, the remelting and outgassing having taken place in brief local heating events, such as electric discharges (Whipple 1966; Cameron 1966). With ~10^5 times the size of matrix grains, they would be far less efficient collectors of volatiles, having 10^{-10} the specific surface area and 10^{10} the diffusional equilibration time. Thus Fig. 1*a* is pertinent to chondrules, and Fig. 1*b*, to matrix.

Ordinary chondrites show a more variable pattern (Larimer and Anders 1967; Anders 1971*a*). Elements fairly high in the condensation sequence (Cu, Ge, Sn, Ga, S) again are depleted by nearly constant factors, averaging 0.24. This suggests a matrix content of 24%. But the last 4 metals in the condensation sequence (Pb, Bi, Tl, In) are depleted by much more variable factors, down to ~10^{-3}. Several other volatile elements whose condensation behavior is less well understood (Cl, Br, I, Hg, Cs, Cd, etc.) show variations of similar magnitude. The depletion generally increases with petrologic type. Apparently these meteorites accreted at so high a temperature that not even the matrix acquired its full complement of volatile metals. And each meteorite seems to have its own individual kind of matrix, reflecting its P, T history up to the time of accretion.

Accretion Temperatures

One can use volatile metals as "cosmothermometers", as first attempted by Urey (1952*a*, 1954). The temperature corresponding to a given metal content can be read off the condensation curve (Fig. 1*b*). Keays et al. (1971) have reported values for 11 L-chondrites, based on Bi, Tl, and In contents. Results ranged from 460 K for L3's to 560 K for L6's (all at an assumed P of 10^{-4} atm) and were reasonably concordant for the 3 thermometers. The event recorded by these thermometers is the *chemical* isolation of the dust grains from the gas, when they ceased to take up volatile metals. However, this probably coincided with their *physical* isolation from the gas (=accretion). Estimates of diffusion rates show that grains of $\leqslant 10^{-5}$ cm will remain in equilibrium with Pb, Bi, Tl vapor in the gas down to 500 K, at cooling times as short as a few years (Larimer 1967). Thus the temperatures calculated can in fact be interpreted as accretion temperatures.

Similar temperature estimates have been attempted for the Earth, Moon, and several classes of achondritic meteorites (Anders et al. 1971; Laul et al. 1972). The procedure is less straightforward because the effects of planetary differentiation are superimposed on the original nebular fractionation and must be separated from it. Moreover, for a differentiated planet one can at best obtain only a planet-wide average abundance of a volatile element. This does not give the true average temperature during accretion, but a value biased toward lower temperatures, because abundance is a logarithmic, not a linear function of temperature (Fig. 1). The error thus introduced seems to be of the order of 30 K (Anders et al. 1971).

The mean values obtained are shown in Table 1. These values apply to an assumed total pressure of 10^{-4} atm, except 10^{-2} atm for the Earth and Moon. The results are moderately sensitive to pressure: at 10^{-4} atm, the values for Earth and Moon become 456 and 502 K. For the L chondrites at least, P can be bracketed between 10^{-2} and 10^{-6} atm by chemical arguments. At pressures of 10^{-2} or 10^{-6} atm, calculated Bi-Tl correlation curves fit the data noticeably less well than at 10^{-4} atm. Moreover, all chondrites contain FeS, whose formation temperature (680 K) is pressure independent. At a pressure of 10^{-2} atm, the condensation curves of Bi and Tl are shifted upward sufficiently to imply accretion temperatures *greater* than 680 K for the more strongly depleted type 6's. Such meteorites then should contain no FeS, contrary to ob-

Table 1. *Accretion temperatures from volatile metal content*

Object	Low-T fraction (%)	Accretion T (K)	P (atm)	Ref.[b]
L Chondrites	24	513[a]	10^{-4}	1
Eucrites	0.8	460	10^{-4}	2
Shergottites	28	471	10^{-4}	2
Nakhlites	38	478	10^{-4}	2
Earth	11	544	10^{-2}	3
Moon	2.4	610	10^{-2}	3

[a] This is an average value, weighted in such a manner as to make it consistent with the temperatures of the remaining, chemically differentiated objects in this table (Anders et al. 1971). The actual range for L chondrites is 460 to 560 K.
[b] 1. Keays et al. (1971)
2. Laul et al. (1972)
3. Anders et al. (1971)

servation. Similarly, a lower limit of 10^{-6} to 10^{-7} atm can be inferred from the absence of Fe_3O_4 (formation temperature, 400 K) from all but the least recrystallized and least depleted type 3's. Thus the pressure in the region of the L chondrites apparently was $10^{-4\pm2}$ atm. Somewhat more restrictive limits will be set in § VIII.

It is remarkable that temperatures as high as 500 K prevailed during accretion of the meteorite parent bodies, because the present black body temperature at 2.8 AU is only 170 K. Thus a powerful, transient heat source must have been present. One possibility is the Hayashi stage of the Sun; another is the collapse stage of the nebula which would release vast amounts of gravitational potential energy over a short span of time (Cameron 1962, 1963, 1969). Indeed, the temperatures inferred for the dust shells or disks surrounding protostars are of the right order: 500 K for ε Aurigae, $\leqslant$1 850 to 350 K for VY Canis Majoris and 850 K for R Monocerotis and other infrared stars (see Kopal, 1971 and Anders 1971*a* for references). Perhaps it will soon be possible to compare time, temperature, and pressure estimates from meteorites with observational and theoretical data for protostars.

There is some indication that the bulk of the accretion in the inner solar system took place between 500 and 700 K. Only 5–10% of the ordinary chondrites are type 3's which accreted below 500 K, and no known meteorites have the FeS-free, but otherwise chondritic composition expected for material accreted above 680 K. The Earth, too, seems to contain substantial amounts of FeS (Anderson et al. 1971), and resembles O4–O6 chondrites in the abundance of volatile elements (Anders 1968; Larimer 1971). There are two limiting interpretations for this pattern: accretion at a constant temperature of $\sim$540 K (at 10^{-2} atm) or at falling temperatures from $>$700 to $\sim$350 K. In the latter case, a large part of the volatiles may have been brought in as a thin veneer of carbonaceous-chondrite-like material, swept up by the Earth in the final stage of accretion (Turekian and Clark 1969).

The Moon seems to be depleted in volatiles by another 2 orders of magnitude relative to the Earth (Ganapathy et al. 1970; Anders 1970; Anders et al. 1971). In terms of the preceding 2 models, this may imply either accretion at a higher temperature (610 K at 10^{-2} atm) or a much lower accretion efficiency in the final stages of formation, when volatile-rich material was being swept up.

Material Balance

The fate of the lost volatiles remains largely undetermined. A few of the most volatile elements (In, Bi, Tl, Cs, and Br) are occasionally enriched in O3 chondrites beyond cosmic levels. Such enrichment might be expected in the final stages of accretion when the gas phase had become enriched in volatiles left behind by the previously accreted O4's–O6's. An even more striking enrichment is shown by Hg, thought to be the most volatile of all metals (Urey 1952*a*, 1954). It is enriched in C chondrites by 1–2 orders of magnitude above cosmic levels and is only slightly depleted in O chondrites (see Anders 1971*a* for references). Larimer and Anders (1967) have tried to explain the "mercury paradox" by postulating the existence of a relatively involatile Hg compound, but Arrhenius and Alfvén (1971) have proposed an attractive alternative. If condensation took place from a partially ionized plasma, Hg, because of its high ionization potential, would be neutralized ahead of most other elements, thus becoming available for condensation at an earlier stage.

Some part of the volatiles may have been lost with the gases during dissipation of the solar nebula. Hence it seems unlikely that the average compositon of all meteorites is close to cosmic. Meteorites apparently did not evolve in a closed system.

Other Cosmothermometers. Fe^{2+} and O^{18}

If the dust remains in equilibrium with the gas on cooling, metallic Fe will be oxidized to Fe^{2+} which enters magnesium silicates and forms the solid solutions $(Mg, Fe)_2SiO_4$ (=olivine) and $(Mg, Fe)SiO_3$ (=orthopyroxene). The Fe^{2+} content of olivine in O chondrites, ranging from 16 mol percent in H's to 31 mol percent in LL's, corresponds to equilibrium at 550–500 K (Grossman 1972). These temperatures are independent of pressure, but depend on H_2/H_2O ratio. The above values apply to a gas of solar composition, $H_2/H_2O = 640$; they are uncertain by some 50° owing to uncertainties in the thermodynamic data. In terms of the two-component model this was the temperature range of the dust when chondrule formation began. It is close to the accretion temperatures of 460–560 K estimated from volatile metals. Thus chondrule formation and accretion may have proceeded simultaneously (Larimer and Anders 1970).

Very similar temperatures have been estimated by Onuma, Clayton, and Mayeda (1972) from the O^{18} content of chondrites. They note that most of the oxygen in a solar nebula will be contained in gaseous species: CO and H_2O at high temperature; H_2O alone at low temperature. If the dust maintains isotopic equilibrium with the gas, its O^{18} content will vary with temperature. The temperature dependence can be predicted from the relevant chemical and isotopic-exchange equilibria.

For the O chondrites, Onuma et al. obtain chondrule formation temperatures of 450 to 470 K. These temperatures again refer to the temperature of the dust at the onset of chondrule formation. They are independent of pressure above 10^{-6} atm, but depend on the O^{18}/O^{16} ratio in the nebula and isotopic fractionation factors for the relevant minerals. The combined uncertainty is probably a few tens of degrees.

The O^{18} temperatures are somewhat lower than the accretion temperatures based on volatile metals, 460 to 560 K. The latter are pressure dependent, however, and can be brought into approximate concordancy with the O^{18} temperatures by lowering the pressure to 10^{-5} atm, well within the quoted limits of $10^{-4\pm2}$ atm (§ VIII). The O^{18} temperatures also are slightly lower than the chondrule formation temperatures estimated from Fe^{2+} (500–550 K), but within the margin of error.

For carbonaceous chondrites, higher chondrule formation temperatures were found, 530 to 620 K at 10^{-4} atm (Onuma et al. 1972). The higher of these values are pressure dependent, because they fall in a range where the O reservoir in the nebula changes from CO to H_2O, by the pressure dependent CO/CH_4 equilibrium. At pressures of 10^{-5} to 10^{-6} atm, these temperatures would be lowered by $\leqslant 50$ and $\leqslant 100$ K.

The higher temperatures (or lower pressures) for C chondrites are consistent with the fact that these meteorites are depleted in Na, K, and Mn, elements of only moderate volatility which are not depleted in O chondrites (Fig. 1).

Onuma et al. also obtained formation temperatures for 2 of the low-temperature minerals in Cl, C2 chondrites: hydrated silicates (probably serpentine and montmorillonite, Bass 1971) and dolomite [(Ca, Mg)CO_3]. The temperatures for Cl's were 360 ± 15 and 360 ± 5 K; for C2's (silicate only): 380 ± 15 K. These minerals probably formed from anhydrous silicates in the primary condensate, by reactions such as:

$$12(Mg, Fe)_2SiO_4 + 14H_2O \rightarrow 2Fe_3O_4 + 2H_2 + 3(Mg, Fe)_6(OH)_8Si_4O_{10}$$

$$4(Mg, Fe)_2SiO_4 + 4H_2O + 2CO_2 \rightarrow 2(Mg, Fe)CO_3 + (Mg, Fe)_6(OH)_8Si_4O_{10}$$

Roughly similar temperatures had previously been estimated from the presence of Fe_3O_4 ($\leqslant 400$ K) and hydrated silicates (300 to 350 K; Larimer and Anders 1967; Anders 1968).

V. *Organic Matter*

Carbonaceous chondrites contain up to 4% carbon. Most of it is in the form of an insoluble, aromatic polymer with –OH and –COOH groups, somewhat similar to coal or to humic acids in soils. The remainder consists of Mg, Fe, and Ca carbonates and a variety of organic compounds. O chondrites contain smaller amounts of C (0.01 to 2%) in an ill-defined chemical form, presumably again an aromatic polymer. The literature up to late 1966 has been reviewed by Hayes (1967) and Vdovykin (1967).

Origin of Organic Compounds

The existence of organic compounds in meteorites is a paradox under the equilibrium condensation model. At equilibrium in a solar gas of 10^{-4} atm, C should be present as small molecules of high volatility: CO at high temperatures (>600 K), CH_4 at low temperatures. Two alternatives may be considered. One view, widely held but never developed in detail, is that organic compounds were produced from $CH_4+NH_3+H_2O$ by Miller–Urey reactions. In these reactions energy is supplied from an external source (UV, charged particles, electric discharges) to produce free radicals which combine to more complex compounds lying thermodynamically "uphill" from the reactants. The other alternative is that the organic compounds formed spontaneously as metastable products, during hydrogenation of CO on cooling of the nebula. Under equilibrium conditions, methane should result by a hydrogenation reaction:

$$CO+3H_2 \rightarrow CH_4+H_2O$$

But hydrogenation of CO proceeds rapidly only on solid catalyst surfaces, in which case partially hydrogenated hydrocarbons with $H/C<4$ are formed in preference to CH_4, e.g.

$$10CO+21H_2 \rightarrow C_{10}H_{22}+10H_2O$$

This reaction, the Fischer–Tropsch synthesis, is used industrially for the production of gasoline.

Some of the minerals of meteorites (Fe–Ni, silicates, and especially Fe_3O_4) are good catalysts for this reaction. One may thus expect it to have taken place on the surfaces of dust grains in the solar nebula. Indeed, meteoritic hydrocarbons show a striking resemblance to Fischer–Tropsch hydrocarbons (Studier, Hayatsu, and Anders 1968). Normal (=straight chain) molecules predominate, followed by 4–5 slightly branched (mono- and dimethyl) isomers. This resemblance is highly significant if one considers that some 10^3 to 10^5 structural isomers exist for hydrocarbons with 15 to 20 C atoms. The Miller–Urey reaction shows no such selectivity, producing all possible isomers in comparable abundance.

A systematic study of the Fischer–Tropsch reaction, as applied to meteorites, was begun by M. H. Studier, R. Hayatsu, and the writer 6 years ago. It appears that all compound classes reliably identified in meteorites can be made by the Fischer–Tropsch reaction, or some variant thereof (Table 2). In the presence of NH_3, even compounds of biological interest are produced: amino acids, including heterocyclic or aromatic ones such as histidine and tyrosine; nucleotide bases such as adenine and guanine; porphyrins, etc. Contamination was precluded by use of deuterium in the synthesis. The Miller–Urey reaction is

Table 2. *Organic compounds in meteorites*[a]

Compound Class	Fischer–Tropsch-Type Reactions[b]
Alkanes (normal, 2-Me, 3-Me, 4,5-DiMe)	+
Alkenes	+
Isoprenoids ($<C_{15}$)	+
Aromatic Hydrocarbons (benzene, alkylbenzenes, polycyclics)	+
Fatty Acids (mainly normal)	?
Nitrogen Bases (adenine, guanine, etc.)	+
Amino Acids (protein and non-protein)	+
Porphyrin-like Compounds	+
Cl- and S-compounds (∅–Cl, thiophenes)	?
Polymer (aromatic, with –OH and –COOH groups)	?

[a] For references see Hayes (1967); Anders (1971*a*); Studier et al. (1971).
[b] + accounts reasonably well for meteoritic feature, ? not yet investigated.

able to account for some of these compounds, but not for others. A few compound classes, including the polymer, yet remain to be investigated, and it is conceivable that other processes will have to be invoked. At present, however, the Fischer–Tropsch reaction adequately accounts for the evidence.

Carbon-Isotope Fractionations

A further argument for the Fischer–Tropsch reaction comes from the isotopic differences between carbonate C (=bonded to O, e.g. $MgCO_3$) and organic C (=bounded to H, e.g. C_xH_y) in carbonaceous chondrites. Clayton (1963) showed that the carbonate C was some 7–8 % richer in C^{13} than the organic C (see Anders 1971*a* for references). Although fractionations of this magnitude are theoretically possible under equilibrium conditions at very low temperatures ($\leq 0°C$), they are not observed in nature. Urey (1967) therefore proposed that the 2 types of C came from 2 unrelated reservoirs while Arrhenius and Alfvén suggested fractionation during carbonate growth from the gas phase, involving multiple desorption or metastable molecules.

It turns out, however, that the Fischer–Tropsch reaction gives an isotopic fractionation of just the right sign and magnitude, owing to a kinetic isotopic effect (Lancet and Anders 1970). From the temperature dependence of the fractionation between 375 and 550 K, the observed fractionations in Cl and C2 chondrites correspond to 358 ± 12 and 392–418 K. These values agree rather well with the O^{18}-based formation temperatures of carbonates and silicates, 360 K for Cl's and 380 K for C2's (Onuma et al. 1972). The Miller–Urey reaction gives a fractionation of only -0.04 ± 0.02 % (Lancet and Anders, to be published.)

It seems likely that the Fischer–Tropsch reaction is also responsible for interstellar molecules (Herbig 1970*a*, *b*). Even C1 chondrites contain only ~6% their cosmic complement of C; for other meteorites and the inner planets, the amount is even smaller. The missing C was probably lost to interstellar space along with H_2, He, H_2O, and other volatiles. Its chemical state may never be known with certainty, but since the retained C appears to show the imprint of the Fischer–Tropsch reaction, it seems likely that the lost C, too, had been involved in this process. Many of the interstellar molecules identified thus far have also been seen in meteorites or in variants of the Fischer–Tropsch synthesis: HCHO, CH_3OH, HCOOH, HCN, HNCO, CH_3CN, COS.

VI. *Primordial Noble Gases*

Two types of "primordial" gas are found in meteorites, termed "solar" and "planetary" (Signer and Suess 1963). We are not concerned with the former which is sporadically distributed and apparently represents solar wind trapped on the surface of the meteorite parent body *after* accretion (Wänke 1965). Our interest is confined to the planetary component which seems to date from a pre-accretional stage. Three properties need to be explained: amounts, elemental ratios, and isotopic ratios.

The *amounts* of planetary gas show a most remarkable correlation with petrologic type, among both the ordinary and carbonaceous chondrites (see Anders 1971*a* for references). Absolute abundances of Ar, Kr, Xe rise by 3 orders of magnitude from O6 to C1 chondrites or ureilites. (He and Ne show a different behavior: they are found only in C1, 2's and a few O3's.) Elemental ratios of the heavier gases (Ar/Kr, Ar/Xe) remain constant within a factor of ≤ 6 over the entire range.

A rather common view is that planetary gases reflect some kind of solubility equilibrium between nebula and solid phase, perhaps modified by adsorption effects. The equilibrium solubility of these gases may be expected to follow Henry's law, the amount of dissolved gas at a given temperature being proportional to its partial pressure. Experimental measurements at the temperatures of interest (~500 K) are difficult, because diffusion rates are too slow to permit equilibration in times comparable to the human lifespan. The problem has been circumvented by growing a mineral directly in a noble-gas atmosphere (Lancet and Anders 1971). Solubility data were obtained for all 5 noble gases in magnetite, between 450 and 700 K. Values were highly reproducible and gave reasonable heats of solution. Henry's law was obeyed below partial pressures of 10^{-2} atm. This suggests that true equilibrium solubilities were being measured.

Data from these experiments may be compared with the noble-gas content of magnetite from the Orgueil C1 chondrite (Jeffery and Anders 1970). Extrapolating the laboratory data to 359 K, the mean formation temperature of Orgueil minerals from O^{18}/O^{16} and C^{12}/C^{13} data (Onuma et al. 1972), one finds that the following partial and total pressures are required to account for the observed gas contents:

	Ar	Kr	Xe
Partial P (atm)	6.8×10^{-11}	9.4×10^{-12}	5.5×10^{-12}
Total P (atm)	4.5×10^{-6}	2.2×10^{-3}	1.16×10^{-2}

The Ar value should be given greatest weight, because the Kr, Xe contents of meteoritic magnetite may be systematically too high owing to retention of a small amount ($<2\%$) of silicate. The total pressure found here, 5×10^{-6} atm, is consistent with previous estimates for the solar nebula. It is higher, however, than the pressure expected for the Arrhenius–Alfvén model.

The *elemental* ratios are not yet fully accounted for. He and Ne have almost equal solubilities in Fe_3O_4 and will therefore appear in nearly cosmic ratio, as observed in meteorites. But solubilities of Ar, Kr, Xe in Fe_3O_4 are nearly equal, in contrast to the decreasing depletion of the heavier gases in meteorites. Perhaps this is beside the point, because more than 99% of the Kr, Xe in Orgueil resides in hydrated silicates, not magnetite (Jeffery and Anders 1970). Thus the meteoritic pattern must be governed by the solubilities in silicate, not magnetite. Measurements on silicates will be needed to settle the matter conclusively, but since hydrated silicates, unlike magnetite, are not close-packed, it seems reasonable to expect higher solubilities, especially for the heavier gases.

The *isotopic* differences between solar and planetary gas have not yet been satisfactorily accounted for. Neither mass-dependent fractionations (gravitational escape, volume diffusion) nor nuclear reactions give comprehensive, quantitatively selfconsistent explanations (Herzog 1972). This is a common shortcoming of all models of the early solar system and hence cannot be held against any one of them.

VII. *Chronology*

This subject has been discussed in my recent review (Anders 1971*a*). In the present context, the most important point is the time scale of the evolution of the solar system. Previous age determinations had shown that all chondrites began to retain Xe^{129} (from extinct I^{129}; $t_{\frac{1}{2}} = 16.4$ My) within 15 My of each other (Podosek 1970). However, all these meteorites had experienced at least 2 heating events, chondrule formation and metamorphism, in the following schematic history.

Initial Condensation	Δt_1	Chondrule Formation	Δt_2	Accretion	Δt_3	End of Metamorphism

Both of these heating events were capable of expelling Xe, I and thus resetting the Xe^{129}–I^{129} clock. The dispersion in I–Xe dates thus gave only the *differences* in $\Delta t_1+\Delta t_2+\Delta t_3$ from meteorite to meteorite, not the *absolute* value of $\Delta t_1+\Delta t_2+\Delta t_3$. Though the total range of 15 My suggested that the time scale was on the order of 10^7 years, a longer time scale, as favored by Arrhenius and Alfvén, could not be ruled out.

Meanwhile some new data have become available which provide an absolute value for $\Delta t_1+\Delta t_2+\Delta t_3$. Alexander et al. (1972) measured Fe_3O_4 from an *unmetamorphosed* meteorite, the C1 chondrite Orgueil, and found it to be 2 ± 2 My younger than the metamorphosed C4 chondrite Karoonda (Podosek 1970). Both meteorites are genetically related by various chemical criteria, and so it is probably safe to assume that Karoonda originally had the same I^{129}/I^{127} ratio as Orgueil. The above age of -2 ± 2 My thus represents $\Delta t_1+\Delta t_2+\Delta t_3$ for Karoonda. A sizable part of the total interval must have been taken up by Δt_3, because cooling times even for small asteroids are on the order of 10^6 to 10^7 yr. Thus the entire interval between condensation and accretion must have been *less* than 2 Myr, perhaps much less.

VIII. *Conditions in the Early Solar System*

Fig. 3 illustrates temperature and pressure estimates for two localities in the early solar system: the source regions of the C1 and ordinary chondrites. Temperatures for the C1 chondrites are very well determined by three isotopic thermometers, all nominally independent of pressure. The pressure, as given by the intercept of the Ar solubility line with the temperature lines, is probably good to within a factor of 2, because Ar contents of meteorites and Ar solubilities are accurately known.

For ordinary chondrites, two pressure-independent thermometers are available, O^{18}/O^{16} and Fe^{2+}. The former, giving temperatures of 450–470 K, is probably more reliable, as discussed in the text. Substituting the O^{18} temperatures in the condensation equation for Bi, we can solve for the pressure corresponding to the observed Bi contents.

$$\log P_t = 9.25 + \log\,[\alpha/(1-\alpha)] - 6\,100/T$$

Here $9.25 = b - \log\,(2\ \mathrm{Bi/H})$, where Bi and H are the atomic abundances of Bi and H, while b is the integration constant of the Clausius–Clapeyron equation, plus the standard entropy of solution. The fraction condensed is α, while $6\,100 = (\Delta H_v - \Delta H_s)/2.303\ R$, where ΔH_v is the heat of vaporization of

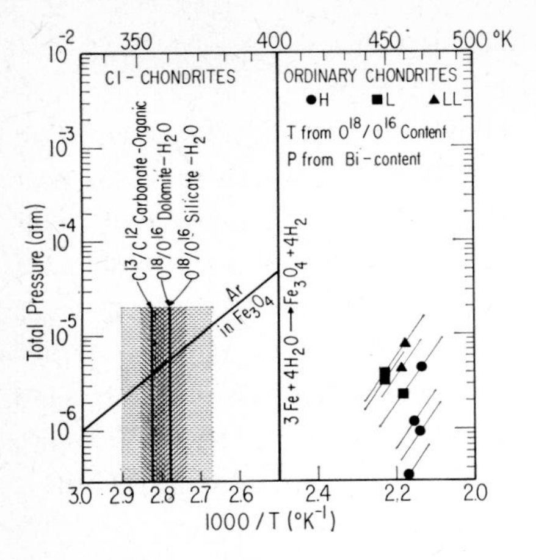

Fig. 3. Temperatures and pressures in the source regions of Cl and O chondrites. Sloping lines represent P,T-dependent equilibria; their intercept with pressure-independent temperature estimates (vertical lines or points) gives both P and T. For Cl chondrites T = 359 K and $P = 5 \times 10^{-6}$ atm is obtained. For the O chondrites, conditions are less well-determined: T = 450–470 K, $P = 10^{-4}$ to 10^{-6} atm. O^{18}/O^{16} data from Onuma et al. (1972) and Reuter et al. (1965); C^{13}/C^{12} and Ar from Lancet and Anders (1971) and unpublished; Bi from Keays et al. (1971) and Laul et al. (unpublished).

Bi, ΔH_s is its heat of solution in nickel-iron, and R is the gas constant (Keays et al. 1971).

Results are plotted in Fig. 3 for 9 meteorites for which both O^{18} and Bi data are available. The points scatter somewhat, but at least some of the scatter may reflect genuine differences among classes. The LL chondrites are highest, while the H chondrites are lowest. The error bars shown (having the slope of the Bi condensation curve) correspond to a nominal uncertainty of $\pm 10°$ in the O^{18} temperatures.

On the face of it, the agreement of the points is not bad. Temperature errors less than 10° would be required to give concordant pressures of $10^{-5} - 10^{-6}$ atm for each chondrite class. However, the actual uncertainty is somewhat greater than that. Neither the vapor pressure equation of Bi nor its heat of solution in nickel-iron are known very accurately. The magnitude of the uncertainty is indicated by the fact that pressures estimated analogously from Tl data are about an order of magnitude higher than those based on Bi data. Thus the pressures in the source region of the ordinary chondrites can only be bracketed between 10^{-4} and 10^{-6} atm. More accurate estimates will be possible when better thermodynamic data become available. But even the present data agree rather well with previous estimates for the solar nebula, and are orders of magnitude higher than those expected for an extensively ionized plasma.

It seems rather significant that the various cosmothermometers and barometers give largely concordant answers. They are based on quite diverse processes: gross chemical transformations, solubilities of trace metals or gases in solids, equilibrium isotope fractionations, kinetic isotope fractionations, etc. One can never be completely sure that this agreement is not fortuitous. Perhaps plasma reactions at much lower pressures can mimic the exact combination of isotopic and elemental ratios corresponding to P, T concordancy at higher pressures. But until such a series of coincidences is actually demonstrated, it seems justified to accept the P, T data in Fig. 3 at face value.

The obvious conclusion thus seems to be that the meteorites and planets formed from a hot solar nebula of ~ 0.1–$1\ M_{\odot}$. Most or all of their properties can be quantitatively accounted for by condensation processes, largely under equilibrium conditions.

This work was supported in part by NASA Grant NGL 14-001-010 and AEC Contract AT(11-1)-382.

References

Alexander, C., et al., 1972, manuscript in preparation.

Anders, E., 1964, Space Sci. Rev., *3*, 583.

Anders, E., 1968, Acc. Chem. Res., *1*, 289.

Anders, E., 1970, Science, *169*, 1309.

Anders, E., 1971*a*, Ann. Rev. Astron. Astrophys., *9*, 1.

Anders, E., 1971*b*, Geochim. Cosmochim. Acta, *35*, 516.

Anders, E., Ganapathy, R., Keays, R. R., Laul, J. C. and Morgan, J. W., 1971, Proc. Second Lunar Sci. Conf., *2*, Geochim. Cosmochim. Acta, Suppl. 2, 1021.

Anderson, D. L., Sammis, C. and Jordan, T., 1971, Science, *171*, 1103.

Arrhenius, G. and Alfvén, H., 1971, Earth Planet. Sci. Lett., *10*, 253.

Bass, M. N., 1971, Geochim. Cosmochim. Acta, *35*, 139.

Cameron, A. G. W., 1962, Icarus, *1*, 13.

Cameron, A. G. W., 1963, Icarus, *1*, 339.

Cameron, A. G. W., 1966, Earth Planet. Sci. Lett., *1*, 93.

Cameron, A. G. W., 1968, in Origin and Distribution of the Elements (ed. L. H. Ahrens). Pergamon: Oxford, p. 125.

Cameron, A. G. W., 1969, in Meteorite Research (ed. P. M. Millman). Dordrecht: D. Reidel, p. 7.

Clayton, R. N., 1963, Science, *140*, 192.

Ganapathy, R., Keays, R. R., Laul, J. C. and Anders, E., 1970, Proc. Apollo 11 Lunar Sci. Conf., *2*, Geochim. Cosmochim. Acta, Suppl. 1, 1117.

Grossman, L., 1972, Geochim. Cosmochim. Acta, in press.

Hayes, J. M., 1967, Geochim. Cosmochim. Acta, *31*, 1395.

Herbig, G. H., 1970*a*, paper presented at American Astronomical Society meeting, Boulder, Colorado.

Herbig, G. H., 1970*b*, Mém. Soc. Sci. Liège [8°], 5e sér., *19*, 13.

Herzog, G. F., 1972, submitted to J. Geophys. Res.

Jeffery, P. M. and Anders, E., 1970, Geochim. Cosmochim. Acta, *34*, 1175.

Keays, R. R., Ganapathy, R. and Anders, E., 1971, Geochim. Cosmochim. Acta, *35*, 337.

Kopal, Z., 1971, This volume.

Lancet, M. S. and Anders, E., 1970, Science, *170*, 980.

Lancet, M. S. and Anders, E., 1971, Meteoritics, *6* 286.
Larimer, J. W., 1967, Geochim. Cosmochim. Acta, *31*, 1215.
Larimer, J. W., 1968, Geochim. Cosmochim. Acta, *32*, 1187.
Larimer, J. W., 1971, Geochim. Cosmochim. Acta, *35*, 769.
Larimer, J. W. and Anders, E., 1967, Geochim. Cosmochim. Acta, *31*, 1239.
Larimer, J. W., and Anders, E., 1970, Geochim. Cosmochim. Acta, *34*, 367.
Laul, J. C., Morgan, J. W., Ganapathy, R. and Anders, E., 1972, Proc. Second Lunar Sci. Conf., *2*, Geochim. Cosmochim. Acta, Suppl. 2, 1139.
Lord, H. C. III, 1965, Icarus, *4*, 279.
Onuma, N., Clayton, R. N. and Mayeda, T. K., 1972, Geochim. Cosmochim. Acta, in press.
Podosek, F. A., 1970, Geochim. Cosmochim. Acta, *34*, 341.
Reuter, J. H., Epstein, S. and Taylor, H. P., Jr, 1965, Geochim. Cosmochim. Acta, *29*, 481.
Signer, P. and Suess, H. E., 1963, in Earth Science and Meteoritics (eds. J. Geiss and E. D. Goldberg). Amsterdam: North Holland, p. 241.
Studier, M. H., Hayatsu, R. and Anders, E., 1968, Geochim. Cosmochim. Acta, *32*, 151.
Studier, M. H., Hayatsu, R. and Anders, E., 1972, Geochim. Cosmochim. Acta, in press.
Turekian, K. K. and Clark, S. P., Jr, 1969, Earth Planet. Sci. Lett., *6*, 346.
Urey, H. C., 1952*a*, The Planets. New Haven: Yale University Press.
Urey, H. C., 1952*b*, Geochim. Cosmochim. Acta, *2*, 269.
Urey, H. C., 1954, Astrophys. J., Suppl. , *1*, 147.
Urey, H. C., 1967, Quart. J. Roy. Astron. Soc., *8*, 23.
Urey, H. C. and Craig, H., 1953, Geochim. Cosmochim. Acta, *4*, 36.
Van Schmus, W. R. and Wood, J. A., 1967, Geochim. Cosmochim. Acta, *31*, 747.
Vdovykin, G. P., 1967, Carbon Matter of Meteorites (Organic Compounds, Diamonds, Graphite), Moscow: Nauka Publishing Office.
Wänke, H., 1965, Z. Naturforsch., *20a*, 946.
Whipple, F. L., 1966, Science, *153*, 54.
Wood, J. A., 1962, Nature, *194*, 127.
Wood, J. A., 1963, Icarus, *2*, 152.

Discussion

H. Alfvén

There is no principal difference between "gas" and "plasma". With present terminology a gas is called a plasma with a very low degree of ionization.

G. Arrhenius

Professor Anders deserves great credit for having so thoroughly documented the notion that the grain temperature in condensation must have been far below the melting points or melting intervals of the solids.

My major reservation against your scheme is on your assumption that the surrounding gas could have been kept at the same temperature as the grains and that it could avoid being partially ionized. The Sun, particularly in the superluminous state invoked, would partially ionize any gas in the solar system. If you talk about blowing components around with the aid of solar wind, how can you use this fully ionized plasma without charge transfer? And even if you were to assume a dark Sun or no Sun, how can you avoid electron flow and ionization in gas moving in a magnetic field? Already the high primordial radioactivity invoked as an early heat source, is a source of ionization. And any time that you speak of a gas at several thousand degrees, this is by necessity partially ionized.

E. Anders

Theories of plasma condensation are in an early stage of development. You have succeeded in explaining, qualitatively, a remarkable number of the meteoritic properties observed. As your model becomes more quantitative it will be easier to see whether one model is better than the other. The neutral gas condensation model can account, more or less quantitatively, for most of the evidence known at present. It is entirely possible that certain features will some day be explained very much better by a quantitative version of the plasma condensation model.

As for the heat source, the gas condensation model is in no way tied to the existence of a superluminous sun. Cameron (1962, 1963) has shown that the collapse of the nebula would heat it to temperatures above 2 000 K, so no external heat source may be required. In fact, his calculations suggest that the Sun formed late, after sizeable bodies had accreted in the nebula.

I agree that a small number of ions will be present in the gas, even if the Sun is not yet radiating. But while these ions may have important magnetohydrodynamic effects, they will have little influence on the condensation path. It is the neutral atoms, not the ions, which condense, and thus a slight degree of ionization of an element can at most retard its condensation to a very minor extent. Only at temperatures above 4 000 K will ionization be appreciable. It is not clear how the gas at 2–3 AU can be maintained at this temperature, expecially if the Sun is not yet radiating.

Our models differ in one very fundamental respect, pressure. According to the gas condensation model, the pressure in the region of the ordinary and carbonaceous chondrites must have been at least 10^{-6} atm, or $>2\times10^{13}$ molecules/cm^3. All kinds of paradoxes and inconsistencies develop if pressures are assumed to be lower, because some of the relevant chemical equilibria are pressure-dependent while others are not. At pressures below 10^{-6} atm, various chemical changes would occur in the wrong order. In the plasma condensation model, on the other hand, densities must be much lower than 10^{13}/cm^3, to prevent ion recombination.

Finally, there are the organic compounds. They look as if they had been made in a neutral gas, not a plasma.

Cameron A. G. W., Icarus, 1, 13 (1962); 339 (1963).

B. A. Lindblad

You are discussing the formation of grains at temperatures of the order at a few hundred degrees. When these grains are sizable—say 10 to 50 microns—they will be moving in Keplerian orbits if there is a central attracting mass. Are you assuming essentially circular orbits in your discussion? If the grains are moving in eccentric orbits a large variation in the temperature will occur. How does this affect your theory of grain formation?

E. Anders

You have brought up a very important point which we had neglected in our earlier papers. The grains we are considering are less than 1 μm in diameter, and so they probably move in circular orbits, owing to the damping effect of the gas. But the asteroid on which they are accreting is moving in an eccentric orbit, and hence sweeps out a fair range of a. Judging from our results to date, the temperature distribution in the nebula was rather flat, and so the composition of grains probably did not vary too much with a. But if the asteroid moved in an eccentric orbit, it would also sweep the region above and below the median plane. Pressures are lower there, and hence condensation of volatiles is less complete. I hope to discuss this problem in one of my next papers.

F. L. Whipple

There is another factor for which we must have internal consistency and which is calculable, but I do not think it has been done well yet. It is based upon the growth rates. The time scales that we have, or the degree that we are limited to 2 million years, are not really critical. But if you assume a density you then, at a given solar distance, do determine the time scale in which you can accumulate these bodies. That time rate of growth is very important I think. If you go to densities very much below 10^{-6} atmospheres, it is difficult to grow these bodies rapidly enough. So the time scale of growth which can be calculated will answer this question because there is always a relationship between the time scale of growth and the amount of material, the density you have there. And then there is an accumulation problem and I really don't think you can accumulate these bodies into the small asteroids and have them to grow to large sizes unless you have a certain limit of dust density. I rather like this density of 10^{-4}. As soon as you specify density and temperature, you specify growth rate and this is going to limit the problem. The key lies in the growth rates and we must get these data and put them to work. With all the data that have been accumulated here, we may solve the problem of the densities and then we can see whether the time scales fall within the appropriate limits or not, as a check.

G. Arrhenius

Calculating the growth rates you arrive at the order of magnitude of 10–100 μm/day. The growth is accelerated by high temperature, particularly when low grain temperature is maintained. The time scale of growth of particles in the primordial circumsolar gas hence could have been at the level of hours, days and weeks, provided the supply of source material was not the limiting factor. The actual time it took probably depends critically on the total time and rate of addition of the matter to the condensing system.

F. L. Whipple

You have to accumulate them as well as grow them. If you have a hot plasma at a low density you blow away everything and I dont see how you accumulate them. At the stage you consider, I think that the accretion is coming to an end.

G. Arrhenius

This question may be discussed further in connection with accretion later in the program.

P. Pellas

I have the following questions and remarks:

1) As volatile properties of the elements is the major parameter used in your model, let me recall that the most volatile elements are the noble gases, which are also easy to measure with high accuracy. Hence your model has to integrate the elemental *and* isotopic ratios measured in meteorites: in the different classes (carbonaceous, enstatite, ordinary) of chondrites, and for the different petrographic types in each of these classes. Your model does not fit, at this moment, the noble gas data.

2) The notion of "matrix" you use, appears to me a very vague one. In the case of carbonaceous chondrites (especially for the C1 and C2) it is possible to speak about a "matrix". But what is excactly the "matrix" in petrographic types 4, 5 or 6? The "matrix" in these latter cases may well be only a smaller grain-size fraction of *broken* chondrules and crystals. If that is the case, why should the volatile elements be enriched in the "matrix"?

3) It appears to me that the enstatite chondrites (of different petrographic types, your types I and II) do not fit your model very well. More important, the enstatite achondrites do not fit your model, either on a fractionated heavy rare gas basis or on a volatile element basis (I put forward the enstatite achondrite case, only because you have tentatively explained within the framework of your model the "accretion temperatures" of the eucrites, shergottites and nakhlites). If you consider that the "low-temperature fraction" corresponds to

the "matrix", where is the matrix in the nakhlites, in the shergottites, in the enstatite achondrites or even in the eucrites?

4) Finally, what do you mean exactly by "metamorphism"? Perhaps it is worthwile to quote Ramdohr's statement (1967): "*High initial temperatures, and long cooling times would create just those textures which many authors ascribe to metamorphism, e.g. indistinct chondrule form and cementation of chondrules and matrix.*"

E. Anders

I very much appreciate your comments, because they bring up points that must have troubled other participants.

1) I do not think we have neglected the noble-gas data. In a paper written more than a year ago but published only recently (Keays et al., 1971) we showed that the heats of solution of Ar, Kr and Xe in meteoritic minerals must be -15 to -25 kcal/mole, if the correlation of noble gases with volatile metals was to be explained. And the solubilities at ~ 500 K would have to be $\sim 10^5$ times higher than the values measured by Kirsten at 1 773 K. Both of these predictions have been confirmed for at least one mineral, magnetite. The heats of solution are -15 to -16 kcal/mole, and the solubilities are indeed $10^5 - 10^6$ times higher than Kirsten's (Lancet and Anders 1971). Other minerals must be measured, of course, but in view of the trend thus far, we can perhaps be forgiven our optimism.

2) In my 1964 paper I referred to the volatile-rich and volatile-poor fractions as "A" and "B" rather than "matrix" and "chondrules", to emphasize the independence of this chemical model from petrographic models. Meanwhile, some evidence has accumulated supporting that A equals matrix and B equals chondrules (Larimer and Anders 1967; Anders 1971). But if you do not accept this identification, we can once again revert to the neutral terms A and B. I think I can make a strong argument for the presence of such components at the time of accretion. One of them (A) must originally have been as fine-grained as the matrix in chondrites of types 1, 2 and 3, i.e. less than 1 μm. If types 4–6 show a coarser-grained matrix today, this is very probably due to metamorphic recrystallization.

3) I have not yet had time to write our paper on enstatite chondrites, but I assure you that the data are entirely consistent with our model. We have not yet measured the enstatite achondrites but from the data in the literature I anticipate no contradictions. Concerning the other achondrites: we believe, with Wood (1962), that all bodies in the inner solar system, including the Earth, Moon and achondrite parent bodies, accreted from a mixture of high-and low-temperature material.

4) I do not think this is the place to go into the case for and against meta-

morphism. There is a voluminous literature on this subject, starting with the papers of Merrill and Zavaritskij in the 1920's and 1930's. Let me note, however, that there are many objections to a monotonic cooling history (at constant or declining rate) as proposed by Ramdohr as an alternative to metamorphic reheating. There are a number of thermometers which show consistently that the temperature during accretion was around 450–550 K: Pb, Bi, In, Tl, noble-gas abundances, O^{18}/O^{16} ratio, absence of Fe_3O_4, presence of FeS, Fe^{2+} content of silicates, etc. And there are other thermometers which show that the peak temperature reached after accretion was around 900–1 200 K: structural state of feldspar, distribution of Mg^{2+} and Fe^{2+} between pyroxenes, O^{18}/O^{16} ratios, etc. Thus temperatures went through a maximum after accretion, rather than falling monotonically.

D. Lal

My question concerns your statement that concordant temperatures are obtained using different cosmothermometers, in particular basing on volatile elements and low temperature minerals. If one starts out with a high temperature vapour and cools it and in this process obtains the different components of meteorites, what does one mean by saying that the accretion temperature is of the order of 500 K?

E. Anders

This is the temperature at which the grains ceased to equilibrate with the gas, and hence ceased to take up volatiles. According to Larimer's (1967) diffusion calculations, grains of $\leqslant 10^{-5}$ cm would still equilibrate with the gas at 500 K on the time scale appropriate to the nebula (decades to centuries).

Thus the end of equilibration was probably brought on by the physical isolation of the grains from the gas, that is, accretion.

J. Sherman

In Professor Arrhenius model the grains are charged by the plasma. If the charge is strongly negative, ions impinging on the grains may be accelerated and impact the surface with relatively high energy. What effect will this have on the absorption of rare gases or trace elements in the matrix? Can this influence the discussion of argon adsorption given in Professor Anders' paper?

G. Arrhenius

At high energies such as in solar flares noble gas atoms penetrate to such depths in the target crystal that they become permanently embedded.

At the much lower energies that we are mainly concerned with here, however, the interaction is limited to the surface atomic layer of the solid. The energy state of the impinging atoms would indeed be expected to influence the desorption rate and the bond formation with the surface atoms in the way discussed in my lecture and also in Professor Massey's presentation. This is true also for the noble gas atoms.

Cometary Meteoroids

By Peter M. Millman

National Research Council of Canada, Ottawa, Ontario

Introduction

Most of the particulate material in interplanetary space encountered by the Earth is associated with comets, both in origin and in physical nature. This first became clear when, in the nineteenth century, a number of orbits followed by streams of meteoroids were found to be identical with those of certain comets. We now know of some dozen meteoroid streams (Jacchia 1963) with associated comets, and we have good reason to believe that all similar streams had a parent comet at some time in the past. The orbits followed by these meteoroids range all the way from those with long periods and high eccentricities, like the Perseids, where the particles spend most of their time further from the Sun than the planet Uranus, down to much smaller orbits, like the Geminids, where the entire path lies well within that of Jupiter. In addition to the recognized streams there is a background complex of meteoritic material in orbit about the Sun. Dynamically, this appears to have dispersed from the more concentrated meteoroid streams; and physically, the particles seem to consist of the same material as in these streams. All of the small particles noted above will be considered together as the cometary meteoroids. They account for roughly 99 per cent of the material run into by the Earth in its orbit round the Sun.

Densities

It is significant that no cometary meteoroid seems to have fallen to Earth as a unit larger than a micrometeorite. This means that the physical structure of the material is so fragile that even the largest particles are fragmented into dust and vapour on entering the Earth's atmosphere. Confirmation of this comes from the physical theory of meteor luminosity (Öpik 1958). Thanks to the firing of a number of artificial meteors from multiple-stage rockets, the luminous efficiency of meteoritic matter is now fairly well known, and using this we have a good idea of the approximate masses involved. When meteor masses are combined with the decelerations recorded by photography, we find bulk densities for the cometary meteoroids that are much lower than the typical

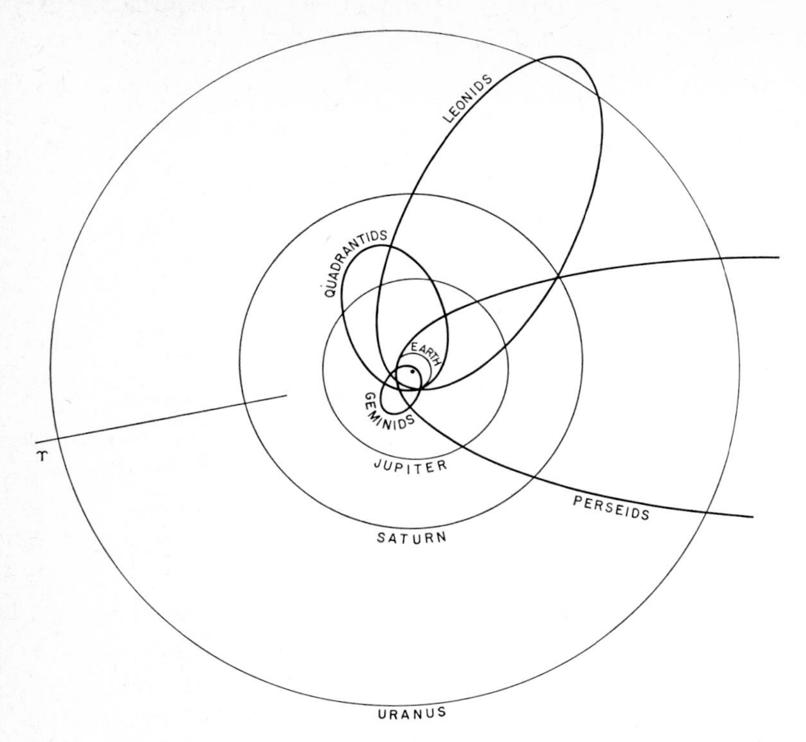

Fig. 1. Typical orbits of several well-known meteor streams, plotted in the plane of the ecliptic. In actual fact these meteor orbit planes have a wide range of high inclinations to the ecliptic (Jacchia 1963).

densities of the stone meteorites which fall to Earth. Verniani (1967, 1969) has summarized the information on meteoroid densities, as derived from 324 precisely reduced meteors photographed with the Super-Schmidt cameras at the field stations of the Harvard College Observatory. These densities are listed in Table 1. The majority of cometary meteoroids have bulk densities well under 0.5 g/cm^3, but a small group, about 15 per cent of those studied, have higher densities, in the range of 1 to 2 g/cm^3. These higher density meteors, both shower and background, have orbits lying within the orbit of Jupiter. Although some investigators have suggested somewhat higher densities than those given by Verniani, there seems no doubt that the actual densities are well under those of the stony meteorites. The generally low densities of all cometary meteoroids indicate a loosely-bound structure containing significant quantities of the lighter elements, probably in the form of ices. Where the matrix of ices has melted or evaporated the remainder will be a porous conglomerate of small particles of the heavier common elements. This picture

Table 1. *Densities of cometary meteoroids after Verniani (1967)*

	g/cm³	Number
Non-shower (sporadic)		
low density	0.21	189
high density	1.38	31
Shower		
Geminid	1.06	20
Virginid	0.7	1
N ι Aquarid	0.63	2
Leonid	0.6	1
η Aquarid	0.6	1
σ Hydrid	0.40	3
Lyrid	0.39	3
S ι Aquarid	0.30	5
Perseid	0.29	8
S Taurid	0.28	18
δ Aquarid	0.27	7
N Taurid	0.26	4
Orionid	0.25	4
Quadrantid	0.20	9
$\varkappa$ Cygnid	0.17	4
α Capricornid	0.14	12
Giacobinid	$<$0.01	2
Overall mean density	0.30	324

is consistent with the fact that all cometary meteoroids break up before reaching the Earth's surface. The extreme example of this was the Tunguska event of 1908, where a cometary mass some hundred-thousand tons in weight was completely pulverized in the atmosphere (Tsikulin 1961).

Mass Distributions

In any discussion of meteoroid origin and evolution the particle mass distribution is an important parameter. Since the mass of the meteoroid is related to the brightness of the meteor, the mass distribution can be determined observationally by a statistical study of the distribution of meteor luminosities as determined photographically or by visual recording. An alternative method is to measure some feature of the meteor radar echo that is related directly to the quantity of ionization per unit path length. This in turn depends on the original mass loss. The mass distribution for a small range in mass is generally defined by a quantity we can call the differential mass index, s. The increment in counted numbers of particles, dN, varies as $m^{-s}dm$, where m is mass and dm the increment in mass limit for counting. The index s has been determined repeatedly by counts of meteors made visually, photographically, and by radar, and has also been estimated by a theoretical extension of the particle-penetration records from the Pegasus and the Explorer satellites. As might be expected the observational values of s cover a broad range, which results

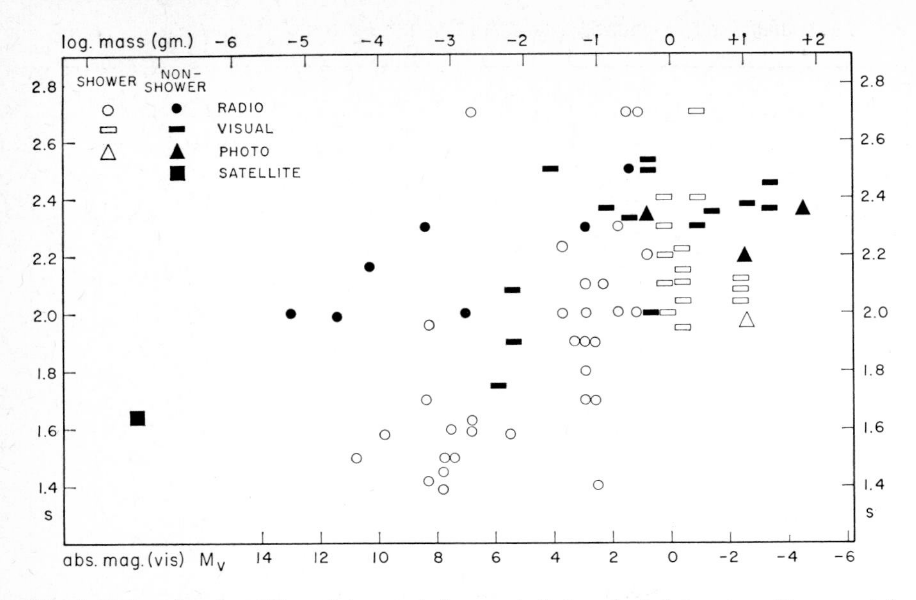

Fig. 2. Values for the differential mass index *s*, plotted against the meteoroid mass at the centre of the mass range over which *s* was determined. Material for this plot was taken from: Clifton and Naumann, 1966; Elford, 1967; Hawkins and Upton, 1958; Kresáková, 1966; Kresáková and Kresák, 1955; Levin, 1955; Lindblad, 1967; McIntosh and Šimek, 1969; Millman, 1935, 1967*b*, 1970*a*, *b*, unpublished visual data on Giacobinids; Millman and Mc Kinley, 1963; Naumann, 1966.

partly from real differences in various parts of the meteoritic complex, and partly from certain systematic errors and selection effects inherent in the particular technique employed. However, the *s*-values, based on the counts involving very large numbers of meteors, approach a general mean for a given mass range. Fig. 2 reproduces a representative selection of *s*-values, found by various observational methods, and plotted against the centre of the mass range used in finding *s*. It will be noted that the trend is to smaller *s*-values for smaller particle masses and, in general, *s* for shower meteors is smaller than *s* for the sporadic background or non-shower meteors.

Over long periods of time collisions among the particles of the meteoritic complex itself will have the effect of gradually fragmenting and pulverizing the material. This action will steadily increase the numbers of small particles and will be especially effective where, as in this case, the cohesive strength within each particle is small. Dohnanyi (1970) has treated the subject theoretically, taking into account the erosive effects of collisions, and finds that, for his model, the observed mass distribution of the sporadic meteoroids is unstable and can only be maintained by being continually fed from some source—the dispersing meteor streams. He finds that to maintain a background complex with $s = 13/6$ requires dispersing meteor streams whose small particles have reached a steady-state distribution with $s = 1.5$. It is of interest that in Figure 2 the non-shower meteors of small mass have s in the range 2.0 to 2.3, while the corresponding shower meteors have a lower limit of s in the range 1.4 to 1.6. This gives further support to the hypothesis that the background meteoritic complex is maintained by the dispersing meteor streams, which in turn have their origin in disintegrating comets. Whipple (1967) has estimated that the addition of some 10^{14} to 10^{15} g/yr of cometary material to the meteoritic complex would maintain it in a steady state. This is about one-tenth of one per cent, or less, of the original mass of a large comet like Halley's (Whipple 1963).

Chemical Composition

Since we cannot study a reliably identified cometary meteoroid in the laboratory, the chemical composition must be determined by less direct methods. On the assumption that all cometary meteoroids originate in comets we can infer the chemical elements present by noting those observed in the spectra of comets. Table 2 lists these under two headings—elements found in the well-known molecular spectra of the coma and tail, and additional elements that show up when the nucleus of the comet approaches close to the sun. Good spectroscopic observations in the second category are not numerous and it is very likely that this list of elements will be lengthened when the next sun-grazing comet is observed.

Additional, and more direct, information comes from the observational data on the spectra of cometary meteors. The elements identified in meteor spectra have also been listed in Table 2. Taking the lists for both comets and meteors we find the common elements we might expect to appear in the range of wavelengths accessible to astronomical observation, and under conditions of fairly low excitation potentials.

The determination of the relative quantities of the various elements in cometary meteoroids is beset with many difficulties and problems. Meteor luminosity comes from a mixture of both atmospheric and meteoritic atoms and molecules, and the proportions of each in the radiating gas are not known. Unknown also is the exact stage of fragmentation of the meteoroid at any given point along the meteor path, and hence the nature of the radiating mech-

Table 2.

Elements identified in spectra of comets (*O'Dell 1971; Slaughter 1969; Wurm 1963*)
Coma and tail: H, He? C, N, O
Nucleus near sun: Na, K, Ca, V, Cr, Mn, Fe, Co, Ni, Cu.

Elements identified in spectra of meteors (*Ceplecha 1971; Halliday 1961; Millman 1963*)
H, Li?, C, N, O, Na, Mg, Al, Si, K?, Ca, Ti, Cr, Mn, Fe, Co, Ni, Sr, Ba?

anism is often in doubt. Even if we do know the various radiating mechanisms operative at any given instant we are still in trouble. The numerical values necessary to convert from the quantity of radiation in a line or band to the numbers of radiating atoms and molecules are missing in most cases. Attempts to calculate meteoroid compositions on the classical basis by finding an effective temperature are unsound, as the radiating gas is very far from conditions approaching thermodynamical equilibrium.

Some recent work at the Ames Research Laboratory of NASA, California (Boitnott and Savage 1970, 1971; Savage and Boitnott 1971), has provided the first opportunity for a reliable determination of percentage composition in meteoroids. In these papers the absolute luminous efficiencies for Na, Mg, Ca and Fe atoms have been measured under simulated meteor conditions. The results are applicable to absolute luminosity measures in meteor spectra if we take the upper portions of the meteor trajectories, where the assumption of free molecular flow is valid. Extensive work on Fe, under conditions of collisional excitation, has also been reported by Friichtenicht and Becker (1971).

Accurate photometry, giving absolute values for the luminosity of various eatures in meteor spectra, has been carried out in only a few cases (Cook and Millman 1955, 1956; Millman and Cook 1959; Cook et al. 1971*a*). Some of

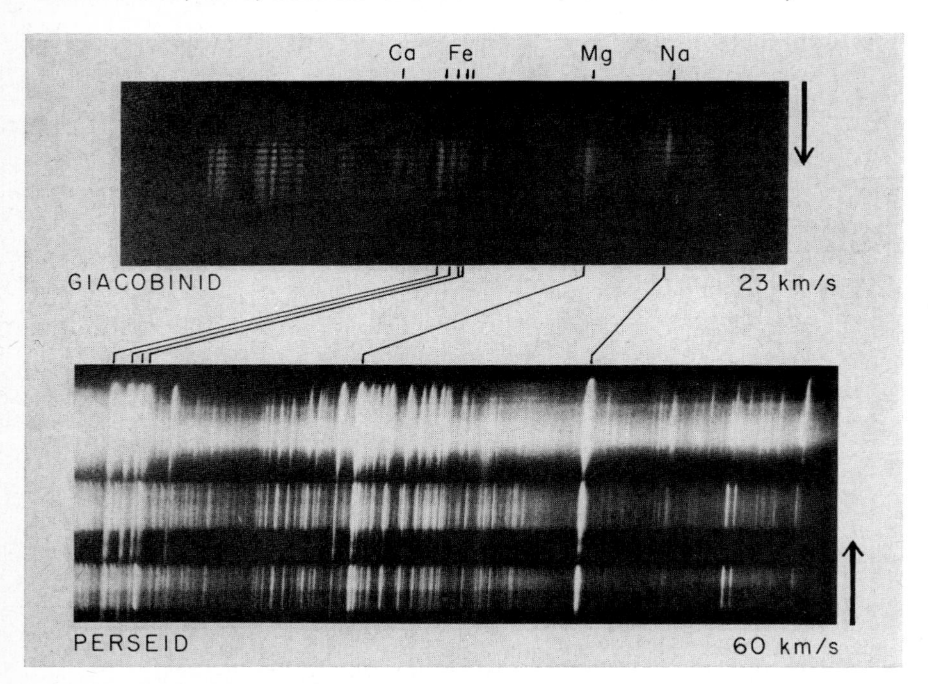

Fig. 3. Examples of the spectra of two cometary meteors, photographed on the Ottawa meteor program. Direction of meteor motion is indicated by the arrows. Positions of some of the neutral atomic lines of Na, Mg, Ca and Fe are indicated.

good lines in the copper that come out under similar conditions and I think it simply means that there is not very much copper in cometary meteoroids. We have some very detailed spectra now. A meteor spectrum with over 900 lines in it was reported at Albany in June, but no copper was found. So this may be significant. Let me point out, however, that the data I have listed for comets have been collected quickly from the literature. The meteor data have a higher significance.

V. Vanýsek

Can the Hα-line be due to hydrogen from the outer parts of the terrestrial atmosphere?

P. M. Millman

No. The hydrogen lines are strongest at the bottom of the trails near a height of 85 km. We generally have hydrogen lines in the spectra of fast meteors.

E. Anders

I want to congratulate you on a beautiful piece of work which has great cosmogonic significance. From your data thus far it seems that comets consist of relatively unfractionated solar matter, not of material lost from the inner planets. This was not really obvious until now. Is there any prospect of extending your work to another 10 or so showers?

P. M. Millman

Yes. We have a lot of material now in the process of reduction, the Geminids for example, and we are building up more. We hope to improve the photometry and expand the material in a year or two. We have not placed too much emphasis on the absolute photometry measures, but now that we are beginning to get some laboratory values that can be used we will go ahead fast. I hope those here can influence people to do laboratory work in this field. It would be very nice to have the laboratory values for some of the atoms like H and Mn and a few others.

P. Pellas

Taking into account your very fine results on the chemical compositions of cometary meteoroids, have you some personal idea concerning the location(s) where the comets have been formed?

P. M. Millman

I have not considered this problem enough to make my opinion at the present worth very much. I generally think of the comets as an integral part of the solar system, and that they were formed together with the rest of the solar system.

H. Massey

Relative values of cross sections are more reliable than absolute values but they must be determined in the same experiment. Otherwise instrumental effects become too important.

P. M. Millman

I may point out that the laboratory work by Savage and Boitnott, and also our photometric values for meteor spectra are absolute in units of ergs/sec.

M. Wallis

I refer to your figure showing the various values of the differential mass spectral index s. Although it could be taken as support for Dohnanyi's picture, alternative interpretations are possible. In particular one would expect the particle distribution in Alfvén's "jet streams", in which particle sticking is important, to have smaller s than the sporadic meteors in which disintegrating collisions may be dominant.

P. M. Millman

I agree with you. Several models are possible. The model I mentioned is only one attempt to fit theory to observations. My presentation is primarily to summarize the observations. And the average observational values we are getting have some statistical significance.

Physical Parameters of Asteroids and Interrelations with Comets

By T. Gehrels

Lunar and Planetary Laboratory, the University of Arizona, Tucson, Arizona

Following the request made by the Organizing Committee, this is a review of our present knowledge of and recent developments in physical studies of asteroids. 1971 marks the beginning of a new stage of great activity on asteroids and comets because of this Nobel Symposium, the first study meeting for asteroid and comet missions to be held in October at the Marshall Space Flight Center, and a colloquium held in Tucson in March 1971. This review is partly based on the proceedings of the Tucson Colloquium (SP-267, see References). It contains 70 papers about 10 of which are extensive review articles and it will be a much needed reference book where none existed before. In addition, this review uses a recent discussion of the literature (Gehrels 1971).

As for observational work on asteroids, positions are obtained in about a dozen observatories in various parts of the world. Physical observations, that is of size, brightness, colors, spectral scans, and polarization, are made at only a few observatories. At least that was the case before 1971, but the interest in making these observations is increasing. Statistical studies are from the Yerkes–McDonald Survey (Kuiper et al. 1958) and the Palomar–Leiden Survey (van Houten et al. 1970*a*).

At Palomar Mountain there is, in addition to the 5-m Hale reflector, a unique Schmidt telescope with wide field, $6^\circ 3 \times 6^\circ 3$. Asteroids can be photographed with it down to the 21st astronomical magnitude, and about 10^5 asteroids would be found if the whole sky were observed. Actually, one does not need to observe the whole sky, since the asteroids are concentrated towards the plane of the ecliptic; nearly all asteroids occur within a 1.6 AU belt centered on the ecliptic. However, even with limited photography the amount of reduction work is very large, this usually limits the area of the sky that is photographed, and the results suffer from uncertainty in completion corrections.

Of the above 10^5 asteroids, the orbits of about 2 000 are presently known. These are fairly well represented by the eleven in Table 1. Sizes are directly determined only for Ceres and two or three others, although, again after 1971, more sizes may soon be determined. Masses have been determined for Ceres and Vesta only, by Hertz and by Schubart (1971). There is some discussion in NASA SP-267 about the reliability of diameter measurements, but my conclusion is that those summarized by Dollfus (1971) are good to the precision indicated. The densities of both Ceres and Vesta then may be 5 g cm^{-3} which should mean at least a partially metallic composition. The total mass in the asteroid ring is 0.0004 that of the Earth.

The other radii in Table 1 are based on the assumption of, approximately, Ceres' reflectivity for all asteroids and on the astronomical magnitude in the B(1,0) column. In some of the present literature one still sees the usage of the poorly defined absolute magnitude g; it is better to discontinue that usage and to use instead B(1,0) which is for blue light, at 1 AU from Earth and Sun, and at 0° phase angle [Table I of Gehrels (1970) gives B(1,0) values for 1683 asteroids].

Photometric studies have been reviewed by Gehrels (1970); the need for further work is so great that a special telescope (Kaufman 1971) is proposed. The asteroids appear to have a dusty regolith, roughened by particle impact, by irradiation, and by sputtering (see the paper by Lal, in this book). A sequence of roughness–size appears to emerge: the large asteroids and Trojans have phase effects similar to that of the Moon, while the small asteroids appear to be much rougher. The exception may be Geographos, with smooth lightcurves, but the radar work of Goldstein (1969) showed great roughness for Icarus, while also a stony-metallic composition was found after combination with photometric data. Unravelling of differences in texture and composition may be possible with additional studies of the opposition effect (Gehrels 1956) and polarization-phase functions (Veverka 1971), including those of comparisons on laboratory samples. The range of phase angles can be extended with measurements made from spacecraft during fly-by and rendezvous missions. A spectrophotometric study in an M.I.T. dissertation by Chapman is about

Table 1. *Orbital characteristics of representative asteroids*

	Radius km	B (1, 0) mag	Mean Motion arcsec/day	Perihelion AU	Aphelion AU	Eccen-tricity	Incli-nation
944 Hidalgo	10	12.1	253	1.98	9.7	0.66	43°
624 Hektor	40 × 110[a]	8.7	306	5.02	5.2	.02	18
279 Thule	30	9.8	400	4.15	4.4	.03	2
153 Hilda	46	8.9	448	3.38	4.6	.15	8
108 Hecuba	37	9.3	614	2.96	3.5	.08	4
1 Ceres	385	4.1	771	2.55	3.0	.08	11
93 Minerva	48	8.8	776	2.36	3.1	.14	9
46 Hestia	37	9.3	885	2.09	3.0	.17	2
434 Hungaria	10	12.1	1 309	1.80	2.1	.07	23
1221 Amor	0.4	19.2	1 333	1.08	2.8	.44	12
1932 HA Apollo	2	15.6	1 959	0.65	2.3	0.56	6

[a] See Dunlap and Gehrels 1969.

completed; an absorption band near 0.94 μm, ascribed to Fe^{2+} and again indicating a metallic composition, is found for Vesta and a few other asteroids.

Lightcurves, that show the brightness variation as a function of time, have been observed for some 50 asteroids. The shapes vary from nearly spherical to elongated with a factor $3\frac{1}{2}$ between longest and shortest axes, but for the regular asteroids the mean deviation from sphericity is only 10% in radius. The periods of rotation are summarized by Taylor (1971), they range from 2 to 18 hours, the majority from 4 to 10 hours. Isochronism of spins, from asteroids to planets, was first discussed by Alfvén (see the paper by Arrhenius in this book), and this suggests that lightcurves should be observed of asteroids fainter than $B(1,0) \sim 15$. The frequency–size relation obtained from the Palomar–Leiden Survey, shows somewhat of a non-linearity, a "hump", between absolute magnitudes 9 and 14, and this appears to indicate that the bright ones are original accretions and the fainter ones collisional fragments (Anders 1965). If so, the faint asteroids might show larger and more rapid variations than the above 50 bright asteroids. This, however, is a suggestion for photographic lightcurve work, for instance with your Schmidt telescope here in Saltsjöbaden (see p. 326 of Gehrels 1970), as these objects are generally too faint for the present photoelectric procedures.

The fourth column of Table 1 shows that it is organized by increasing mean motion, the reciprocal of the orbital period, which is to be compared to 300 arcseconds per day for Jupiter; it is seen that some of these asteroids are in resonance with Jupiter: Thule makes four revolutions against Jupiter's three, Hestia nearly three revolutions for every one of Jupiter's, etc.

The last four columns in Table 1 give the orbital characteristics of the elliptic orbits, where the aphelion distance is the greatest distance from the Sun. The asteroid belt appears to be defined at about 3.3 AU, excepting Hidalgo, the Trojan asteroids, Thule, and the Hilda asteroids. Especially for the fainter asteroids, observed in the Palomar–Leiden Survey, the boundary near 3.3 AU is clearly seen and that drop in frequency appears important for the study of asteroidal origin. At the same time, Marsden (1971) notes that no comet has aphelion distance smaller than 4.1 AU. Table 2 lists a few of the short period

Table 2. *A few short-period comets*

	Perihelion AU	Aphelion AU	Eccen-tricity	Period yr	Incli-nation	Radius km
Encke	0.34	4.1	0.85	3.3	12°	0.6–3.5
Grigg-Skjellerup	1.00	4.9	.66	5.1	21	0.7–4.2
Tuttle-Giacobini-Kresák	1.12	5.1	.64	5.5	14	0.1–0.8
d'Arrest	1.37	5.7	.61	6.7	18	0.6–3.3
Ashbrook-Jackson	2.31	5.3	0.40	7.5	13	1.6–9.3

comets; the orbital elements are taken from Marsden (1970) and the radii from Roemer (1966) for assumed reflectivities of 70 and 2%. These short-period comets with aphelia near 5 AU are usually referred to as the Jupiter family.

For the distinction of asteroids and comets it is recognized that Jupiter has a strong influence such that five types of objects appear dynamically intermixed at present: (1) Ceres, Pallas, and Vesta (see below); (2) other original condensations; (3) collisional fragments; (4) extinct cometary nuclei; (5) comets showing a coma (see below). Originally, however, the asteroids may have formed from a plasma between 2.2 and 3.3 AU. [These numbers should not be taken too literally: as pointed out above there are exceptions, including objects with high eccentricity. In addition, Mars has had a dynamical effect to cause the present positions and groupings of the asteroids at the inner side of this belt (see Williams 1971)]. Inside of the Martian orbit the asteroids all have highly eccentric orbits; the discovery of one with a nearly circular orbit, that might have accreted there, would be spectacular.

The 2.2–3.3 AU plasma belt had a curious density: so low that no major planet formed, and yet high enough for asteroidal conglomerates. Ices and volatile materials apparently were not abundant.

At large distance from the Sun, conglomerates apparently have a mixture of dust and volatiles (see Whipple's paper in this book). Their evaporation effects, when they are perturbed to come closer to the sun, cause particle expulsion observed as a coma and the observer identifies the object as a comet. (Even if at a following apparition the coma is not observed, the object remains identified as a comet.)

The cores of comets appear to be quite small. Roemer (1968; a sample of her radii is in Table 2) found 0.6–10 km (albedo 0.7; 3.5–60 km for albedo 0.02) for the radii, and some increase of size with increasing distance from the Sun. Roemer (1971, personal communication) believes this to be an effect of observational selection. A paper by Marsden and Sekanina (1971) shows a possible connection of comets and the 10-m interplanetary boulders of Harwit (1967). I wonder if these cometary boulders are observed, after entry in the Earth's atmosphere, as fireballs. Wetherill (1971) has found a connection of orbital characteristics of comet Encke and other comets with those of fireballs observed by the Prairie Meteorite Network of the Smithsonian Astrophysical Observatory. Until now it appeared there was a discrepancy of Wetherill's conclusion on cometary origin of the meteorites with that of Anders (1971) who connects meteorites with 7 selected families of asteroids that have eccentricities and inclinations high enough to allow their collision fragments to cross the orbit of Mars.

Next we return to Table 1 to discuss specific asteroids. Hidalgo, by the

2.2–3.3 AU criterion, appears to be an extinct cometary object. Only one Trojan is listed in Table 1, Hektor, and it occurs in the preceding equilaterial region of Sun and Jupiter. The Trojan regions appear to be additions to the above 2.2–3.3 AU rule and here also the density of matter was large enough to form fairly large solid objects.

The Trojans have been thought of as escaped satellites of Jupiter but that idea must be abandoned because van Houten et al. (1970*b*) found evidence, with the Palomar Schmidt, of 700 Trojans in the preceding region (that is to a limit of about the 21st magnitude). Their statistics and rotation rates seem to be similar to those of the asteroids.

The following region, L_4, was photographed in March 1971, again at Palomar, and the van Houtens have a preliminary indication that the number there is less than in the preceding region. Interpretations of the two different densities are surmised either in the adoption of a chance occurrence of higher density of the solar nebula preceding Jupiter, or on a consideration of the systematic dynamical effects that might have caused the asymmetry as well as causing mass concentration in both equilateral regions.

At this stage of discussing the outer asteroids in Table 1, a few remarks ought to be made about the present completion of discovery of asteroids. A beginning has been made to look for Trojans of Neptune, and none have been found, but even at Palomar we could not have observed objects smaller than about 190 km in radius; smaller ones are beyond the limit of the telescope because of the large distance. Trojans of Saturn have not been found, at least not larger than 30 km in radius. Asteroids between Jupiter and Saturn have not been found, to a limit of 10 km (these radius limits are for Ceres' reflectivity).

From the outside of the solar system going inward we are then left with asteroidal objects in the preceding and following equilaterial regions of Jupiter and an asteroid belt starting rather suddenly near 3.3 AU. Hidalgo, Thule, and even the Hilda objects (see Table 1) are suspected to be extinct cometary nuclei. In the asteroid belt itself and among the Amor and Apollo objects cometary nuclei may occur (see below).

Hilda, Hecuba, Minerva and Hestia represent groups of asteroids that are dynamically associated and usually called after these representative ones: the Hilda asteroids, etc.

Ceres is an exceptional case of size. In the accretion process, Ceres, Pallas, and Vesta apparently became large enough to let their own gravitation play a role whereby their capture cross section increased (Hartmann 1968). They subsequently avoided major impacts while small impacts could not destroy them (perhaps partially because they are of unusual metallic composition and density).

Next in Table 1, we come to the inner boundary of the asteroid belt which is determined by Mars and that is where the curious Hungaria grouping occurs. Because of their high inclination, while most photographic plates are taken near the ecliptic, the Hungaria asteroids are incompletely known but the Palomar–Leiden Survey showed that there may be a large number of them.

The Amor and Apollo asteroids have been listed in greater detail in Table 3. Table 3 has been provided for this Symposium by Marsden as a complete listing of all asteroids with perihelion distance smaller than 1.5 AU, excepting approximately a dozen for which the orbit was based on three observations only and these appear hopelessly lost. Table 3 is entirely due to Marsden but for the column "radius" which I added similarly to the one for Table 1. Notice that some are lost forever and for some, such as Adonis and Apollo,

Table 3. *Minor planets with perihelion distance $q \leqslant 1.50$ AU (Marsden 1971, personal communication)*

	Perihelion AU	Aphelion AU	Eccentricity	Period yr	Inclination	B(1, 0)	Radius km	Status[a]	Nature[b]
1566 Icarus	0.19	2.0	0.83	1.1	23°	17.6	0.8	S	???
1936 CA (Adonis)	0.44	3.3	.76	2.6	1	18.6	0.5	R?	C?
1971 FA	0.56	2.4	.62	1.8	22	16.3	1.5	S?	P
1937 UB (Hermes)	0.62	2.7	.62	2.1	6	18.1	0.6	L	
1932 HA (Apollo)	0.65	2.3	.56	1.8	6	15.6	2	R?	
1685 Toro	0.77	2.0	.44	1.6	9	16.3	1.5	S	
1620 Geographos	0.83	1.7	.34	1.4	13	16.0	1.7	S	P
1959 LM	0.83	1.9	.38	1.6	3	14	4	L	
1950 DA	0.84	2.5	.50	2.2	12	16.1	1.6	L?	
1948 EA	0.89	3.6	.60	3.4	18	16.6	1.3	R?	C??
1953 EA	1.03	4.0	.59	4.0	21	19.6	0.3	R?	C?
1960 UA	1.05	3.5	.54	3.4	4	18.1	0.6	R	C??
1968 AA	1.06	3.2	.51	3.2	24	16.6	1.3	S	
1950 LA	1.08	2.3	.36	2.2	26	14	4	L?	
1221 Amor	1.08	2.8	.44	2.7	12	19.2	0.4	S	
1580 Betulia	1.11	3.3	.49	3.2	52	15.7	2	S	
1627 Ivar	1.12	2.6	.40	2.6	8	14.2	4	S	
433 Eros	1.13	1.8	.22	1.8	11	12.4	9	S	P
887 Alinda	1.15	3.9	.54	4.0	9	16.3	1.5	S	C?
719 Albert	1.19	4.0	.54	4.1	11	16.9	1.1	L	C?
1036 Ganymed	1.22	4.1	.54	4.3	26	10.9	17	S	C?
1963 UA	1.24	4.0	.53	4.3	11	15.6	2	R?	C?
1953 RA	1.26	3.3	.45	3.4	13	15.7	2	R?	
1474 Beira	1.39	4.1	.49	4.5	27	13.0	7	S	C?
1134 Kepler	1.43	3.9	.47	4.4	15	15.4	2	S	C?
1009 Sirene	1.44	3.8	.45	4.3	16	16.9	1.1	L?	C?
1139 Atami	1.45	2.4	.26	2.7	13	14.3	3	S	
1963 RH	1.48	3.3	.38	3.7	21	14.3	3	L	
1198 Atlantis	1.50	3.0	0.34	3.4	3	16.8	1.1	L?	

[a] Status: S = secure, R = recoverable, L = lost.
[b] Nature: P = planetary, C = cometary.

it will take an appreciable effort to recover them, if that is at all possible (Roemer 1971).

The asteroids in Table 3 are listed in order of increasing perihelion distance and the first ten are the Apollo asteroids. The Amor asteroids are defined to be the ones with perihelion distance between 1.00 AU and 1.38 AU (the perihelion distance of Mars). It is seen that there is a drop in frequency of occurrence beyond 1.26 AU so that a cutoff near 1.38 AU for the definition of Amor asteroids appears logical. Marsden (personal communication) has pointed out that Betulia, 1953 EA and probably also Alinda may in the not too distant future have perihelion distances smaller than 1.00 AU and then become Apollos.

Marsden bases the cometary classification, in the last column of Table 3, on the work of Sekanina regarding the connections between meteor streams and minor planets. On the other hand, Marsden uses the shape, apparent from the photometric lightcurves, to identify planetary origin with a P. It is noted that the criterion of geocentric velocity mentioned by Anders in the discussion below would result in a C for Icarus and Adonis, and a P for Apollo, Hermes, Geographos, 1948 EA and 1950DA (see Table 2 of Anders and Arnold 1965).

1971 FA incidentally is the new Apollo found last March (see Gehrels, Roemer, and Marsden 1971); it will probably be observed again this fall and shall then be named Daedalus. The experience with 1971 FA showed that a special search for Apollo and Amor asteroids should be made. Until now these objects have been found only because long trails were noticed on plates taken for other purposes, while the trails of 1971 FA were not longer than two or three times that of ordinary asteroids. Öpik (1963), Hawkins (1963), Baldwin (1964), Hills (1971), and Williams (1971) have theoretically predicted rather large numbers of them and the Palomar–Leiden Survey has confirmed that. Improved statistics and understanding of these objects are important to the study of interrelations of asteroids and comets.

In conclusion I thank B. G. Marsden for his contribution to this paper and the National Aeronautics and Space Administration for financial support.

References

The abbreviation "NASA SP-267" is used for the proceedings of a colloquium "Physical Studies of Minor Planets", edited by T. Gehrels, NASA SP-267 (US Government Printing Office, Washington, D.C.), 1971.

Anders, E., 1965, Icarus, *4*, 399.
Anders, E., 1971, NASA SP-267.
Anders, E. and Arnold, J. R., 1965, Science, *149*, 1494.
Baldwin, R. B., 1964, Astron. J., *69*, 377.
Dollfus, A., 1971, NASA SP-267.

Gehrels, T., 1956, Astrophys. J., *123*, 331.
Gehrels, T., 1970, Chap. 6 in "Surfaces and Interiors of Planets and Satellites" (ed. A. Dollfus). Academic Press, London.
Gehrels, T., 1971, Trans. Amer. Geophys. Union, *52*, 453.
Gehrels, T., Roemer, E. and Marsden, B. G., 1971, Astron. J., *76*, 607.
Goldstein, R. M., 1969, Icarus, *10*, 430.
Hartmann, W. K., 1968, Astrophys. J., *152*, 337.
Harwit, M., 1967, Chap. 44 in "The Zodiacal Light and the Interplanetary Medium" (ed. J. L. Weinberg). National Aeronautics and Space Administration, Washington, D. C, NASA SP-150.
Hawkins, G. S., 1963, Nature, *197*, 781.
Hills, J. G., 1971, NASA SP-267.
Houten, C. J. van, Houten-Groeneveld, I. van, Herget, P. and Gehrels, T., 1970*a*, Astron. Astrophys. Suppl., *2*, 339.
Houten, C. J. van, Houten-Groeneveld, I. van, and Gehrels, T., 1970*b*, Astron. J., *75*, 659.
Kaufman, M., 1971, Sky and Telescope, *42*, 170.
Kuiper, G. P., Fujita, Y., Gehrels, T., Groeneveld, I., Kent, J., Van Biesbroeck, G. and van Houten, C. J., 1958, Astrophys. J. Suppl., *3*, 289.
Marsden, B. G., 1970, Astron. J., *75*, 206.
Marsden, B. G., 1971, NASA SP-267.
Marsden, B. G. and Sekanina, Z., 1971, Astron. J., *76*, 1135.
Öpik, E., 1963, Advan. Astron. Astrophys., *2*, 219.
Roemer, E., 1968, Astron. J., *73*, 533.
Roemer, E., 1971, NASA SP-267.
Schubart, J., 1971, NASA SP-267.
Taylor, R. C., 1971, NASA SP-267.
Veverka, J., 1971, NASA SP-267.
Wetherill, G. W., 1971, NASA SP-267.
Williams, J. G., 1971, NASA SP-267.

Discussion

E. Anders

Another criterion is available for distinguishing cometary from planetary Apollo asteroids: geocentric velocity. Arnold and I (1965) pointed out that the Apollo asteroids fell into 2 groups on the basis of geocentric velocity: 12–18 km/sec and 24–30 km/sec. We called them "asteroidal" and "cometary", because we were unable, in our Monte Carlo calculations, to produce the high-velocity objects from any of the known asteroids. We could, however, derive such objects from Encke's comet. In the planetary perturbations treated by Arnold's Monte Carlo method, planetocentric velocity is not strictly an invariant; first, because the object meets the planet in different parts of its orbit on successive encounters, and the non-circularity of the planet's motion causes a slight acceleration (or deceleration) on each encounter; second, because the object usually interacts with more than one planet. Nonetheless, at geocentric velocities greater than $\sim$6 km/sec, the changes are not very great.

Our assignments have been checked in two cases thus far. 1620 Geographos, one of our "asteroidal" objects, has a strongly elongated shape as expected for a fragment from the asteroid belt. 1566 Icarus, classified by us as "cometary", is spherical within ~10%, as expected for an object stored for ~4.5 AE in a collision-free environment such as Oort's comet belt. And according to Gehrels' color measurements, Icarus falls well outside the asteroid field in a UBV diagram.

Anders, E., and Arnold, J. R., 1965, Science, *149*, 1494.

T. Gehrels

We struggled with the cometary/planetary identification of Icarus (p. 194 of Gehrels et al. 1970), and I have further questions: Why could a cometary nucleus not be elongated? Why does Trojan Hektor have an elongated shape, 40 × 110 km, alike that of Geographos? Hektor is so large; is that also a fragment? These are topics for further study.

Gehrels, T., Roemer, E., Taylor, R. C., and Zellner, B. H., 1970, Astron. J., *75*, 186.

S. K. Runcorn

The distribution in space of the axes of rotation of asteroids seems of great interest. It seems possible to obtain this information if observation of the amplitudes of the lightcurves is made from different points in the relative orbits of Earth and asteroid. Do the results show a great scatter in the directions of these axes? One would suppose that the small scatter of the periods of rotation of the asteroids is some consequence of the accretion process. But steady accretion by small grains will produce spherical bodies—the elongated asteroids must therefore be formed by the coalescence or collision of bodies of comparable size. This would, one supposes, have the effect both of dispersing the periods of rotation and the directions of the axes of rotation. How can the close grouping of the periods be explained?

T. Gehrels

Hektor, for instance, has a very steep lightcurve in the one part of the orbit, and little variation in another. The axis of rotation lies near the plane of the ecliptic. One can get a precise determination of the period and of the direction of the axis of rotation provided lightcurves are observed in many points of the orbit. This requires a large amount of work and reliable axis orientations are presently available only for a few asteroids—not enough to make a statement about the scatter.

V. Vanýsek

Can some non-gravitational forces make asteroids jump from one "family" to another in spite of the strong perturbations caused by Jupiter?

T. Gehrels
Indeed, changes in the orbital elements, e.g. concerning the classification between Amor and Apollo asteroids, are possible.

Theory of Jet Streams

By Jan Trulsen

Auroral Observatory, University of Tromsø, Tromsø

Introduction

In the following a jet stream will denote a collection of grains moving in similar elliptical orbits around a central gravitating body, collisions between the individual grains being responsible for forming the stream and keeping it together against external perturbations. Such streams are postulated to play an important role in the theory for the formation and evolution of the Solar System, by professors Alfvén and Arrhenius (1970). We will here be concerned with the question whether such material streams can actually form out of a more uniform distribution of grains. The above authors gave the first qualitative arguments for their development, based on the following two facts:

1) A particle will return to the place of its last collision. Thus particles are not easily lost from a stream.

2) Two particles colliding partially inelastic will have orbits more similar after the collision than before. Thus inelastic collisions will tend to focus the individual orbits.

Two attempts have been made to construct a quantitative theory of the dynamics of jet streams. Baxter and Thompson (1971) consider a two-dimensional configuration, the collisions being completely inelastic, but number conserving. This author (1971) studied a three-dimensional model with collisions being only partially inelastic. In both theories the effect of accretional and fragmentational processes were neglected. Neglected are also effects from self-gravitation and rotational degrees of freedom for the individual grain. Since the mass of the stream would be much smaller than the mass of the central body, self gravitational effects will be rather small. The neglect of rotational degrees of freedom seems justified on the ground that the rotational energy stored in small grains can only be a very small fraction of their translational kinetic energy or even of the energy associated with the internal motions in typical stream configurations.

After having reviewed these two theories the results from a numerical simulation of jet stream dynamics will be discussed.

The Three Dimensional Model

Starting with the three dimensional model the individual collision process is required to satisfy, besides momentum conservation, the relation

$$\mathbf{v}_{1b}^2+\mathbf{v}_{2b}^2+\frac{1}{2}((\beta-1)^2-1)((\mathbf{v}_{1b}-\mathbf{v}_{2b})\cdot\mathbf{k})^2=\mathbf{v}_{1a}^2+\mathbf{v}_{2a}^2$$

where $\mathbf{v}_{1,\,2b}$ and $\mathbf{v}_{1,\,2a}$ are the velocities of the two colliding particles before and after the contact collision. $\boldsymbol{k}$ is the impact vector which is a unit vector parallel to the line connecting the centers of the two particles at impact. β is a parameter describing varying degrees of inelasticity, $\beta=2$ corresponding to an elastic collision, whereas for the lower limit $\beta=1$, a head-on collision would be completely inelastic. Note that whatever the value of β all grazing collisions are elastic.

With collision processes of this kind the following kinetic equation of Boltzmann type can be derived

$$\frac{\partial f}{\partial t}+\mathbf{v}\cdot\frac{\partial f}{\partial \mathbf{r}}-\frac{\mathbf{r}}{r^3}\cdot\frac{\partial f}{\partial \mathbf{v}}=D^2\int d\mathbf{v}^1\int d\mathbf{k}(\mathbf{v}^1-\mathbf{v})\cdot\mathbf{k}\left\{\frac{f(\bar{\mathbf{v}}^1)f(\bar{\mathbf{v}})}{(\beta-1)^2}-f(\mathbf{v}^1)f(\mathbf{v})\right\}$$

$$(\mathbf{v}^1-\mathbf{v})\cdot\mathbf{k}>0$$

where

$$\bar{\mathbf{v}}=\mathbf{v}+\frac{1}{2}\frac{\beta}{\beta-1}(\mathbf{v}^1-\mathbf{v})\cdot\mathbf{kk}$$

$$\bar{\mathbf{v}}^1=\mathbf{v}^1-\frac{1}{2}\frac{\beta}{\beta-1}(\mathbf{v}^1-\mathbf{v})\cdot\mathbf{kk}.$$

D is the grain diameter.

For $\beta=2$ the collision operator reduces to that originally derived by Boltzmann. The collision operator looks singular in the limit $\beta=1$. This is, however, not so. A simple transformation will displace the apparent singularity to $\beta=2$, the collision operator taking the form

$$C[f,f]=D^2\int d\mathbf{v}^1\int d\mathbf{k}\,(\mathbf{v}^1-\mathbf{v})\cdot\mathbf{k}f(\mathbf{v}^1)\left\{\frac{4f(\bar{\bar{\mathbf{v}}})}{(\beta-2)^2}-f(\mathbf{v})\right\}.$$

$$(\mathbf{v}^1-\mathbf{v})\cdot\mathbf{k}>0$$

with

$$\bar{\bar{\mathbf{v}}}=\mathbf{v}+\frac{\beta}{\beta-2}(\mathbf{v}^1-\mathbf{v})\cdot\mathbf{kk}.$$

Both forms of the collision operator satisfy particle number and momentum conservation.

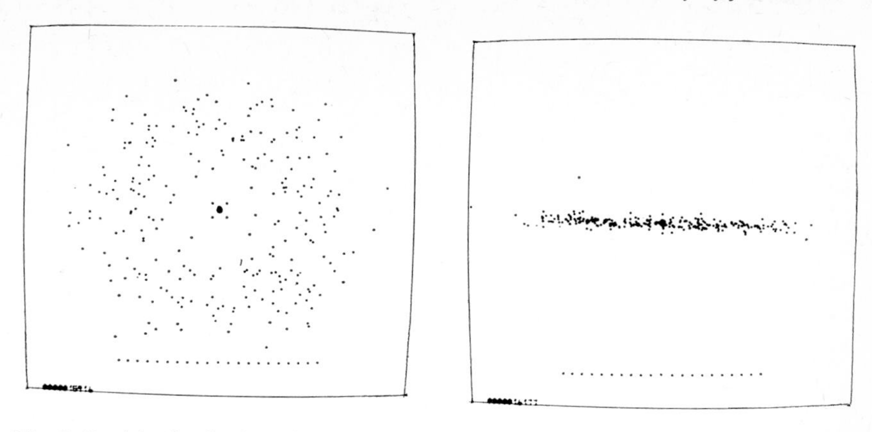

Fig. 3. Particle distribution after approximately 19 collisions/particle (t = 1 000) for $\beta = 1.4$.

for

$$a_1 < a < a_2, e < e_0 \quad \text{and} \quad i < i_0 \quad \text{with}$$

$$a_1 = 0.8, a_2 = 1.2, e_0 = 0.1 \quad \text{and} \quad i_0 = 30^\circ.$$

Some snapshots of the actual distribution of particles from the ecliptic pole and in the ecliptic plane respectively are shown in Figs. 1–5. These snapshots give a feeling for the extension of the system in the radial and polar directions initially and at a later stage for 4 different values of the inelasticity coefficient, $\beta = 1.1$, 1.4, 1.7 and 2.0. An impression for the distribution of eccentricity can be gained by adding animation. Thus a computer movie was prepared, the distribution of particles being seen from an observer in a spaceship following a circular orbit outside the stream and inclined 90° to the ecliptic plane.

Some simple statistics of the particle distribution as a function of time are

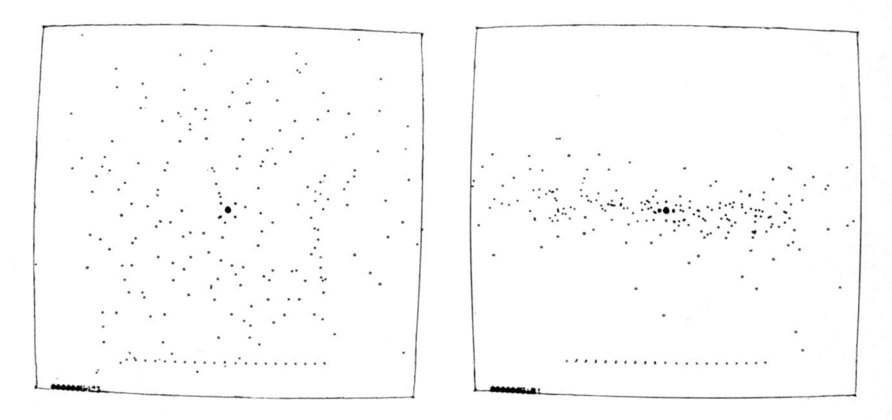

Fig. 4. Particle distribution after approximately 12 collisions/particle (t = 720) for $\beta = 1.7$.

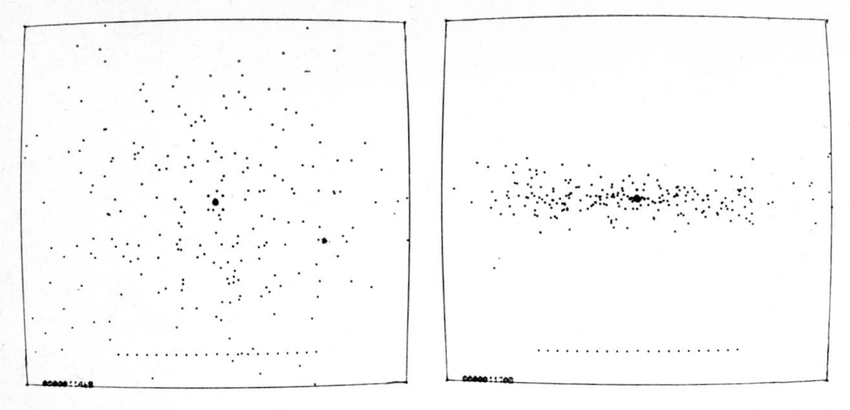

Fig. 5. Particle distribution after approximately 4 collisions/particle (t = 340) for $\beta = 2.0$.

given in Figs. 6–14. Here smoothed-out histograms of semimajor axis, eccentricity and inclination are given at different times. The main aspects of the evolution referred to above are clearly visible, e.g. the rapid initial phase and the extreme importance of the degree of inelasticity on the development of the distributions of semimajor axis and eccentricity.

Discussion

The tendency of the system to approach a Saturnian ring configuration is only one part of the formation of more proper jetstreams. To this end also a focusing of the particle distribution in the radial direction is required. This is equivalent to a focusing of the distribution of semimajor axis around some central value. This effect was not demonstrated by the numerical simulation, rather the opposite effect was observed. For the most inelastic case studied, $\beta = 1.1$ this outward diffusion from the maximum density in the radial direction is rather

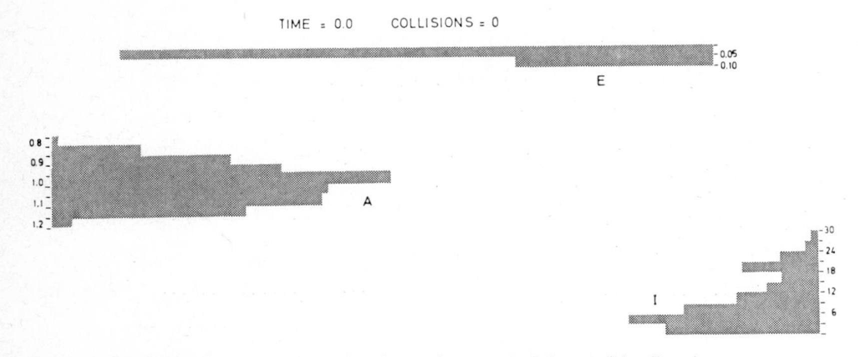

Fig. 6. Initial distributions of semimajor axis, eccentricity and inclination.

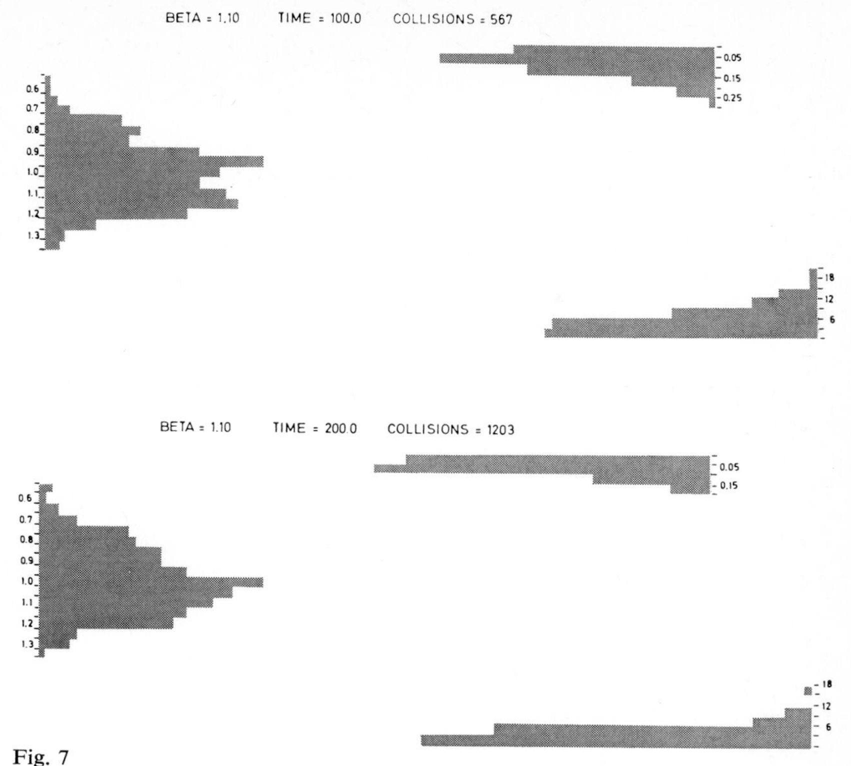

Fig. 7

Fig. 7–14. Evolution of the distributions of semimajor axis, eccentricity and inclination for different values of β. The histograms are taken at times t = 100, 200, 500 and 900.

small after the more violent initial stage. This diffusion rate would be still less or may be slightly reversed letting $\beta \rightarrow 1.0$.

Several remarks could be made in this connection. The tendency towards a complete scattering of the system for elastic collisions is rather strong. Only after a few collisions per particle an almost uniform distribution of semimajor axis is produced. For the particular collision model chosen here all near grazing collisions will be near elastic without regard to the value of β. Thus the observed radial diffusion rate could be due to the numerical dominance of near grazing over head-on collisions to a ratio of approximately 3 to 1. This is also closely related to the establishment of pressure balance in the initial stage. This depends on the ability of the system to transfer excess internal energy in one transverse direction to the other one. The collisions contributing most to this exchange rate are those taking place with impact vectors making an angle of approximately 45° with the relative velocity vector. In such a collision only half of the available internal kinetic energy is lost for the extreme case of $\beta = 1.0$. In this respect the two- and three-dimensional models discussed above differ drastically, this exchange rate necessarily

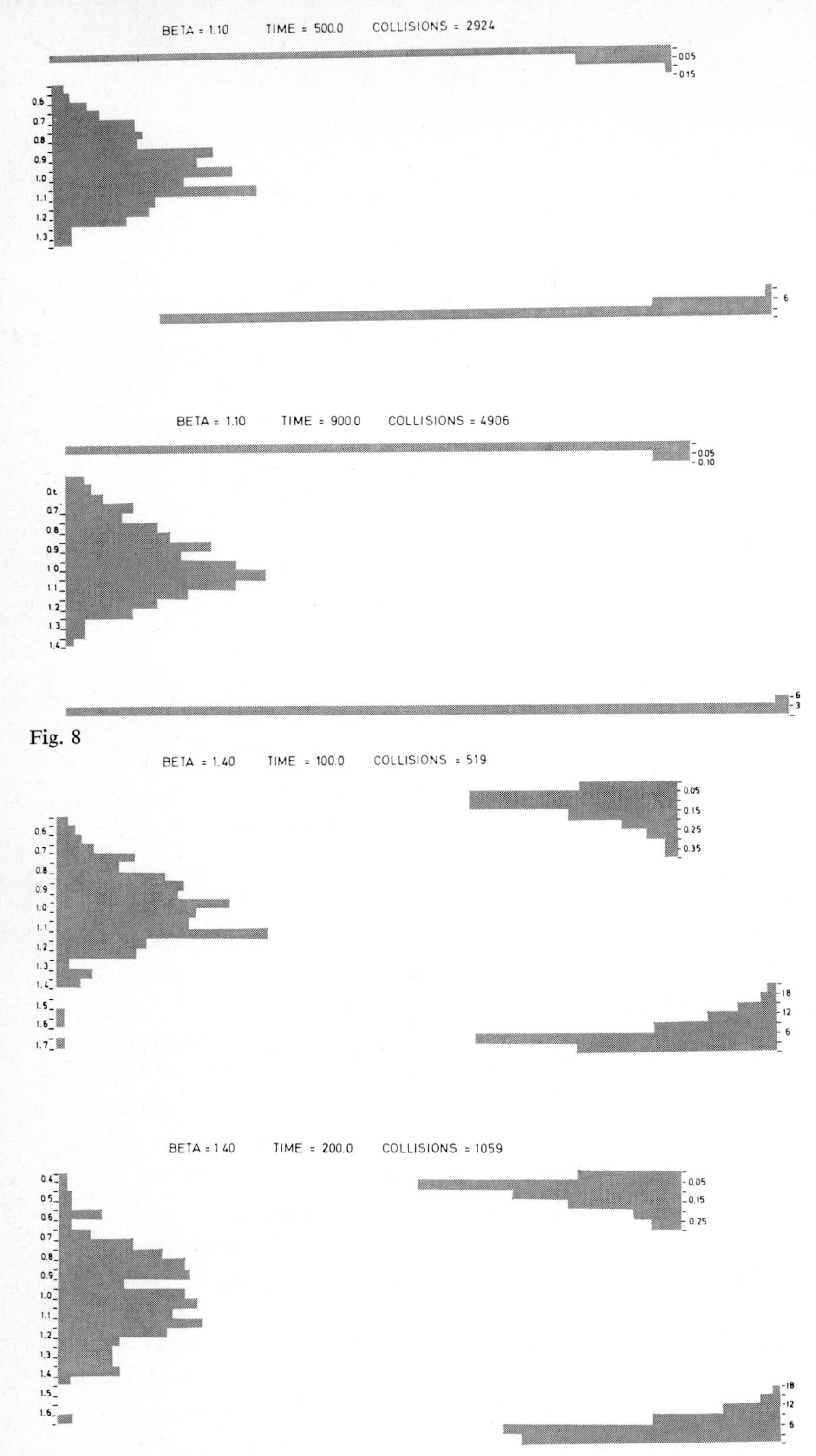

BETA = 1.10 TIME = 500.0 COLLISIONS = 2924
BETA = 1.10 TIME = 900.0 COLLISIONS = 4906
BETA = 1.40 TIME = 100.0 COLLISIONS = 519
BETA = 1.40 TIME = 200.0 COLLISIONS = 1059

Fig. 8

Fig. 9

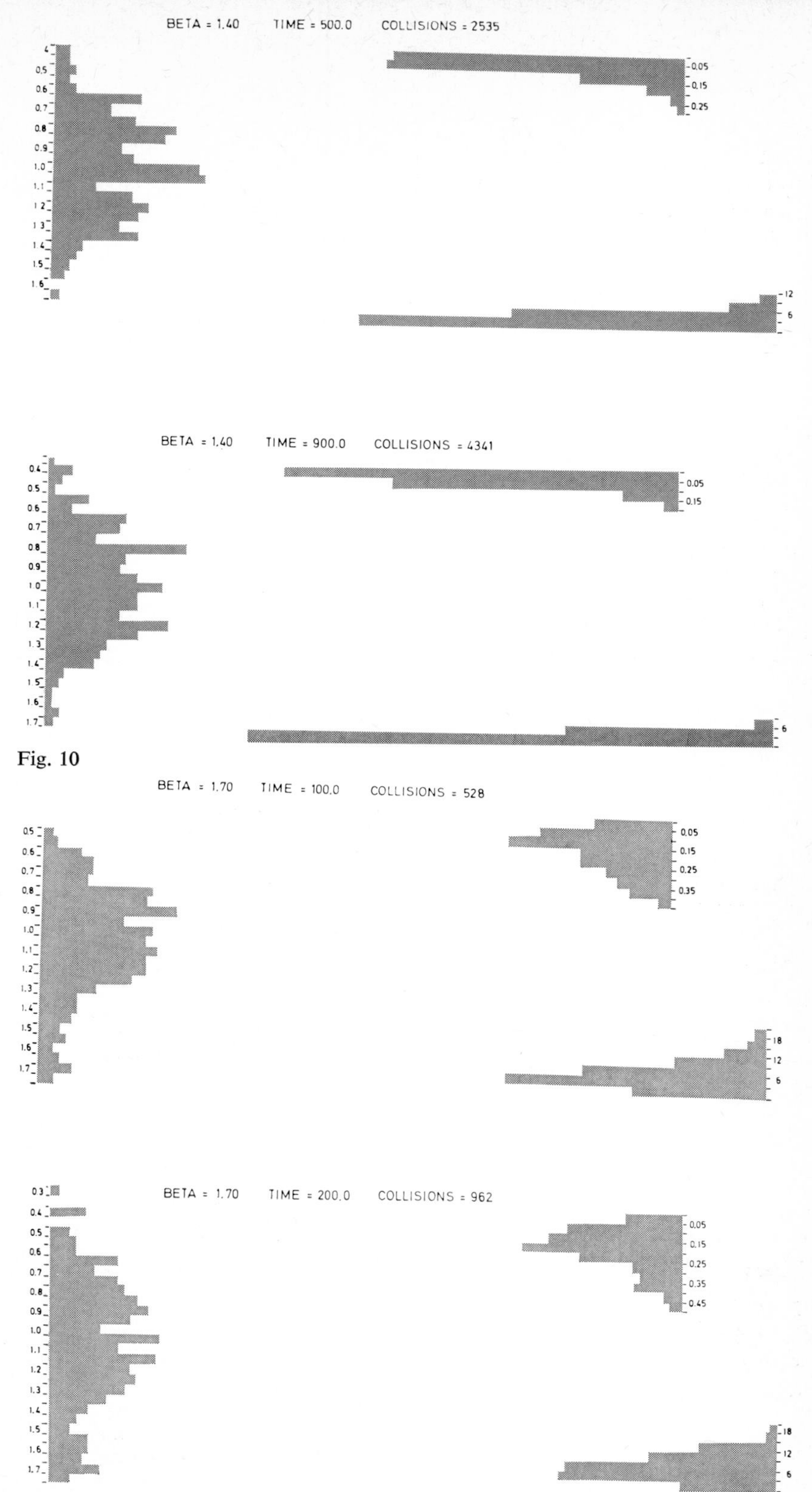
BETA = 1.40 TIME = 500.0 COLLISIONS = 2535
BETA = 1.40 TIME = 900.0 COLLISIONS = 4341
BETA = 1.70 TIME = 100.0 COLLISIONS = 528
BETA = 1.70 TIME = 200.0 COLLISIONS = 962

Fig. 10

Fig. 11

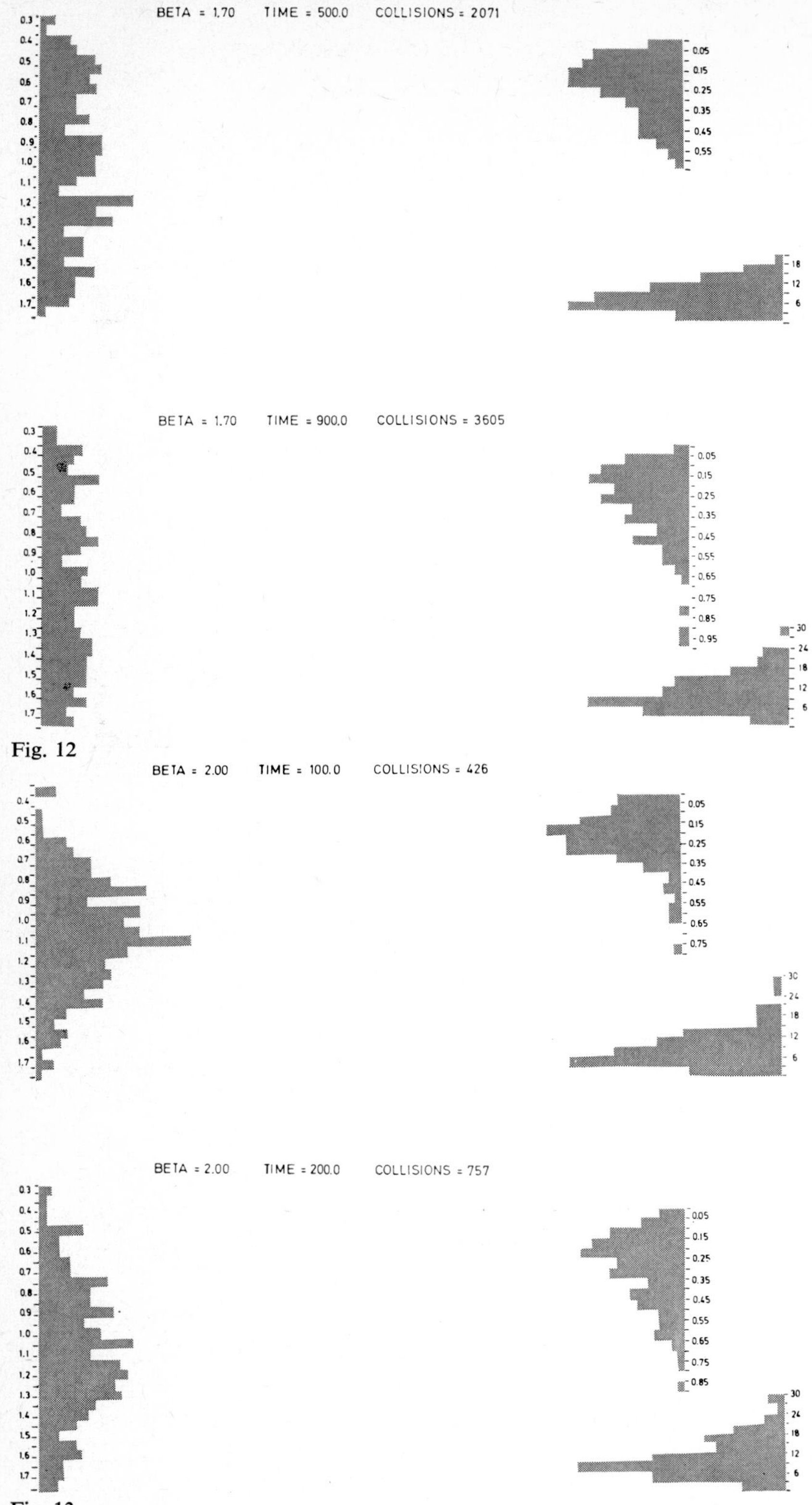
BETA = 1.70 TIME = 500.0 COLLISIONS = 2071
0.3
0.4
0.5
0.6
0.7
0.8
0.9
1.0
1.1
1.2
1.3
1.4
1.5
1.6
1.7
0.05
0.15
0.25
0.35
0.45
0.55
18
12
6
BETA = 1.70 TIME = 900.0 COLLISIONS = 3605
0.3
0.4
0.5
0.6
0.7
0.8
0.9
1.0
1.1
1.2
1.3
1.4
1.5
1.6
1.7
0.05
0.15
0.25
0.35
0.45
0.55
0.65
0.75
0.85
0.95
30
24
18
12
6
BETA = 2.00 TIME = 100.0 COLLISIONS = 426
0.4
0.5
0.6
0.7
0.8
0.9
1.0
1.1
1.2
1.3
1.4
1.5
1.6
1.7
0.05
0.15
0.25
0.35
0.45
0.55
0.65
0.75
30
24
18
12
6
BETA = 2.00 TIME = 200.0 COLLISIONS = 757
0.3
0.4
0.5
0.6
0.7
0.8
0.9
1.0
1.1
1.2
1.3
1.4
1.5
1.6
1.7
0.05
0.15
0.25
0.35
0.45
0.55
0.65
0.75
0.85
30
24
18
12
6

Fig. 12

Fig. 13

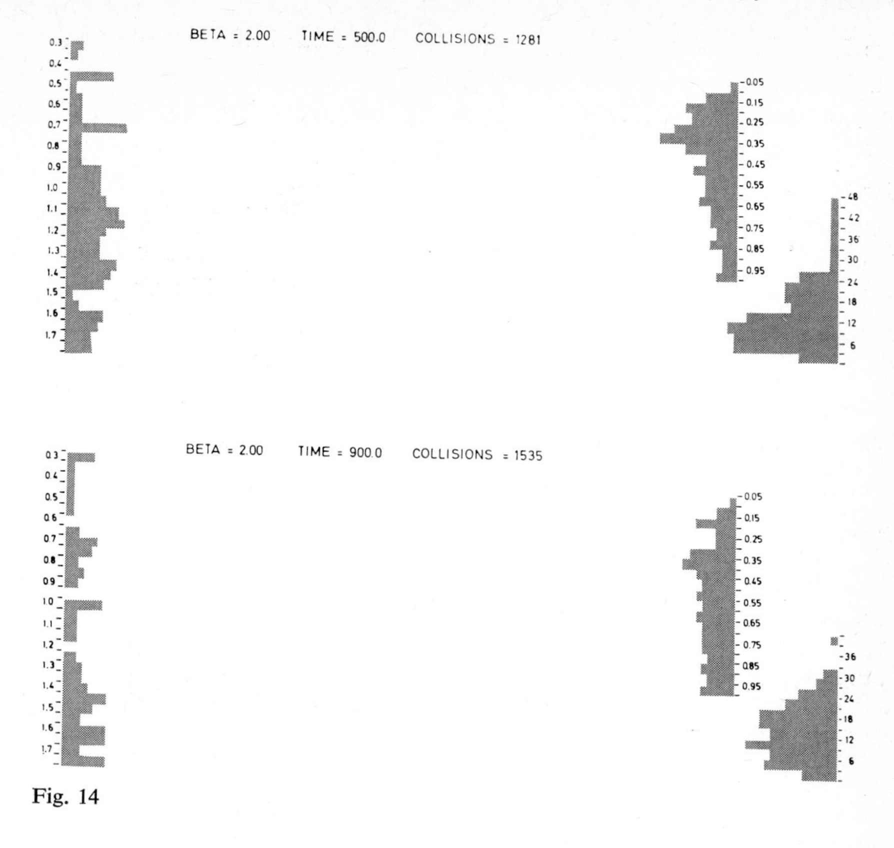

Fig. 14

vanishing in the former case. It is a matter of discussion what rate is the most physical one. The two-dimensional model clearly underestimates this effect, the three-dimensional one probably overestimating it.

We have been studying an initial value problem. The formation of jet streams would take place under conditions more similar to a steady state problem with a continuous source of new grains. Thus the rapid diffusion rate in the initial stage should not bother us too much.

The conclusion of the simulation is therefore that the question whether jet stream configurations can arise still remains very much an open one. To decide this problem more information of actual collision processes is needed. We would thus like to know the degree of inelasticity for a typical collision as a function of impact vector and relative velocity.

It would seem like a rather modest number of particles were used in this simulation. Thus only very rough statistics can be obtained with 250 particles distributed in a three-dimensional space. Finding possible collision processes between particles moving in elliptic orbits are, however, rather time consuming, computation time for a typical run on the IBM 360/91 lying in the one hour range for 250 particles and increasing with the square of the particle number for a fixed number of collisions per particle.

The numerical simulation shows that the time expansion method in the three-dimensional model usually has a rather limited range of validity. In order to extend the validity beyond the rapid initial stage one should at least be extremely careful to use an initial distribution satisfying the pressure balance condition. In the two-dimensional model this pressure balance problem is absent. It is, however, not clear whether there will still be a rapid initial stage, that is, if some kind of balance will be established between the distributions of L and e, and whether the imposed initial distribution satisfies such a requirement. It would be very helpful to perform the numerical simulation corresponding to this model.

The author would like to thank Professor Hannes Alfvén for his interest and encouragement during this work.

References

Alder, B. J. and Wainwright, T. E., 1959, J. Chem. Phys., *31*, 459.
Alder, B. J. and Wainwright, T. E., 1960, J. Chem. Phys., *32*, 1439.
Alfvén, H. and Arrhenius, G., 1970, Astrophys. Space Sci., *8*, 338.
Baxter, D. C. and Thompson, W. B., 1971, 12th Colloquium of IAU, Tucson, Arizona, March 8–10.
Bogolyobov, N. N., 1946, Problems of a Dynamical Theory in Statistical Physics. State Technical Press, Moscow.
Trulsen, J., 1971, Astrophys. Space Sci., *12*, 789.
Trulsen, J., 1972, Numerical Simulation of Jet Streams, under preparation.

Discussion

H. Alfvén

In the two-dimensional case, there is no doubt that collisions between particles in Kepler orbits will lead to the formation of jet streams. Thompson and Baxter have shown that for periodic orbits the "diffusion coefficient" is negative, which means that collisions make the orbits more similar.

The question whether a three-dimensional model gives the same result depends upon the assumptions about the nature of the collision. In the cases treated by Trulsen so far, a collision can transfer radial oscillation energy into axial oscillation, which eventually will cause a radial spread. Trulsen's results illustrate for example the spread of a gas. It is very likely that a more realistic model will show jet stream formation also for the three-dimensional case.

Baxter, D. C., 1971, IAU Colloquium No 12 "Physical Properties of Asteroids." Tucson, Arizona.

J. Trulsen

As Alfvén mentioned the broadening of the distribution of semimajor axis comes about to a large degree due to a larger transfer of kinetic energy between the two transverse directions. For the two-dimensional model this transfer rate necessarily vanishes. In the real case there will be a transfer of energy. at what rate is a matter of discussion. One would hope that laboratory experiments will shed some light on this problem. The Baxter–Thompson two-dimensional result should therefore be treated with care, this even more so because their completely inelastic collisions are probably rather far from reality. As a comparison, for the most inelastic case in my model, $\beta=1$, only one half of the kinetic energy in the center of mass system is lost on the average in a collision.

The ideal gas case corresponds to $\beta=2$ for which the spreading was most drastic. To conclude, the existence of a radial collisional focusing mechanism is still an open but very interesting problem.

Z. Kopal

If we accept that collisions of one kind or another tend to focus the orbital elements of meteor particles, this will lead to exchange not only of energy but also momenta. Is this consistent with an alignment of the axes of rotation reported by Gehrels?

J. Trulsen

The rotation of the particles has not been considered in my work. Inclusion of the rotation of small particles would probably not change my results as the energy stored in such rotation is very much smaller than the kinetic energy of the translational motion.

E. Anders

How do your values of β compare with those observed in laboratory studies of hypersonic and subsonic impacts? In hypervelocity impacts about one-half the kinetic energy is spent on crushing and heating the rock.

J. Trulsen

I would prefer not to stress one particular value of β as the most physical one. It would depend on composition and structure of the grains involved. Equally important is to decide what dependence on impact vector is closest to reality. Fractionation was not included. The energy loss associated with particle break-up could be very important. Particle break-up would also increase the collision frequency for the system, thereby increasing the effect of collisions on the dynamics.

T. Gehrels

Faint asteroids are more concentrated towards the plane of the ecliptic than are the brighter ones. I wonder if this is not an observational confirmation of a fair degree of inelasticity in the collisions among asteroids. Are the rings of Saturn similarly explained? Is it reasonable to predict that the smallest particles in the asteroid belt occur in a thin sheet in the invariable plane? Dr R. Roosen, author of a Ph.D. dissertation on the counterglow (see Roosen 1970) has warned me that no such effect has been observed in zodiacal light nor in the counterglow.

Roosen, R. G., 1970, Icarus, *13*, 184.

J. Trulsen

For the evolution of the asteroid belt the gravitational perturbations from the planets are very important. I have not studied the effects of including both collisions and gravitational perturbative forces in the evolution of particle distributions and will therefore hesitate to give any answer to Gehrels' question off hand. One would here need to make an estimate of the relative importance of collisional and gravitational forces.

One might argue as does Gehrels that the collisional effects would show up strongest in the distribution of the smaller bodies. To give an answer to this question is certainly a challenging problem.

Meteor and Asteroid Streams

By B. A. Lindblad

Lund Observatory, Lund, Sweden

Meteor Streams

Introduction

A meteor stream may be defined as an assembly of interplanetary particles moving in similar orbits. The meteors in a given stream enter the atmosphere of the Earth along roughly parallel paths. Their apparent point or area of divergence on the celestial sphere is known as the meteor stream radiant. Meteors which cannot readily be associated with a known radiant are called sporadic meteors.

Catalogues of meteor stream radiants have been published by Denning (1899), Olivier (1911, 1929), McIntosh (1935), Hoffmeister (1948), Astapovich (1956) and others. These visual observing programmes have given valuable information about radiants, hourly rates and dates of occurrence of meteor streams, but have yielded only limited information about meteor velocities and orbits. Owing to the relatively large observational errors inherent in visual work there is the possibility that a number of fictitious radiants have found their way into the catalogues.

With improvement in photographic techniques more exact methods of recording meteors become possible. The photographic double station technique has made it possible to obtain precise radiants, velocities and heliocentric orbits of single meteors. The Harvard photographic meteor programme has produced the largest body of homogeneous orbital data to date—but also the contributions from the USSR stations in Odessa and Dushanbe have substantially widened our knowledge of the orbital characteristics of meteors.

The distinction between stream meteors and sporadic meteors is to a certain extent arbitrary. The classifications used in the past were based primarily on geocentric quantities such as radiants, velocities and dates of occurrence, with the orbital elements being used as a secondary criterion for stream identification. The meteor streams detected in the Harvard programme have been summarized by Jacchia (1963) and McCrosky and Posen (1959).

Computer techniques for the detection of meteor streams have been introduced by Southworth and Hawkins (1963) and Southworth (1968). This approach, based directly on the orbital elements, is particularly useful for detecting minor streams, as only a few, well determined orbits are necessary for delineating a stream. The computer technique has proved very useful

Table 1. *Photographic meteor catalogues*

Type of data	Station	No. of orbits	No. used in search	Reference
Small camera	Mass, NM	144	136	Whipple (1954)
Super Schmidt	NM	413	411	Jacchia et al. (1961)
Super Schmidt	NM	313[a]	311	Hawkins et al. (1961)
Wide angle camera	Prairie	100	100	McCrosky (1968)
Super Schmidt	NM	353	346	Posen et al. (1967)
Small camera	D	49	49	Katasev (1957)
Small camera	O	225	221	Babadjanov et al. (1963, 1967)
Small camera	D	253	253	Babadjanov et al. (1963, 1967)
Total		1 845	1 827	
Super Schmidt	NM	1 790[b]	—	McCrosky et al. 1961
Small camera	D	77[c]	—	Babadjanov et al. 1969

[a] Meteors common to Jacchia et al. (1961) have been excluded.
[b] Graphical reduction procedure. Meteors common to references Jacchia et al. (1961) and Hawkins et al. (1961) have been excluded. Data sample not used in present study.
[c] Data sample received too late for inclusion in stream search.

Mass = Massachusetts; NM = New Mexico; Prairie = Prairie network; D = Dushanbe (Stalinabad); O = Odessa.

also in the study of diffuse streams with a poorly defined radiant area and a long period of activity. Extensive searches in the photographic catalogues based on the Southworth and Hawkins *D*-criterion have been made by Lindblad (1971 *a*, 1971 *b*).

This paper summarizes some of our knowledge of meteor streams, as obtained from the photographic data, and analyses in some detail the orbital dispersion within a meteor stream. The data sample used in the study is detailed in Table 1. It consists of 1304 precisely reduced Harvard orbits and 523 precisely reduced Odessa and Dushanbe orbits.

Velocity-Elongation Diagram

The mean geocentric velocity and the mean elongation λ of the apparent radiant from the apex of the Earth's motion were computed for all streams listed in Lindblad (1971*a*). Datum points for 25 well recognized photographic meteor streams are plotted in the velocity-elongation diagram Fig. 1. Streams are denoted by customary symbols. Two closely spaced points in the diagram indicate northern and southern branches, respectively. The continuous curve in the diagram gives the theoretical relation between geocentric velocity and elongation assuming parabolic velocity at the Earth's perihelion.

Comparatively few datum points appear in the lower right-hand part of Fig. 1. These points cluster very near to the parabolic limiting curve and represent meteor streams in retrograde orbits whose members make head-on collisions with the Earth's atmosphere. All of these streams have orbital

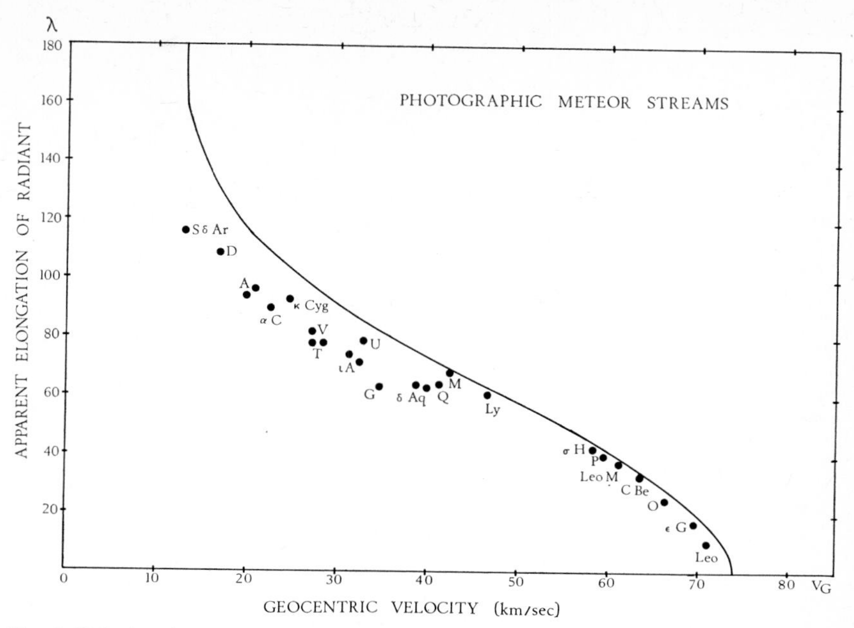

Fig. 1. Velocity-elongation diagram for well-studied photographic streams.

periods of more than 30 years. They will hereinafter be referred to as long-period streams. Because of their high geocentric velocities members of these streams produce detectable luminosity even for fairly small masses. These meteor showers were among the first to be studied by visual observers.

Most of the points in the velocity-elongation diagram cluster along a curve branching off from the parabolic limiting curve at approximately $\lambda = 60°$ and displaced by about 5 km/sec with respect to the curve. Members of these streams are moving in direct, low-inclination orbits, with orbital periods of less than 10 years. These streams will be referred to here as the short-period meteor streams.

New Photographic Streams

A number of new photographic streams have been detected in the computer searches by Southworth and Hawkins (1963) and Lindblad (1971 *a*, 1971 *b*). Most of the newly detected photographic meteor streams are concentrated in the upper left-hand part of the velocity-elongation diagram. These meteors are approaching the Earth from the anti-apex direction and are catching up with us at relatively low velocities. It is obviously difficult to observe members of these streams since only fairly large meteor masses will produce enough luminosity for detection. For the visual observer it is difficult to establish their shower identification, since the low geocentric velocity implies a diffuse radiant.

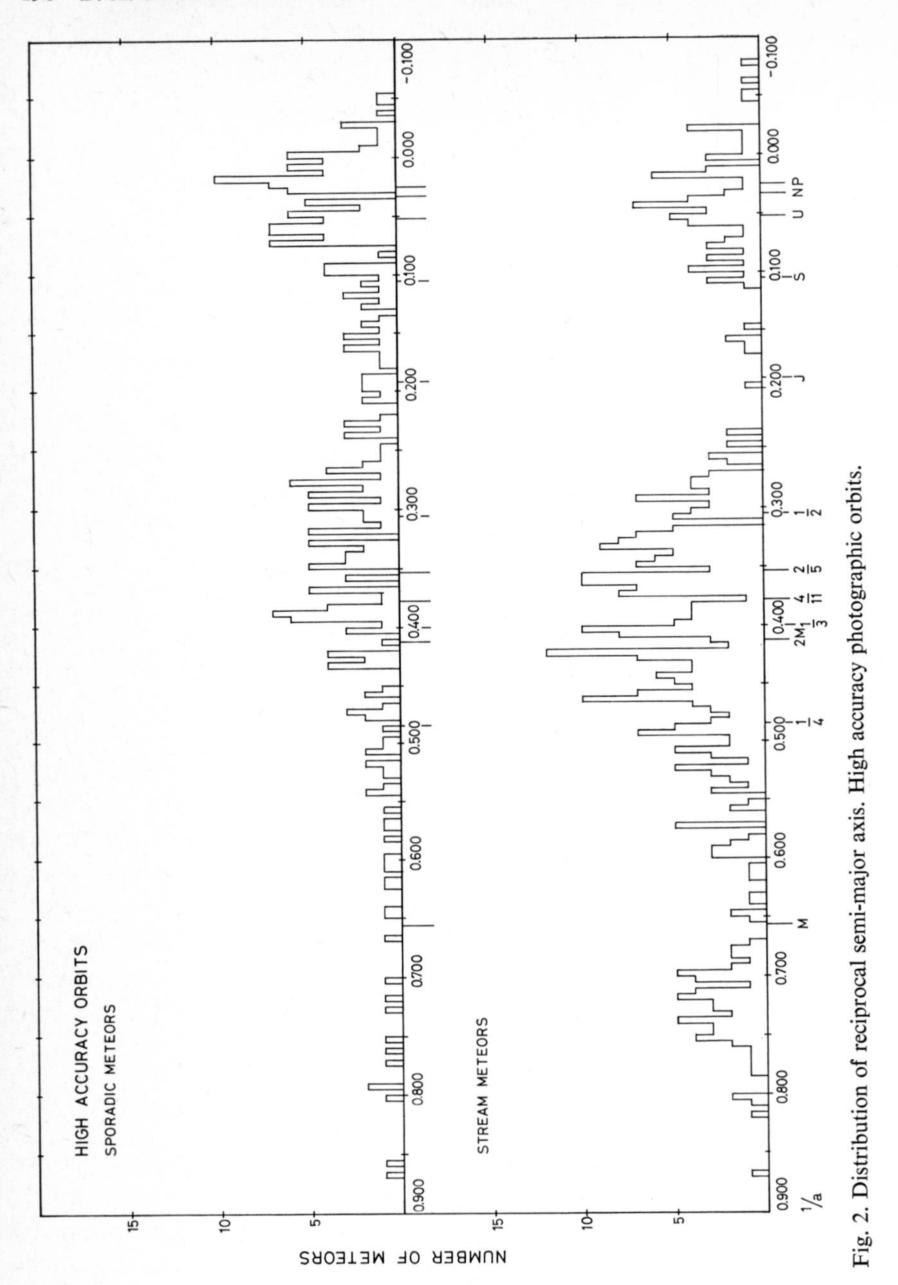

Fig. 2. Distribution of reciprocal semi-major axis. High accuracy photographic orbits.

Distribution of 1/*a*

The difference between short-period and long-period meteor streams is clearly brought out in the distribution of reciprocal semi-major axis $1/a$. Lindblad (in press) has recently analysed the available, precisely reduced photographic orbits according to quality. Fig. 2 depicts the $1/a$-distribution for the most accurate stream and non-stream orbits. The division of the stream component into two main groups, the long-period orbits with orbital periods in excess of Saturn's and the short-period orbits with periods between those of Jupiter and Mars is clearly shown. For the purpose of this study the short-period meteor streams are defined by the condition $0.65 > 1/a > 0.20$ $(AU)^{-1}$.

Dispersion in Orbital Elements of Short-Period Meteor Streams

The large quantity of precisely reduced photographic data now available has enabled a fairly detailed study of the stream structure among the short-period meteor orbits. Table 2 lists pertinent data for 18 short-period photographic meteor streams for which 7 or more members were detected in the computer search, Lindblad (1971*a*). The Draconid stream has been added as a fiducial point. The Geminid shower has been excluded since this stream has a mean $1/a > 0.70$ and thus represents an extreme case of a stream with specific orbital energy less than that of Mars. The evolutionary history of this stream is obviously quite different from that of the other short-period streams.

The following new, or less well studied, photographic streams have been included in the data analysis: The δ Arietids, ϱ Geminids, Northern σ Leonids, β Piscids, Northern and Southern Andromedid-Piscids, and δ Piscids. These short-period streams are detailed in Lindblad (1971*a*, 1971*b*). The δ Arietids were discovered by McCrosky and Posen (1959), the ϱ Geminids and σ Leonids by Southworth and Hawkins (1963), the β and δ Piscids by Lindblad (1971*a*).

The photographic Andromedid-Piscid stream was first thought to be associated with the great Andromedid shower and thus with Bielas comet. It now appears likely that this stream has no association with the comet and that it is the same stream as the visual Piscid shower reported by Hoffmeister (1948). The Northern σ Leonid stream is identical with the Virginid stream listed by Hoffmeister. Hoffmeister coined the term "ecliptical currents" for the short-period streams, and suggested that they were related to the asteroids. In the literature Hoffmeister's important work on the short-period meteor streams has not received the attention it deserves, possibly because of the author's strong belief in the meteor stream-minor planet relationship.

Scatter in Orbital Energy

Let us assume that a meteor stream is initially formed as a dense group of similar orbits. After formation a number of perturbing forces act on the individual particles. With time these forces, both gravitational and non-gravitational, will produce substantial changes in the orbits. During the stream dispersion process, the orbital energy of the individual members will

Table 2. *Short-period photographic meteor streams*

Stream	Duration	No. of members	V_G	λ	a	e	i	ω	Ω	π	$\sigma\left(\frac{1}{a}\right)$	$\sigma(\pi)$	$\sigma(\Omega)$
Quadrantids	2–4 Jan.	14	41.2	63°7	3.037	.677	71.9	171.1	282.4	93.6	.018	3°7	0°4
ϱ Geminids	15–27 Jan. + 9 May	7	17.1	105°5	2.808	.720	3.3	251.8	290.2	182.0	.053	3°8	25°3
Virginids	5 March–1 April	9	27.4	82°2	2.592	.825	5.4	279.1	0.9	280.0	.086	5°3	12°0
N σ Leonids	31 March–13 May	11	17.8	100°8	2.337	.679	0.4	249.3	29.3	278.6	.071	4°8	12°9
α Capricornids	17 July–10 Aug.	18	22.6	90°3	2.421	.758	7.3	270.2	126.9	37.1	.050	7°0	5°7
S ι Aquarids	19 July–8 Aug.	8	32.3	71°6	2.348	.892	4.1	130.9	305.4	76.3	.182	8°0	6°9
N ι Aquarids	23 July–4 Sept.	7	31.2	74°3	2.581	.878	5.7	305.2	145.3	90.5	.159	15°9	16°2
S δ Aquarids	21 July–29 Aug.	24	40.0	63°0	3.094	.967	26.2	149.5	309.6	99.1	.120	3°5	6°1
N δ Aquarids	5–25 Aug.	8	38.8	64°0	2.504	.958	20.2	327.8	140.7	108.5	.124	5°2	5°5
β Piscids	10 Aug–12 Nov. + 15 Dec.	25	15.7	114°6	2.699	.697	3.1	231.6	186.9	58.5	.070	7°0	35°5
$\varkappa$ Cygnids	19–25 Aug.	7	24.7	92°8	4.496	.769	38.2	199.8	147.7	347.5	.046	3°4	1°8
N Andromedids-Piscids	3 Sept.–7 Nov.	30	20.5	96°4	2.661	.751	5.7	258.6	200.2	98.8	.067	13°8	16°0
S Andromedids-Piscids	14 Sept.–14 Nov.	24	20.2	94°2	2.385	.724	3.6	83.4	9.6	93.0	.111	13°0	14°2
N Taurids	8 Oct.–21 Dec.	42	28.3	79°2	2.269	.834	3.3	293.4	228.1	161.5	.070	13°0	19°1
S Taurids	10 Sept.–14 Dec.	98	27.3	79°1	2.153	.815	5.8	113.2	33.4	146.7	.130	16°7	22°8
δ Piscids	16 Sept.–4 Oct.	8	30.5	75°6	2.399	.874	4.3	302.5	179.2	121.7	.135	5°6	6°2
Draconids	9 Oct.	2	17.1	108°6	3.332	.700	24.6	176.9	195.5	12.5	.004	0°0	0°1
S δ Arietids	3–19 Dec.	8	13.0	116°1	2.441	.634	1.0	46.8	78.1	124.9	.077	4°8	4°4

Table 3 *Mean orbital elements of asteroid streams*

Name	No. of mem-bers	a	e	i	ω	Ω	π	Family	Remarks
Rosa	11	3.134	0.149	0.2	279.5	181.5	101.0	1	
Janina	8	3.133	0.180	1.6	268.9	141.3	50.2	1	
Coronis	7	2.865	0.050	0.5	264.4	165.3	69.7	3	
Denone	10	2.861	0.061	0.2	102.7	218.9	321.6	3	
Elvira	8	2.870	0.083	1.7	164.5	200.6	5.1	3	
Lacrimosa	17	2.424	0.126	1.3	298.2	259.6	197.8	3	Loose association
Anahita	19	2.261	0.177	1.0	17.2	310.7	327.9	6, 32	
Nephele	14	3.133	0.183	1.5	278.4	63.8	342.2	1	
Hertha	9	2.423	0.197	2.1	1.8	328.1	329.9	32	
Gisela	10	2.214	0.152	3.0	188.7	189.4	18.1	6, 7	Alfvén A
——	7	2.213	0.201	3.2	172.1	171.1	343.2	6	Alfvén A
Eriphyla	15	2.882	0.079	2.5	241.1	118.3	359.4	3	
Lucretia	8	2.234	0.143	5.5	46.4	1.1	47.5	7, 8	Alfvén C

the problem using the proper elements and added several new families to Hirayama's list. The currently used numbering of the asteroid families is mainly that introduced by Brouwer (1951).

Alfvén (1969, 1970) has drawn attention to the fact that within the Flora family there exist groups of asteroids which exhibit similarity in all five orbital elements. Alfvén has put forward the jet stream hypothesis to explain this grouping. Arnold (1969) used computer techniques to detect families and streams amongst the asteroids. Danielsson (1969, 1971) has studied the statistical significance of asteroid streams.

Lindblad and Southworth (in press) have recently applied the Southworth stream search program to a data sample including both the numbered asteroids and the PLS asteroids listed by van Houten et al. (1970). Streams detected in the search and having more than 7 members are listed in Table 3. These streams have been briefly discussed by Lindblad and Southworth (in press).

Scatter in Orbital Energy

The asteroid streams reported by Lindblad and Southworth are plotted in the $\sigma(1/a)-1/a$ diagram in Fig. 6 (filled circles). There is a tendency for increased scatter in $1/a$ as the value of $1/a$ increases. As indicated in Fig. 6 the streams represent concentrations within the recognized families of Themis, Coronis, Flora and Nysa. It is therefore of interest to study the $\sigma(1/a)-1/a$ diagram for asteroid families as well. Fig. 7 depicts datum points for all Brouwer families with 10 or more members (filled circles). The family classification is that of Brouwer, and families have been denoted by their customary numbers. Fig. 7 includes the families found in the PLS (open circles). The family classification is that of Lindblad and Southworth (in press).

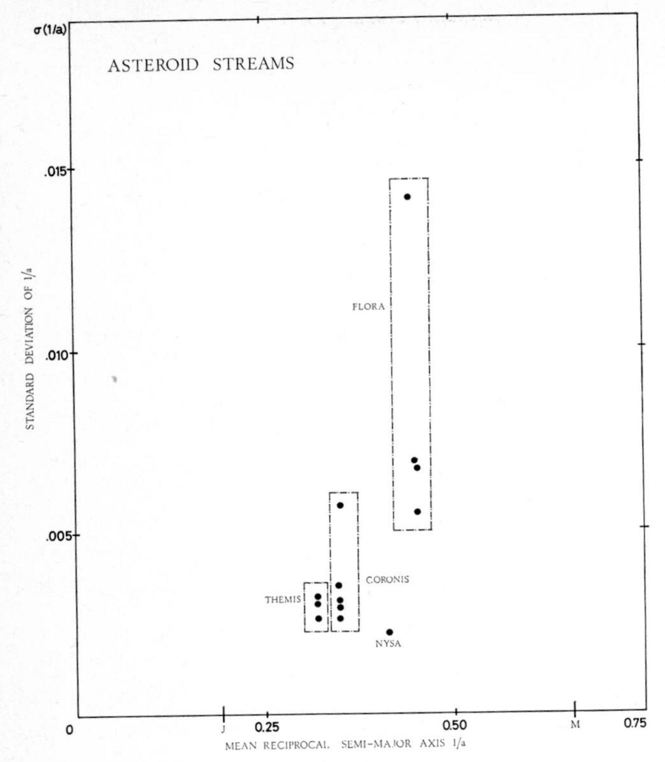

Fig. 6. Scatter in orbital energy $\sigma\left(\frac{1}{a}\right)$ vs $1/a$. Asteroid streams.

Fig. 7 shows that the scatter of $1/a$ within a family increases significantly with increasing mean value of $1/a$. The families of Flora (6–9) and Nysa-Michela (32) exhibit the largest scatter in $1/a$. Families 1, 3 and 6 show a somewhat larger scatter in the PLS sample than in the numbered asteroid sample. Either the family limits have been chosen slightly wider in the PLS sample, or the fainter asteroids of the PLS exhibit larger scatter. This question merits further detailed attention.

Percentage of Asteroids in Streams

The PLS survey assigns to each individual orbit an index of quality. A special search was made in this sample to find out how the number of asteroids in streams varied with orbit quality. Two searches at the same rejection level D_s were made, the first included 980 PL orbits of highest quality (type 1); the second 856 PL orbits of lower quality (types 2–4). The percentage of asteroids belonging to streams in the two data samples was 41 and 31, respectively. The stream percentage thus decreases with decreasing accuracy of the orbits. Similar results have been obtained in the meteor

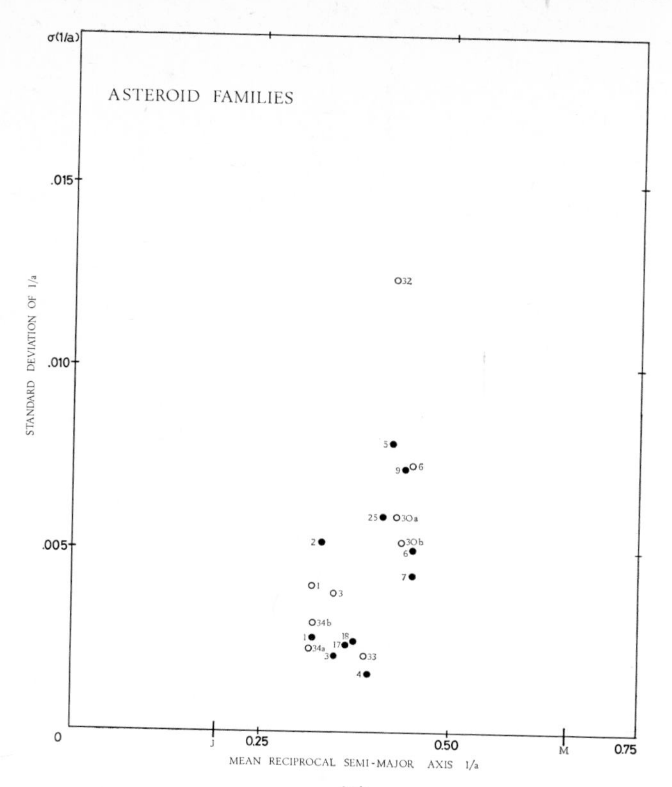

Fig. 7. Scatter in orbital energy $\sigma\left(\frac{1}{a}\right)$ vs $1/a$. Asteroid families. Numbered asteroids ●. Palomar-Leiden asteroids ○.

studies. If asteroid streams only represented spurious groupings in the data no dependence on orbital accuracy would be detected.

The percentage of stream members varies markedly with the degree of concentration of the orbits to the ecliptic plane. The results of a search in the combined numbered asteroid–PL asteroid sample (NASA SP-267) are typical. The percentage of stream members in the numbered asteroid sample was 16.5 compared with 30.0 for the PLS data. The high percentage of stream objects in the PLS is a direct consequence of the concentration of the orbits to the ecliptic plane resulting from a high inclination cut-off in the sample.

Significance of Asteroid Streams

The statistical significance of asteroid streams was assessed by computer searches in random samples. Four independent but similar sized samples ($N=2\,282$) were constructed and searched. The results of the four searches were averaged and summarized in Fig. 8. The diagram shows that one would

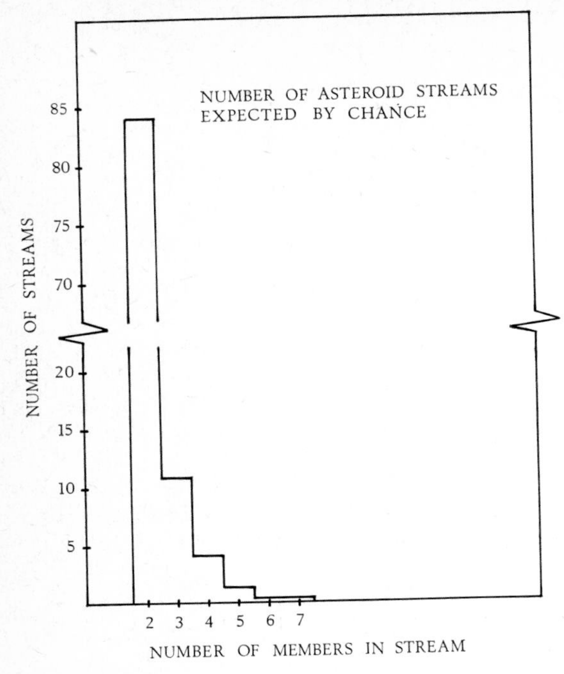

Fig. 8. Number of asteroid streams in random sample.

expect about 85 two-member streams, 11 three-member, 5 four-member and less than 2 five-member and 1 six- or seven-member stream by chance in a sample of 2 282 orbits. The real asteroid sample studied for streams contained 2 929 orbits. Scaling up the previous results one finds that not more than one of the streams listed in Table 3 could be a chance association of orbits.

References

Alfvén, H., 1969, Astrophys. Space Sci., *4*, 84.

Alfvén, H., 1970, Astrophys. Space Sci., *6*, 161 .

Arnold, J., 1969, Astron. Journ., *74*, 1235.

Astapovich, I. S., 1956, Main Catalogue of 20th Centrury Meteor Radiants. Ashabad, Akad. Nauk Turkman SSR.

Babadjanov, P. and Kramer, E., 1963, Ionosphere and Meteors. Section V of IGY Program, No. 12, Moscow.

Babadjanov, P. B. and Kramer, E., 1967, Smithsonian Contr. Astrophys., *11*, 67.

Babadjanov, P. B., Getman, T. I., Zausayev, A. F. and Karaselnikova, S. A., 1969, Bull. Inst. Astrophys., Akad Nauk Tadzjikistan SSR No. 49.

Brouwer, D., 1951, Astron. Journ., *56*, 9.

Danielsson, L., 1969, Astrophys. Space Sci., *5*, 53.

Danielsson, L., 1971, Physical Studies of Minor Planets, NASA SP-267.

Denning, W. F., 1899, Mem. Roy. Astron. Soc., *53*, 203.

Hawkins, G. and Southworth, R., 1961, Smithsonian Contr. Astrophys., *4*, 85.

Hirayama, K., 1928, Japanese Journ. Astr. and Geophys., *5*, 137.

Hirayama, K., 1933, Proc. Imp. Acad. Japan, *9*, 482.

Hoffmeister, C., 1948, Meteorströme. Verlag Johann Ambrosius Barth, Leipzig.
van Houten, C. J., van Houten-Groeneveld, I., Herget, P. and Gehrels, T., 1970, Astr. Astrophys. Suppl., *2*, 339.
Jacchia, L., 1963, Meteors, Meteorites and Comets: Interrelations, in The Moon, Meteorites and Comets, p. 774. The Solar System, vol. 4 (eds. Middlehurst and Kuiper), Univ. of Chicago Press, Chicago.
Jacchia, L. and Whipple, F., 1961, Smithsonian Contr. Astrophys., *4*, 97.
Katasev, L., 1957, Photographic Methods in Meteor Astronomy. Gosudarstvennoe Izdatel'stvo, Tekhniko Teoreticheskoi Literatury, Moskva. (Engl. Transl. Monsun Press, Jerusalem 1964).
Kirkwood, D., 1890, Publ. Astron. Soc. Pac., *2*, 48.
Kirkwood, D., 1891, Publ. Astron. Soc. Pac., *3*, 95.
Lindblad, B. A., 1971*a*, Space Res., *11*, 287, Akademie-Verlag, Berlin.
Lindblad, B. A., 1971*b*, Smithsonian Contr. Astrophys., *12*.
Lindblad, B. A., The Distribution of 1/a in Photographic Meteor Orbits. Evolutionary and Physical Properties of Meteoroids. Proc. IAU Colloq. No. 13 (in preparation).
Lindblad, B. A. and Southworth, R., A Study of Asteroid Families and Streams by Computer Techniques. Physical Studies of Minor Planets, NASA SP-267 (in press).
McCrosky, R., 1968, Orbits of Photographic Meteors, in Physics and Dynamics of Meteors, p. 265, Reidel Publ. Co., Dordrecht.
McCrosky, R. and Posen, A., 1959, Astron. Journ., *64*, 25.
McCrosky, R. and Posen, A., 1961, Smithsonian Contr. Astrophys., *4*, 15.
McIntosh, R. A., 1935, Mon. Not. Roy. Astron. Soc., *95*, 709.
Olivier, C., 1911, Trans. Am. Phil. Soc., N. S., *22*, Part 1.
Olivier, C., 1929, Publ. Leander McCormick Obs., *5*, 1.
Posen, A. and McCrosky, R., 1967, NASA Contr. Rep. CR-862.
Southworth, R., 1968, Discrimination of Stream and Sporadic Meteors, in Physics and Dynamics of Meteors, p. 404, Reidel Publ. Co., Dordrecht.
Southworth, R. and Hawkins, G., 1963, Smithsonian Contr. Astrophys., *7*, 261.
Whipple, F., 1954, Astron. Journ., *59*, 201.
Whipple, F. and Wright, F., 1954, Mon. Not. Roy. Astron. Soc., *114*, 229.

Discussion

F. L. Whipple

There is no question that meteoroids are dispersed in orbital elements much faster than perturbations, the Poynting-Robertson effect, and collisions can explain. I have tried for many years to find other physical effects that can produce the dispersion. I hope that some of you present will look for the sources. In 1940 I tried electrical and magnetic effects to no avail. Perhaps plasma magnetic fields can do something. I suspect it is some phenomenon of light pressure on spinning grains as Öpik has suggested.

B. Lehnert

I may possibly be wrong by orders of magnitude, but it just occurs to me that one should perhaps look into the interaction between the meteor bodies and a surrounding plasma to see if there may exist some additional forces determining the dynamics of these obstacles. In particular, various types of plasma waves may be excited by the motion of the latter when they become electrically charged, being similar to the plasma "brakes" on artificial satellites discussed earlier by myself in 1956, and later by others. (See Tellus *8*, 408, 1956.)

B. A. Lindblad

It is obvious that we have a scatter in the orbits and it is important that we find out what physical mechanism is causing the scatter because if we know this we will be able to put a time scale on the diagram, Fig. 3.

And a general comment to Professor Lehnert: The sizes of these meteoroid particles are of the order of 1 cm in diameter, so they are fairly large particles.

P. M. Millman

I have just inserted some densities from Table 1 of my paper into Figure 3 of Lindblad's paper. Although by no means a perfect correlation there appears to be a definite tendency for the low meteoroid densities to be at the bottom of the diagram and the higher meteoroid densities to be at the top. In other words the new streams tend to have lower mean densities than the old streams.

B. A. Lindblad

I have also noticed this correlation and I find it very suggestive.

On Certain Aerodynamic Processes for Asteroids and Comets

By Fred L. Whipple

Smithsonian Astrophysical Observatory and Harvard College Observatory

1. *Introduction*

Let us assume that during planetary formation in our solar system, gas and dust grains constitute a Laplacian-type nebular disk rotating about a central mass, roughly that of the Sun today. The grains or planetesimals are generally accreting in all size ranges from atomic and molecular to large bodies. For smaller bodies, the gas, presumed to be approximately a solar mix, acts as a buffer against high velocities of encounter that might cause collisional destruction. The motions are fairly circular, controlled by a nearly inverse-square law of central force, while the gas and, at first, the grains are spread about the fundamental plane, held by the gas pressure against the surface gravity in the plane and against the perpendicular component of the central force. The effects of turbulence are neglected, as are plasma effects. The interaction of the accreting bodies with the gas and dust thus follows the laws of aerodynamics. For simplicity, the finite bodies are assumed to be spheres.

The general relations applicable to the aerodynamics are presented in the Appendix so that the narrative need not be encumbered with details. The Appendix furthermore contains the basic physical data, the equations relating gas pressure, density, etc. to position in the nebula, and numerical data relevant to the processes discussed in the paper.

In such a system, the gas pressure varies from point to point, depending on gravity and possible magnetohydrodynamic effects if a plasma is present. Radial variations in gas pressure affect the net central radial acceleration on the gas, causing the gas to deviate from the motion appropriate to the local gravitational acceleration and therefore appropriate to the motions of the solid grains, as shown in Fig. 1. The grains thus meet a resisting medium and tend to move in toward the protosun or are accelerated radially outward, depending on the sense of the effective radial acceleration acting on the gas. Under favorable circumstances in the nebula, these radial effects can produce striking systematic motions of the particles—for example, concentrating icy material toward proto-Neptune from the outer portion of the nebula. The effect of radial pressure on the motions of planetesimals is discussed in Section 2.

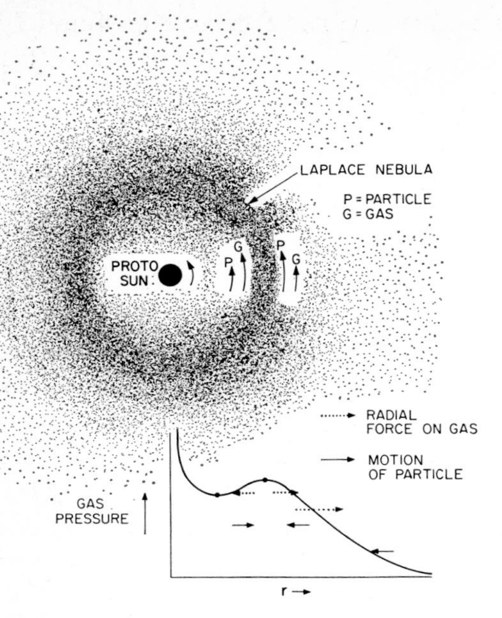

Fig. 1.

The origin of chondrules, millimeter-sized glassy spherules in meteorites, remains unsolved. However, aerodynamic processes could cause asteroids selectively to accrete chondrules, as is discussed in Section 3. New implications of these results as bearing on the origin of chondrules are also presented.

2. *Effects of Radial Pressure*

Let Δg be the deviation from the central gravitational acceleration acting on the gas, whether caused by spiraling magnetic fields or by a radial pressure gradient in neutral gas, dP/dr, where P is the total gas pressure and r is the radial distance to the center of the protosun. For a gas of density ϱ, the radial correction to the central acceleration is, by the classical formula,

$$\Delta g = +\frac{1}{\varrho}\frac{dP}{dr}. \tag{1}$$

Note that in equation (1) the sign is reversed from that for stellar interiors because the gas in the nebula is held in radial equilibrium by its motion in the central gravitational field for $dP/dr = 0$. A positive pressure gradient outward from the center increases the effective central acceleration from that of gravity alone and causes the gas in equilibrium motion to rotate more rapidly. In that case, the gas accelerates the grains in their near-circular motion and causes them to move radially outward. The opposite is true for $dP/dr < 0$.

To determine the quantitative motion of the grains, note that the near-circular velocity change ΔV in the gas under a central acceleration g follows from the circular velocity V law

$$V = g^{\frac{1}{2}} r^{\frac{1}{2}}, \tag{2}$$

by differentiation, to the form

$$\Delta V = \frac{r^{1/2}\,\Delta g}{2g^{1/2}}, \text{ and } \frac{\Delta V}{V} = \frac{1}{2}\frac{\Delta g}{g}. \tag{3}$$

The motion of the spherical grain or planetesimal in the solar nebula under the influence of the gas "wind" is generally complicated. The present paper will deal in detail with only two extreme and simple illustrative cases: (I) where the "drag" force F for $v \sim \Delta V$ is large so that the grain at velocity v with respect to the gas would be stopped in a time t_e (see equations (A4) in the Appendix), a small fraction of the period of revolution T_p, i.e., $t_e \ll T_p/2\pi$; and (II) where the grain or planetesimal at relative velocity $v = \Delta V$ would not be stopped for time comparable to or greater than the period of revolution, i.e., $t_e \gg T_p/2\pi$.

In case I, where $t_e \ll T_p/2\pi$, the grain will largely follow the rotational motion of the gas and thus experience a central acceleration reduced or increased by Δg. Hence, it will execute a radial motion at velocity v with respect to the gas derived from $F(v) \simeq m\Delta g$, where m is the mass of the grain. The velocity v will be small compared to the mean molecular velocity $\bar{v}$ in the gas.

In case II, where $t_e \gg T_p/2\pi$, the much larger grain or planetesimal will experience a tangential drag or an accelerating force at velocity $\sim \Delta V$ and slowly spiral radially in or out with respect to the rotating system of gas according to the modified equations of motion in a central force.

Case I. Grains carried along with the gas, $t_e \ll T_p/2\pi$. The relative velocity v of the grain in the gas is smaller than ΔV and very much smaller than the speed of sound. Thus, case I divides into only two subcases: case I a, when the mean free path L is larger than the particle diameter $2s$, requiring Epstein's law (equations (A3d), (A4d), and (A5)); and case I b, when $L \leqslant 2s$ and the Reynolds' number, R_e, is small, requiring Stokes' law (equations (A4a) and (A5)).

Case I a. $v \ll \Delta V$, $v \ll \bar{v}$, $L > 2s$, and $t_e \ll T_p/2\pi$. For a grain (radius s, mass m, density ϱ_s) the terminal radial speed in the cloud under the acceleration Δg becomes from Epstein's law (equations (A4d) and (A5))

$$\frac{dr}{dt} = t_e\,\Delta g = \frac{\varrho_s\, s}{1.4\,\varrho\bar{v}}\,\Delta g, \tag{4}$$

where Δg is given by equation (1) for a radial pressure gradient in the gas or by the effect of a radial acceleration on the gas from some other source such as a magnetic field.

Case Ib. $v \ll \Delta V$, $v \ll \bar{v}$, $L \leqslant 2s$, and $t_e \ll T_p/2\pi$. Here the terminal radial speed of the grain is derived from Stokes' law (equations (A4a) and (A5)), so that

$$\frac{dr}{dt} = t_e \Delta g = \frac{2\varrho_s s^2 \Delta g}{9\eta}, \tag{5}$$

where η is the viscosity of the gas.

Equation (5) is independent of the gas density so long as $L \leqslant 2s$ and holds fairly well to $R_e = 10$ and within a factor of 3 to $R_e = 10^2$.

Case II. Grains and planetesimals meeting a resisting medium at $v \simeq \Delta V \ll V$. Here, $t_e \geqslant T_p$ and generally $L < 2s$, $v \ll \bar{v}$, while R_e may become appreciable for large bodies, although in most cases of interest R_e is small. Tisserand (1890) solved this two-body problem for small orbital eccentricity e and constant $v = \Delta V \ll V$. To the first order in e, $de/dt = 0$, so that a small orbital eccentricity is retained either under resistance or acceleration for gas in nearly circular motion.

For circular velocity V of mass m at distance r from the effective center of mass M, and a resisting or accelerating force F, the classical two-body solution takes the form

$$\frac{1}{r}\frac{dr}{dt} = \frac{2F(\Delta V)}{mV} = \frac{2\Delta V}{t_e V} = \frac{1}{t_e}\frac{\Delta g}{g}, \tag{6}$$

by equations (2) and (3).

The appropriate expression for t_e can be chosen from equations (A4a, b, or c) depending on the value of R_e for $v = \Delta V$ (equation (A2)). Since the spiraling rate is extremely small for large bodies with large values of R_e, only the Stokes' law (equation (A4a)) will be presented explicitly, viz.:

$$\frac{dr}{dt} = \frac{9\eta r}{2\varrho_s s^2}\frac{\Delta g}{g}. \tag{7}$$

The intermediate case between I and II, when $t_e \sim T_p/2\pi$, has been solved by F. Franklin (private communication) for Epstein's and Stokes' laws, the drag being proportional to v. The maximum rate of orbital change is $dr/dt = 0.5\Delta V$ at $t_e = 0.8T_p/2\pi$. When $t_e = T_p$, dr/dt equals 0.8 the value given by equation (11), approaching equation (11) as a limit for larger values of t_e. For $t_e < T_p/30$, equations (4) and (5) overestimate dr/dt by about 30% but improve for smaller t_e. The intermediate case thus gives values of dr/dt varying from $0.3\Delta V$ at $t_e = T_p/30$ to a maximum of $0.5\Delta V$ at $t_e = T_p/8$ and dropping to $0.3\Delta V$ at $t_e = T_p$. Near but outside these limits, equations (4) and (5) or equation (6) give fairly satisfactory values for dr/dt.

Application in the "Cometary Region"

As shown in the Appendix in Tables A1 and A2, we define the "cometary region" at $r=25$ AU, with a typical density in the plane of $10^{-11.4}$ g cm^{-3}, $T=55$ K, $P=10^{-2.1}$ dyne cm^{-2}, giving a total areal density across the plane of $10^{2.3}$ g cm^{-2} (2×10^{-5} solar mass per square AU) for a solar mixture by weight of gases (H, He, Ne, Ar) 0.9803, ices (C, N, O plus H) 0.0175, and earthy materials (heavy elements) 0.0022. The corresponding viscosity will be $\eta=10^{-4.5}$ cgs, $\bar{v}=10^{4.86}$ cm sec^{-1} for a mean molecular weight of 2.34, and $\bar{v}_i=10^{4.41}$ cm sec^{-1} for the ices-plus-earthy material of mean molecular weight 18.4 and a mass fraction $f_i=0.0197$.

The peak of the gas pressure and density will have been farther out, near $r=30$ AU at Neptune's present solar distance, but the order-of-magnitude calculations at $r=25$ AU will illustrate the nature of the spiraling phenomena caused by a negative pressure gradient outward near the edge of the nebula. Suppose the pressure falls linearly with r to 0 at 50 AU Then $dP/dr=-10^{-16.7}$ cgs, while $\Delta g=10^{-5.3}$ cm sec^{-2} by equation (1) and $\Delta V=10^{3.2}$ cm sec^{-1} by equation (3) for a central solar mass. A forming icy grain of radius s and density 0.1 g cm^{-3} thus meets a resisting velocity of 16 m sec^{-1}, small compared to molecular velocities. The Reynolds number for equation (5) becomes $R_e=10^{-3.6}\ s$ (cm), so that we are in the realm of Epstein's and Stokes' laws of drag up to "cometesimals" of radius 100 m or larger at great solar distances. Correspondingly, the mean free path of the gas molecules (Table A2) is of the order of 1 m or more so that little error is made in applying Epstein's law up to perhaps $s=10$ m, while to about the same limit the diffusion slowing of the accretion rate by the second term of equation (A15) can be neglected. Hence, by equation (A17), the time t_1 for a tiny grain to grow to a radius s_1 (cm) becomes

$$t_1=6.3\ s_1\ \text{yr}, \tag{8}$$

if we accept an accommodation coefficient $a_i=1.0$.

For the Epstein case, by equations (4) and (A17) and an integration, the change in radial distance from r_0 at $t=0$ to r_1 at t_1 and s_1 becomes

$$r_1-r_0=\frac{a_i f_i \bar{v}_i \Delta g}{11.2\bar{v}}\,t_1^2=\frac{1.43\,\varrho_s^2\,\Delta g\, s_1^2}{a_i f_i \bar{v}_i \bar{v}\varrho^2} \tag{9}$$

applicable when $v \ll \bar{v}$, $t_e \ll T_p/2\pi$, and $L \leqslant 2s$ and when changes in ϱ and g with r are neglected. Note that any exhaustion of condensable gas slows the growth rates and thereby increases the total amount of spiraling.

Thus, our cometesimal loses solar distance from r_0 to r, as given by equation (9)

$$r_0 - r_1 = 0.8 \times 10^{-5}\, s_1^2 \text{ AU}, \tag{10}$$

while it grows to radius s_1.

But equation (8) shows that it will grow to a radius $s_1 \sim 10$ m in some 10^4 years, during which it spirals in a nominal 8 AU. We neglect here significant changes in density and other quantities depending on solar distance. The spiraling rate slows, of course, as the pressure gradient falls off.

Note that the time varies inversely as the nebular density ϱ and that the amount of spiraling varies as ϱ^{-2}, so that the inward spiraling for a constant $\Delta g = \varrho^{-1} dP/dr$ is increased for cometesimals of a given size at greater solar distances or at lower solar densities, even though the time scale varies as ϱ^{-1}. Reduction in growth rates by any cause increases the total amount of spiraling for a given gas density and gradient.

The sizable magnitude of the spiraling rate for cometesimals growing at the edge of the solar nebula may explain three facets of the present solar system: (*a*) the comparable masses of Neptune and Uranus, (*b*) the reduction of Neptune's solar distance from Bode's "law", and (*c*) the unobservability of a "comet belt" beyond Neptune, expected by the writer (Whipple 1964).

If we assume (Kuiper 1951; Cameron 1962; Whipple 1964) that Uranus and Neptune are aggregates of cometesimals, then it is otherwise surprising that there should have been enough material at Neptune's distance to make

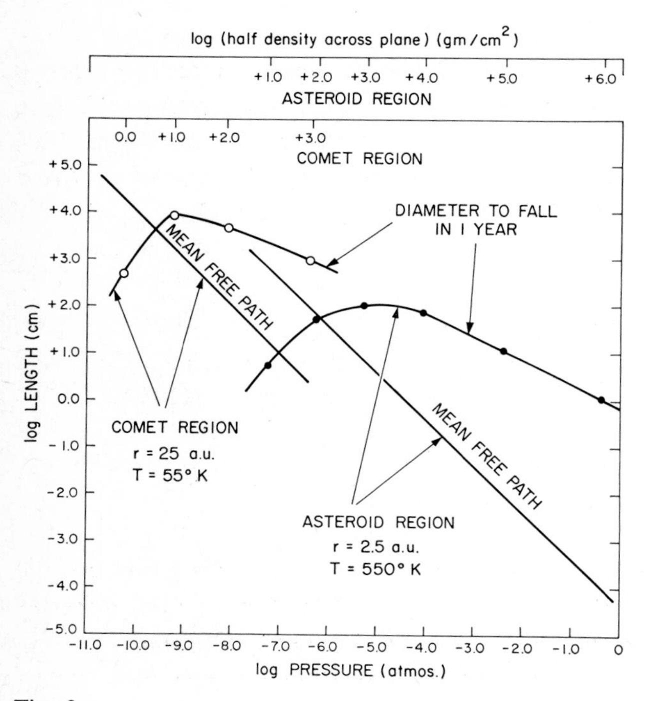

Fig. 2.

a planet as large as Uranus. The spiraling effect, however, could bring in the cometesimals to a solar distance where the pressure gradient was smaller and pile up the cometary material from greater distances for accretion at Neptune, producing no sizable planet (Pluto?) beyond. Neptune's mean solar distance may also have been reduced. Thus, my expectation of a comet belt beyond Neptune is perhaps unwarranted. In fact, the observations indicate that less than one Earth mass exists in a ring beyond Neptune to a solar distance of 50 AU (Hamid, Marsden, and Whipple 1968).

Cameron (1971) suggests that the cometary region may never have been heated sufficiently to destroy the original interstellar grains present in the uncontracted nebula. Hence, centers for grain growth may have been extremely numerous, leading to a rapid exhaustion of the ices in growing relatively small cometesimals. Further growth to cometary and planetary masses would have been difficult or perhaps impossible without concentration of the small bodies toward regions of small radial pressure gradient. Further concentration toward the plane of the nebula by gravity could also have aided in the accumulation process. Fig. 2 plots the minimum diameters of cometesimals that could fall to $1/e$ of their height above the plane in 1 year at the terminal velocity of free fall. The numbers show that once cometesimals reach meter dimensions, they concentrate quickly toward the plane.

Application in the Asteroidal Region

As discussed in the Appendix and shown in Tables A1 and A2, we may take a "typical" asteroidal region at $r=2.5$ AU, with density in the plane of $10^{-8.4}$ g cm^{-3} or a pressure of $10^{-4.1}$ atm, $T=550$ K, $P=10^{+1.9}$ dyne cm^{-2}, giving a total areal density across the plane of $10^{4.3}$ g cm^{-2} (2×10^{-3} solar mass per square AU), $\eta=10^{-3.8}$ cgs, $\bar{v}=10^{5.35}$ cm sec^{-1}, and $\bar{v}_i=10^{4.78}$ cm sec^{-1} for earthy molecules of mean molecular weight 33.5 and a mass fraction $f_i=0.0022$.

Suppose, for illustration, that the gas pressure doubles in 1 AU toward the Sun so that $dP/dr=-10^{-11.3}$ cgs, $\Delta g=-10^{-2.9}$ cm sec^{-2} by equation (1), and $\Delta V=10^{4.1}$ cm sec^{-1} by equation (3). In the gas the mean free path is 0.5 cm and $R_e=0.64\ s$. This puts us squarely in the Newtonian range of drag (equation (A4c)) and the range of slow grain growth (equation (A16)) for sizable planetesimals of $s\gtrsim10$ m. Growth to such sizes for $\varrho_s=3$ occurs rapidly, ~1.4 cm yr^{-1}, for a short time and then slows. As we have seen, in the transition range of s from case I to case II the radial spiraling rate for a change approximately fifty fold in radius by Stokes' law averages about 0.3 ΔV or $dr/dt=10^{-2.1}$ AU yr^{-1}. Thus, the reduction in r is a fraction of 1 AU while the planetesimal is growing to meter dimensions.

For larger bodies, $s>10$ m, equation (7) can be used if applicable, and we will introduce the equations for large Reynolds numbers (t_e from equation

(A4c) and dr/dt from equation (6)) and larger bodies, even though the relative velocity ΔV (equation (3)) may be small. On this basis,

$$\frac{1}{r}\frac{dr}{dt}=\frac{\varrho(\Delta V)^2}{3\varrho_s V}\frac{1}{s}, \tag{11}$$

where the sign of dr/dt is that of ΔV, given by equation (3), $\Delta V=2^{-1}r^{\frac{1}{2}}g^{-\frac{1}{2}}\Delta g$.

For our adopted $\Delta V=10^{4.1}$ cm sec^{-1} at $r=2.5$ AU, neglecting growth, equation (11) becomes numerically (with Δt in years, s in cm)

$$\frac{\Delta r}{r}=\frac{1.22}{s}\Delta t. \tag{12}$$

Appreciable change of solar distance occurs when Δt in years becomes comparable to the radius of the planetesimal measured in centimeters and varies directly as the nebular density (equation (11)). The growth rate, however, is really quite uncertain. The assumed value of ΔV is still small (less than 1 %) compared to the orbital velocity but is some 7 % of the mean molecular velocity. Turbulent motion of the gas might easily produce random relative velocities of the order of 1 km sec^{-1} so that a fast growth rate might be applicable (first term of equation (A15)). In that case, the change of r for a further growth to kilometers might not be significant. The slow growth rate, however, would make a significant change in r quite likely while processes tending to slow the growth rate would augment the radial change.

We may conclude, therefore, that effects of pressure gradients in the asteroid belt could be very important in shifting planetesimals toward Mars or conceivably Jupiter, provided the growth rates of the asteroids are not too rapid or, if so, the pressure gradients are sufficiently large. For the basic development of the asteroids on a short time scale of 10^3 to 10^4 years, the effect would be minor, barring very high-density gradients. However, should the nebular density be 10^{-5} atm or less in the asteroid region and the growth time 10^6 to 10^8 years, pressure gradients could well have decimated the asteroid region by spiraling the planetesimals toward Mars or Jupiter, depending on the distribution of gas and the location of original grain growth. Possibly, this effect contributed to the failure of an Earth-sized planet to develop between Mars and Jupiter.

General Remarks

My attention was first directed to systematic interactions between gas motions and growing planetesimals in a postulated solar nebula by Hoyle's (1960) theory involving the expansion of a centrally condensed nebula by the outward force of spiraling magnetic fields from the Sun. In this process, the plasma would be pressed outward from the Sun, carrying with it the neutral gas and, according to Hoyle, also the planetesimals. In fact (Whipple 1964), the reduced effective gravity on the gas will, following the arguments of this paper, cause the planetesimals to spiral inward toward the Sun rather than outward.

Cameron (1969) mentions the effect of a radial pressure gradient in producing an inward spiraling of planetesimals but does not discuss the alternative possibility, viz., outward spiraling, should a positive outward pressure gradient occur. From a more recent paper (Cameron 1971), the reason becomes apparent. His solutions for conditions in the solar nebula does not permit positive pressure gradients. Generally, however, should it be possible for a toroid of higher density to occur in the solar nebula, the growing planetesimals would be drawn toward it from the inside as well as from the outside, increasing the growth rate of an accreting planet. The importance of the effect, as we have seen above, depends mostly on the nebular density, its distribution, and the time rates of chemical condensation and of accretion. For a short time scale ($\sim 10^3$ to 10^4 years) such as Cameron envisages, the pressure-gradient effect would not be important unless the density gradient were relatively large.

3. *Chondrules*

Chondrules, roughly millimeter spherules found abundantly in many meteorites, have been aptly described by Euchen (1944) as products of a "fiery rain" in a primeval solar-system nebula. Chondrules are clearly mineral droplets that have cooled rapidly; some show evidence of supercooling. On the basis of the quantitative loss of volatile elements, Larimer and Anders (1967) and Keays, Ganapathy, and Anders (1971) deduce that chondrules were formed in an ambient temperature of some 550 K and pressure 10^{-6} to 10^{-2} atm. Since melting temperatures are roughly 1 300° greater, some violent heating mechanism must have been involved. Noteworthy is a suggestion by Wood (1963) that the quick heating was produced by shock waves in a primitive solar nebula. Volcanic and impact processes have been suggested, as has the pinch effect in lightning (Whipple 1966).

Whatever the source of droplet formation, a major evolutionary problem concerns the high abundance of chondrules among several classes of meteorites; in some it exceeds 70% by mass! Accepting the concept that meteorites are broken fragments of asteroids that were originally accumulated from solids in a gaseous solar nebula, one's credulity is taxed by the added assumption that a substantial fraction of the solid material should have been in the form of spherules. Thus, the purpose of this paper is to explore the possibility that chondrules may have been selectively accumulated on some

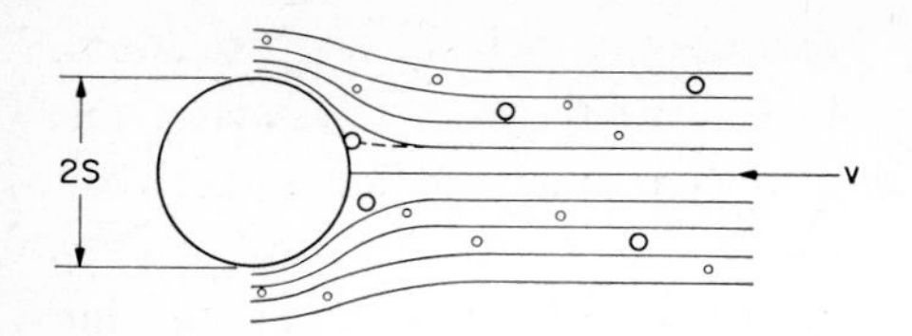

Fig. 3.

asteroidal bodies, thereby eliminating the undesirable supposition that chondrules constituted a major fraction of the dispersed solids in any part of the nebula.

Almost axiomatic is the assumption that the accumulation process for smaller asteroids essentially ceased when the solar nebula was removed, presumably by the effect of the solar wind from the newly formed Sun in its brilliant Hayashi (1966) phase. Possibly, the largest asteroids can still continue to grow in vacuum conditions, but the relative velocities of particle impact on asteroids less than perhaps 100 km in dimension would be generally dissipative rather than accumulative because of the low velocities of escape against gravity.

While the solar nebula was present, however, small bodies moving through the gas would have exhibited aerodynamic characteristics. At a given body velocity and gas density, solid particles having a mass-to-area ratio below a certain value would be carried around such a body by the inertia and viscosity of the gas currents so as not to impinge or accumulate on it (Fig. 3). The physical conditions for certain such accumulation processes will be established in the following sections of this paper.

Impact of Small Particles on a Sphere Moving Through a Gas

Taylor dealt with this basic problem for a cylinder in his paper "Notes on Possible Equipment and Techniques for Experiments on Icing on Aircraft" (1940). Langmuir and Blodgett (1945) derived numerical results by theory and calculation for cylinders, wedges, and spheres moving through air containing water droplets or icy spheres. Fuchs (1964) and Soo (1967) summarized the subject for subsonic flow and included both theoretical and experimental results by various investigators. Probstein and Fassio (1969) investigated "dusty hypersonic flows". The transonic case has apparently not been attacked seriously. The following discussion is based on the presentations by Langmuir and Blodgett, augmented by the summaries by Fuchs and Soo.

A sphere of radius S is assumed to move at velocity v through a gas of density ϱ and viscosity μ, containing in suspension small spheres of radius s and density ϱ_s. Because the flow about the forward surface of the moving body is relatively streamlined at rather high values of the Reynolds number,

R_e, the applicable Reynolds number is reduced greatly from its usual value (equation (A2)) because the small particles will not be thrown violently into the full velocity of the gas flow v, except perhaps near the stagnation point. As will be seen, the relatively small value of the applicable Reynolds number permits the application of the simple Stokes' law of particle drag at values of v far above those for which the law might intuitively appear to be valid. Because the Stokes force F on a sphere of radius s moving at a velocity v_s through a gas (equations (A4a) and (A5)) is independent of the gas density when $L<2s$, the impact equation for particles impinging on the larger sphere of radius S is widely applicable in a solar nebula where the density cannot be accurately specified. The Stokes approximation begins to fail significantly for $R_e>10$, but the drag force is overestimated only by about a factor of 3 at $R_e=10^2$ (see Probstein and Fassio). An inertia parameter ψ is defined as

$$\psi=\frac{s^2\varrho_s v}{9\eta S}, \tag{13}$$

which is the ratio of the inertia force to the viscous force for small particles in the stream. Theory and experiment show (Fig. 3) that particles of radius $<s$ do not impact the sphere for $\sqrt{\psi}<0.2$ for potential flow and 0.8 for viscous flow.

From the nature of Langmuir and Blodgett's definition of ψ, equivalent to $\psi=t_e v/2S$, where t_e is the "stopping time" (see equations (A4a, b, c, d) of the Appendix), I have extended their result to the Epstein region of drag when $L>2s$. The consequent definition of ψ by equation (A4d) becomes

$$\psi=\frac{\varrho_s s v}{2.8\,\varrho\bar{v}S}, \tag{14}$$

where $\bar{v}$, as before, is the mean molecular velocity.

At high values of R_e, when Stokes' law deteriorates, the behavior of the impact changes as a function of another paramter, ϕ, defined by

$$\phi=\frac{R_e^2}{2\psi}=\frac{18\varrho^2 vS}{\varrho_s\eta}. \tag{15}$$

Figure 4 illustrates the changes in impact efficiency for values of ϕ up to $\phi=10^4$. Note that the limiting value of $\sqrt{\psi}$, initiating the impacts, is nearly independent of ϕ, while the efficiency of impact is not greatly dependent on ϕ. Hence, limiting conditions for impaction of particles of radius s, on a sphere of radius S, up to R_e somewhat less than 10^2, can be confidently given by equation (13) when ψ (limit) ~ 0.04 and, although I know of no experimental evidence, probably by equation (14).

Let us then assume that condrules have a radius of 0.05 cm and decide that particles of $\frac{1}{3}$ this radius (diameter 1/30 cm) should not impact our asteroid

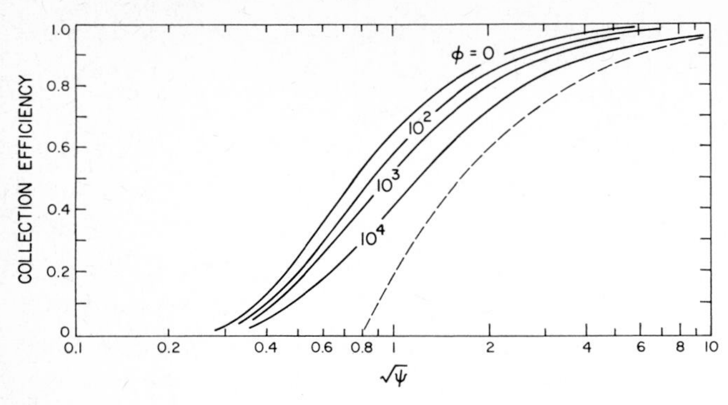

Fig. 4.

of radius S moving at velocity v_1 through the primeval nebula of viscosity η. Then, $s=1/60$ cm, $\sqrt{\psi}=0.2$, and

$$v_1=\frac{0.36\eta}{s^2\varrho_s}S=432\eta S=0.069S, \quad \text{Stokes or} \tag{16 a}$$

$$v_1=\frac{0.11\bar{v}\varrho S}{\varrho_s s}=10^{5.7}\varrho S, \quad \text{Epstein,} \tag{16 b}$$

if we take ϱ_s as 3 g cm^{-3}, employ cgs units, and adopt the "solar mix" (Table A1) of primeval gas. The viscosity becomes approximately $\eta=1.6\times10^4$ cgs and $\bar{v}=2\times10^5$ cm sec^{-1}.

Equations (16 a, b), appropriately applied to the postulated solar nebula at $r=2.5$ AU from the data of Table A2, relate in Figure 5 log v_1 (ordinate) to log pressure (abscissa), for a large range of values of $2S$, the diameter of the protoasteroid. The discontinuities result from discontinuity in the theory from Epstein's to Stokes' laws. The curves are cut off at the top for large S and P by the limit of 10^4 for the parameters ϕ (equation (15)).

The curves in Fig. 5 represent the lower limits of velocity for a body, of diameter $2S$, above which chondrules of radius 1/60 cm would encounter the body and not be swept away by the aerodynamic flow. Near the limiting velocity, very few particles of s as small as 1/60 cm would encounter the protoasteroid, while millimeter chondrules would almost all impinge on it. Near the curves, the velocity of impact of chondrules of radius $s=1/60$ cm would be greatly reduced from the critical velocity of the protoasteroid.

If we confine our attention to the range in pressure from 10^{-6} to 10^{-2} atm and to protoasteroids of diameter ~ 100 km, we see that chondrules and smaller particles will readily be accumulated at low pressures and low velocities, while the higher pressures set realistic velocity limits on the selective mechanism of accumulation. Because velocities ~ 1 km sec^{-1} seem "reasonable" for such bodies at the higher pressures, and even higher velocities at

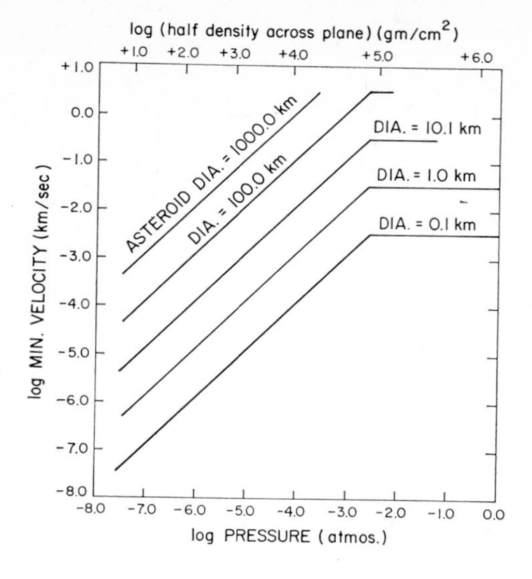

Fig. 5.

lower pressures, Fig. 5 suggests that chondrule production itself might be involved in the ærodynamics of the protoasteroid–nebula interaction.

Note first that unless selective accumulation is very efficient, the chondrites demand a suspiciously high production efficiency of chondrules from dust. Could the chondrules have been concentrated by gravity toward the plane of the nebula? The calculation is simple near the plane, where the gravity, $g(z)$, (equation (A14)) is proportional to the distance, z, from the plane, $g(z)/z$ being constant. Thus, the time of fall, t_f, from z to z/e at the limiting velocity of free fall against gas drag, is given by

$$t_f = \frac{z}{t_e g(z)}, \tag{17}$$

where t_e is the stopping time (equations (A4a, b, c and d)), depending on the law of drag.

For chondrules of density 3 g cm^{-3} and radius 0.05 cm, the Epstein (equation (A4d)) or Stokes (equation (A4a)) approximation is appropriate for the low velocities involved, the choice depending on the gas density. Fig. 6 presents such times of fall, t_f, for the nebular pressures considered in Table A2. Only in the range of density below a pressure of 10^{-5} atm are the times short enough ($\lesssim 3\,000$ years) for a significant settling of chondrules toward the nebular plane in the expected time scales of planetary formation. However, very small protoasteroids can fall quickly to the plane, as shown by Fig. 2, a self-explanatory plot derived from equation (17). Thus, the coagulation of protoasteroids should take place mostly near the plane, while chondrules would likely be well distributed in z.

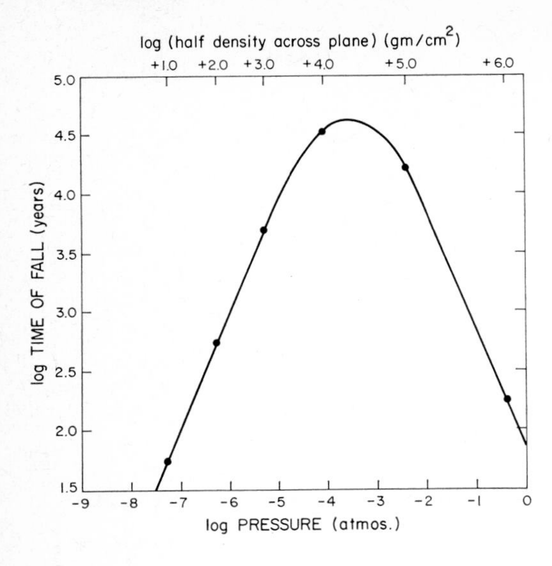

Fig. 6.

But the larger protoasteroids might *produce their own chondrules* and then selectively accumulate them. Two processes are possible:

(*a*) At subsonic velocities, the bow wave could generate a charge differential by friction among dust particles of different sizes so that lightning might occur, producing chondrules according to the author's suggestion (1966).

(*b*) At supersonic velocities, the shockwave in a relatively warm nebula might melt the dust and produce chondrules directly from dust globs or by aggregation of droplets from small dust particles.

No critical discussion of these possible processes will be made here, but the author plans to investigate the matter further. In both cases, however, the chondrules, being formed by the protoasteroid, should be identical in composition with the matrix material in the same layer of the final asteroid, except for volatiles lost by melting. This relation, observed among chondrites, would follow strictly in process (*a*) and also for large (compared to chondrules) globs in process (*b*) that were only superficially heated by the shockwave and then broken on impact. The fine dust in process (*b*) would probably be volatilized and lost.

In conclusion, then, we see that aerodynamical processes may be partially or even entirely responsible for the existence of chondrules. Furthermore, aerodynamic flow must be carefully considered when the growth rates of planetesimals by accumulating processes are being calculated. Gross overestimates of accumulation rates may otherwise be derived.

I am indebted to Fred A. Franklin for his assistance with regard to the mathematical solutions noted in this paper.

Appendix

1. Aerodynamic Relationships

A spherical grain of radius s, density ϱ_s, and mass m will experience a drag ($-$) or accelerating ($+$) force F by interaction with a gas of density ϱ at relative velocity v. The drag equation in a neutral gas is formally stated as

$$F = \frac{C_D}{2} \pi s^2 \varrho v^2, \tag{A1}$$

where C_D is the dimensionless drag coefficient dependent on the Reynolds number R_e if the mean free path, L, of the gas atoms or molecules is small compared to s.

For gases the Reynolds number is defined by

$$R_e = \frac{2\varrho v s}{\eta}, \tag{A2}$$

where η is the viscosity of the gas, given approximately by $\eta = \frac{1}{2}\varrho \bar{v} L$, in which $\bar{v}$ is the mean kinetic speed of the atoms or molecules.

In case $L < 2s$, following the approximations by Probstein and Fassio (1969),

$$C_D = 24/R_e \quad \text{for } R_e < 1, \tag{A3a}$$

$$C_D = 24/R_e^{3/5} \quad \text{for } 1 < R_e < 10^3, \tag{A3b}$$

$$C_D = 0.44 \quad \text{for } R_e > 10^3, \tag{A3c}$$

where equation (A3a) represents Stokes' law of drag and (A3c) the Newtonian case, which holds fairly well for supersonic velocities ($v > \bar{v}$).

For the subsonic case when $L \geqslant 2s$, the drag equation becomes generally complicated, but for $v \ll \bar{v}$, the Epstein approximation has some validity:

$$F\,(\text{Epstein}) = A\bar{v}\varrho s^2 v, \tag{A3d}$$

where $A \simeq 4\pi/3$ if the accommodation coefficient of atoms or molecules on the moving sphere is unity, but is better taken as $5.6\,\pi/3$ (Kennard 1938) as will be adopted here.

For our purposes, the various forms of the drag equation can be simplified by the introduction of the quantity *stopping time* t_e, which is the time for the body's speed to be reduced from v to v/e by drag. Equations (A3), can then be expressed as follows:

$$R_e < 1 \qquad t_e = \frac{2\varrho_s s^2}{9\eta}, \tag{A4a}$$

$$1 < R_e < 10^3 \quad t_e = \frac{2^{3/5}\, \varrho_s\, s^{8/5}}{9\eta^{3/5}\, \varrho^{2/5}\, v^{2/5}}, \tag{A4b}$$

$$R_e > 10^3 \quad t_e = \frac{6\varrho_s\, s}{\varrho v}, \quad \text{and} \tag{A4c}$$

$$\text{Epstein} \quad t_e = \frac{\varrho_s s}{1.4\varrho \bar{v}}, \tag{A4d}$$

where for all, the drag law becomes simply

$$\frac{F(v)}{m} = \frac{v}{t_e}. \tag{A5}$$

In the case of equations (A4b) and (A4c), t_e has been changed by a factor of less than 2 to produce the generality of equation (A5), which is correct for all cases. Stokes' law (equation (A4a)) attains an error of only a factor of 3 up to $R_e = 10^2$. Note that t_e is independent of v for both Stokes' and Epstein's laws.

2. The Assumed Solar Nebula

The Laplace-type solar nebula is assumed to consist of a solar mix of gases rotating under the gravitational attraction of a central mass, $M = 1$ solar mass, with a density $\varrho(z)$, where z is the distance above or below the plane of rotation. The areal density above or below the plane is $\sigma(z)$ integrated from the plane to height z. Thus, the total areal density across the entire nebula is $2\sigma(\infty)$, where $\sigma(\infty)$ is adopted arbitrarily as a function of the distance r from the central mass M.

The gas, of mean molecular weight μ, is assumed to be of temperature T, constant with z, obeying the perfect-gas relationship between pressure P and density ϱ,

$$P = \frac{R_g T \varrho}{\mu}, \tag{A6}$$

where the gas constant, $R_g = 8.317 \times 10^7$ ergs deg^{-1} mole^{-1}. No simple general expression exists for $\varrho(z)$ and $\sigma(z)$ even under the simple assumptions of constant T, a central force, and an infinite plane of gas defined by $\sigma(\infty)$. However, the approximations used by Cook and Franklin (1964) are quite adequate for the accuracy relevant to the present problem. These authors combine the infinite-plane and central-force solutions. By defining z^+ as the distance perpendicular to and across the plane containing one-half the total areal mass, they are able to preserve the form of either the infinite-plane solution or the central-force solution in an approximation including both forces.

Let us solve for

$$z^+ = \frac{2\pi r^3 \sigma(\infty)}{M} \left(\left\{ 1 + \frac{M[E_2^{-1}(1/2)]^2}{\pi \theta_1 G r^3 \sigma^2(\infty)} \right\}^{1/2} - 1 \right), \tag{A7}$$

where G is the constant of gravity, $\theta_1 = \mu/2R_g T$, and $E_2(x)$ is the error function of x.

Then Cook and Franklin show that if

$$\theta_2 = \frac{GM}{r^3} + \frac{4\pi G \sigma(\infty)}{z^+}, \tag{A8}$$

it follows that

$$\varrho(z) = \frac{2\sigma(\infty)}{\pi^{1/2}} \theta_1^{1/2} \theta_2^{1/2} \exp(-\theta_1 \theta_2 z^2), \tag{A9}$$

and

$$\sigma(z) = \sigma(\infty) E_2 \left(\theta_1^{1/2} \theta_2^{1/2} z \right). \tag{A10}$$

Equations (A7) to (A10) apply best when the central force exceeds the infinite-plane vertical acceleration on the gas near the plane, i.e., for a low-density nebula, or near the protosun.

In an alternate solution, when the infinite-plane acceleration exceeds that of the central force, let

$$z^+ = \frac{2\pi r^3 \sigma(\infty)}{M} \left\{ \left[1 + \frac{M \tanh^{-1}(1/2)}{2\pi^2 \theta_1 G r^3 \sigma^2(\infty)} \right]^{1/2} - 1 \right\}. \tag{A11}$$

Then

$$\varrho(z) = \sigma(\infty) \theta_1 \theta_2 z^+ \operatorname{sech}^2(\theta_1 \theta_2 z^+ z), \tag{A12}$$

and

$$\sigma(z) = \sigma(\infty) \tanh(\theta_1 \theta_2 z^+ z), \tag{A13}$$

θ_2 being defined as in equation (A8) with z^+ from equation (A11).

In both cases, for $z \ll z^+$ the vertical acceleration, $g(z)$, near the plane is given by

$$\frac{g(z)}{z} = 4\pi G \varrho(0) + \frac{GM}{r^3}, \tag{A14a}$$

and more generally by

$$g(z) = 4\pi G \sigma(z) + \frac{GM}{r^3} z. \tag{A14b}$$

Table A1. *The solar nebula*

Material	Mass per cent	Mol. wt.
Gases	98.03	2.30
Icy	1.75	17.43
Earthy	0.22	33.5
Earthy & Icy	1.97	18.4
Total	100.00	2.34

The assumed composition of the solar nebula is shown in Table A1. The *gases* are H_2, He, Ne, and Ar. The *icy* elements are assumed to form from H_2O, NH_3, and CH_4, while the *earthy* elements constitute the remainder of the elements in atomic form, following the abundances given by Goles (1969).

Solutions of equations (A7) to (A14) have been carried out at two positions in the solar nebula for various values of $\sigma(\infty)$:

Asteroidal Region: $r=2.5$ AU; $T_{gas}=550$ K and
Cometary Region: $r=25$ AU; $T_{gas}=55$ K.

The temperature in the asteroidal region is based on meteoritic abundances and loss of volatiles given by Larimer and Anders (1967) and is supported by the further conclusions of Keays, Ganapathy, and Anders (1971) that in 13 chondrites at formation, $T=530^{+80}_{-60}$ K and $P=10^{-4\pm 2}$ atm.

The temperature in the "cometary region" is assumed arbitrarily.

Table A2 lists results for calculations as a function $\sigma(\infty)$, or one-half the total areal density across the nebula (g cm^{-2}). In the table, $P(0)$ and $\varrho(0)$ are the total gas pressure (atm) and the gas density (g cm^{-3}), respectively, in the plane; z^+ is distance across the plane, including $\frac{1}{2}$ the total areal density (cm); mass is per (AU)^2 in solar units; $g(z)/z$ is the gravity constant for $z \ll z^+$; and L is the mean free path of molecules in the gas in the plane.

Table A2. *Data concerning the assumed solar nebula*

log $\sigma(\infty)$ g cm^{-2}	log P(0) atm	log ϱ(0) g cm^{-3}	log z^+ cm	Mass Sun (AU)$^{-2}$	log g(z)/z sec^{-2}	log L cm
Asteroidal region, r = 2.5 AU T = 550°K						
+6.0	−0.38	− 4.66	+10.71	1/4.47	−10.74	−4.09
+5.0	−2.38	− 6.66	+11.70	1/44.7	−12.73	−2.10
+4.0	−4.12	− 8.40	+12.44	1/447	−14.23	−0.35
+3.0	−5.25	− 9.54	+12.57	1/4470	−14.55	+0.78
+2.0	−6.26	−10.54	+12.57	1/44700	−14.59	+1.78
+1.0	−7.26	−11.54	+12.57	1/447000	−14.59	+2.78
Cometary region, r = 25 AU T = 55°K						
+3.0	− 6.38	− 9.66	+12.70	1/4470	−15.73	+0.40
+2.0	− 8.14	−11.38	+13.41	1/44700	−17.21	+2.12
+1.0	− 9.24	−12.52	+13.55	1/447000	−17.55	+3.27
0.0	−10.26	−13.54	+13.57	1/4470000	−17.59	+4.28

Grain Growth Rates

We wish to consider stationary grains that are growing by the accretion of atoms and molecules from the nebula. For atoms or molecules that can "freeze" on the grain, constituting a fraction f_i of the total density by weight, with an accommodation coefficient a_i, a mean molecular velocity $\bar{v}_i$, and a mean diffusion coefficient D_i, a spherical grain of radius s and density ϱ_s grows at a rate

$$\frac{ds}{dt}=\frac{a_i f_i \bar{v}_i \varrho}{4\varrho_s}\left(1+\frac{\bar{v}_i s}{4D_i}\right)^{-1}. \tag{A15}$$

The second term becomes significant when the grain has reached such a size that for a velocity v through the gas small compared to $\bar{v}_i$, grain growth is inhibited by gas diffusion of the appropriate atoms and molecules. Approximately, $D_i=1.4\,\eta_i/\varrho$, where η_i is the viscosity appropriate to the "freezing" atoms and molecules. Hence, from the second term of equation (A15), the grain growth rate slows when $s\gg 5.6\,\eta_i/\bar{v}_i\varrho$, given by

$$\frac{ds}{dt}=\frac{1.4a_i f_i \eta_i}{\varrho_s s}, \tag{A16}$$

becoming independent of the general gas density and varying inversely as the radius of the grain.

The transition from rapid grain growth (first term of equation (A15); e.g., Kuiper 1951) to diffusion-limited grain growth (equation (A16)) begins when the mean free path L_i becomes small compared to $2s$, or roughly when the drag law at low grain velocities changes from the Epstein law (equation (A4d)) to Stokes' law (equation (A4a) and (A5)). Hence, for the Epstein case we keep only the first term in equation (A15), so that a very small grain grows to radius s_1 in time t_1 given by

$$t_1=\frac{4\varrho_s}{a_i f_i \bar{v}_i \varrho}s_1 \tag{A17}$$

References

Cameron, A. G. W., 1962, The formation of the sun and planets, Icarus, *1*, 13.

Cameron, A. G. W., 1969, Physical conditions in the primitive solar nebula, in Meteorite Research (ed. P. M. Millman). Springer-Verlag, New York, p. 7.

Cameron, A. G. W., 1971, Presented at I.A.U. Colloquium #13, Albany, N.Y., June. To be published in Evolutionary and Physical Properties of Meteoroids.

Cook, A. F. and Franklin, F. A., 1964, Rediscussion of Maxwell's Adams Prize Essay on the Stability of Saturn's Rings, Astron. J., *69*, 173.

Euchen, A. T., 1944, Über den Zustand des Erdinnern, Naturwiss, *32*, 112.

Fuchs, N. A., 1964, The Mechanics of Aerosols. The Macmillan Co., New York, Ch. 4.

Goles, G. G., 1969, Cosmic abundances, nucleosynthesis and cosmic chronology, in The Handbook of Geochemistry, Part 1, Ch. 5. Springer-Verlag, New York.
Hamid, S. E., Marsden, B. G. and Whipple, F. L., 1968, Influence of a comet belt beyond Neptune on the motions of periodic comets, Astron. J., *73*, 727.
Hayashi, C., 1966, Evolution of protostars, in Annual Review of Astronomy and Astrophysics (ed. Leo Goldberg), *4*, Annual Reviews, Inc., Palo Alto, Calif.
Hoyle, F., 1960, On the origin of the solar nebula, Q. J. R. Astron. Soc., *1*, 28.
Keays, R. R., Ganapathy, R. and Anders, E., 1971, Chemical fractionations in meteorites—IV. Abundances of fourteen trace elements in L-chondrites; implications for cosmothermometry, Geochim. Cosmochim. Acta, *35*, 337.
Kennard, E. H., 1938, Kinetic Theory of Gases. McGraw-Hill Book Co., New York, p. 310.
Kuiper, G. P., 1951, On the origin of the solar system, in Astrophysics (ed. J. A. Hynek). McGraw-Hill Book Co., New York, p. 357.
Larimer, J. W. and Anders, E., 1967, Chemical fractionations in meteorites—II. Abundance patterns and their interpretation, Geochim. Cosmochim. Acta, *31*, 1239.
Langmuir, I. and Blodgett, K. B., 1945, Mathematical investigation of water droplet trajectories, General Electric Res. Lab. Rept. RL-225; also in J. Meteorol., *5*, 175 (1948).
Probstein, R. F. and Fassio, F., 1969, Dusty hypersonic flows, A.I.A.A. Journ, *8*, 4.
Soo, S. L., 1967, Fluid Dynamics of Multiphase Systems. Blaisdell Pub. Co., Waltham, Massachusetts, Ch. 5.
Taylor, G. I., 1940, Notes on possible equipment and techniques for experiments on icing on aircraft, R & M No. 2024. Aeronautical Research Comm., London.
Tisserand, F., 1890, Mécanique Céleste, vol. IV, p. 216.
Whipple, F. L., 1964, The history of the solar system, Proc. nat. Acad. Sci., *52*, 565.
Whipple, F. L., 1966, Chondrules: suggestion concerning the origin, Science, *153*, 54.
Wood, J. A., 1963, On the origin of chondrules and chondrites, Icarus, *2*, 152.

Discussion

H. Sato

If you consider the spin of the small body, you will have a side force. The motion of the body may be modified by the force.

E. Anders

Your scheme, which looks very promising for asteroids, may also be applicable to the Earth and Moon. Our studies on Apollo 11 to 14 rocks have shown consistently that the Moon is depleted in volatile elements by a factor of 10^{-2} relative to the Earth. Now, the volatiles were probably brought in by micron-sized particles, which continued to equilibrate with the nebula down to low temperatures, in contrast to the millimeter-sized chondrules which had too small a specific surface area to condense volatiles. According to your scheme the Moon, being a smaller body, would accrete small particles less efficiently than did the Earth, thus ending up with a smaller abundance of volatiles. The difference might be enhanced by the Moon's higher velocity relative to the gas, if it was already orbiting the Earth at that time.

F. L. Whipple

I have not included any gravitational force here.

G. Arrhenius

You suggested that lightning may have been the process responsible for melting of grains to form chondrules. However, lightning is typically a discharge phenomenon limited to dielectric media at pressures many orders of magnitude above those that we are talking about here. It would seem important to apply your basic idea in the range of environments considered for other reasons, particularly then (*a*) at pressures below 10^{-4} atm, (*b*) at some specified degree of ionization, and (*c*) in those magnetic and electric fields that are most pertinent to the time and place.

If this is done I think that you will find the "lightning" semantically transformed to "quiet discharge", "glow discharge" or "electric current flow with ohmic gas heating". It would seem to be difficult to build up very high potential gradients under these circumstances because of the leakage.

F. L. Whipple

This is especially so if you build it up by turbulence. It is advantageous to build it up quickly by a body moving through the gas. Experiments have been made at the Smithsonian Institution with discharge through low pressure gases, dropping a solar mix through it. They get little things that look much like chondrules but I cannot get the meteoriticists to study them because they are so busy studying the lunar samples.

P. Pellas

Surely, some chondrites are so chondrule-rich that it is a puzzling fact (e.g. Bjurbole). However, the mechanism you propose could only be acceptable, if it obeys some (contradictory!) requirements:

1. The chondrules of a given chondrite (e.g. Bjurbole) must show the chemical fractionation which is observed in each different class of chondrites.

2. Before being collected by the asteroid, they were in space and in such a case they have to be irradiated by solar wind and solar flares (and by galactic cosmic rays, though this latter component was probably negligible due to the relatively strong solar modulation and the small grain-size you assume, 0.05 cm). Now, it appears that the chondrules in Bjurbole are not irradiated as single objects. This observation, which we have made by applying a track-method, is in fact checked by the rare gases measurements on Bjurbole which show the absence of solar-type gases.

3. Contradictorily with observation 1 there exists some chondrules showing very different chemistry among them. That is the case for *Chainpur* (Keil et al.

1964), for *Sharps* and for carbonaceous chondrites of type II, III and even in the case of *Karoonda* (C4). Clearly, in these later cases, mixing processes should have occurred before the chondrules might be collected on the assumed asteroid. It appears to me that, only in this last case, the mechanism you suggest may be a good one. Much better, it can be easily checked either by track-method, or by rare gas measurements on single chondrule basis.

H. Alfvén

From the study of the behaviour of the solar wind around the Moon and from laboratory experiments which we shall be glad to show you tomorrow there is rather much known about how a stream of gas behaves in the neighbourhood of a solid body (Danielsson and Lindberg 1965; Kristoferson 1969). The model you have discussed applies only to a case when we have a very high density which is probably above the densities we should be considering. Normally hydrodynamic phenomena are not so important but we have rather complicated phenomena of a different kind, namely the build-up of electrostatic charges. This was discussed by Dr Lehnert yesterday, and his model is rather far from what happens in a dust storm in a desert.

Danielsson, L., and Lindberg, L., 1965, Arkiv f. Fysik *28*, 1.
Kristoferson, L., 1969, J. Geophys. Res. *74*, 906.

F. L. Whipple

It does depend on the densities you assume.

H. Alfvén

It is necessary to treat the problem quantitatively.

The Structure and Formation of Comets

By V. Vanýsek

Department of Astronomy and Astrophysics, Charles University, Prague

Introduction

The physical processes in comets, beside the distribution and changes in their orbital elements, are the most spectacular phenomena in our solar system. The vast variations in general appearance, brightness and spectra as well as gravitational and non-gravitational effects mirroring in their motions are determined not only by the actual physical state and position in the gravitational field, but were coded in the cometary bodies at the time of their formation. Therefore our advancing knowledge in the cometary physics will contribute significantly to the understanding of the solar system history. There are good reasons to believe that comets are in some sense relicts of the matter which once was dispersed in the primordial nebula.

The following paragraphs are devoted mainly to a few topics which seem to be main problems of the present state of cometary physics: We would like to know a) the composition of the precursors of observed radicals; b) the composition of dust particles surrounding the cometary nucleus; c) if there exist some stratifications of the material in the nucleus.

The problems of the kinematics of ion tails and influence of the solar wind rise up a large complex of problems which are more familiar to other scientists who studied the cometary tails as phenomena developing from the interaction of the interplanetary plasma with the solar wind.

However, the processes which probably are less influenced by corpuscular streams and do not exhibit hydromagnetic phenomena seem to be more suitable for the study of the structure of cometary bodies. The aim of this paper is to review the present state of our knowledge about the behaviour of neutral molecules and dust particles.

Parent Molecules and Ice Particles in Comets

As to the composition of a cometary nucleus, the theoretical as well as observational results and particularly recent extraterrestrial observations of bright comets seem to agree with the idea of Whipple (1950, 1957) that it consists of a conglomerate of ices and earthy material. The measurement of the Lyman-α resonance line made by the OAO-2 and OGO-5 orbital observatories revealed a large H I cloud surrounding the coma of Comet Tago-Sato-Kosaki (1969*g*) and Comet Bennett (1969*i*). Simultaneously OAO-2 measured also the hydroxyl radical. The highly abundant neutral hydrogen and OH indicate that they very probably result from dissociated H_2O molecules and that ordinary ice could be one of the dominant components of the cometary nucleus. Hydrogen would be present in compounds with other elements, mainly oxygen, carbon or nitrogen, because the presence of the atomic or molecular hydrogen in cometary nuclei is highly unlikely.

As far as we can tell, the idea of icy comet nuclei has met no significant difficulties when confronted with observational results. However, little is known about the entire structure of such nuclei. When we speculate about the very early physical processes that took place during the formation of comets, then assuming either a homogeneous or a stratified nucleus means significant differences in the approach to the problem.

The same holds for the microstructure of ice and physical characteristics of dust particles and eventually for hypothetical parent molecules or precursors of the observed molecular radicals.

The formation of observed radicals in cometary atmospheres is still an unsolved problem. According to laboratory results, the photodecomposition of possible precursors of such typical radicals as CN and C_2 in the solar radiation field at a distance of 1 AU is a relatively slow process. For instance, experimental cross-sections for dissociation of unsaturated acetylene molecule C_2H_2, which is one of the supposed parent molecules for C_2, and solar flux integrated over the dissociation continuum lead to a lifetime of about 50 hours; many other relevant compounds have lifetimes nearly 100 hours such as C_2N_2 which is presumed to dissociate into CN. The very common molecule C_2H_4 which could be parent for C_2 with the decay time of about 4–6 hours, produced no C_2 Swan bands, typical of comets, when photodissociated in the laboratory.

In comets, however, the characteristic time of existence for parent particles is probably not longer than 10^4 seconds.

Even if the methods used for the determination of the characteristic time of the decomposition of hypothetical parent particles are not quite accurate it seems to be evident that such processes in cometary atmospheres are relatively fast and require only several hours. This is in disagreement with computed lifetimes obtained for some compounds which have been suggested as parent molecules in comets.

The lifetime τ is defined as a reciprocal value of the dissociation probability

$$\tau^{-1} = \int \sigma_\nu F_\nu \, d\nu, \tag{1}$$

where σ_ν is the photodissociation cross-section and the flux at a frequency ν is defined as $F_\nu = cu_\nu/h\nu$ where u_ν is the density of solar radiation.

The value of σ_ν is about 10^{-18} to 10^{-17} cm^2 for the most common compound and $\tau = 10^5$ to 10^6 seconds.

However, results concerning the prospective parent molecules for cometary radicals (Potter and Del Duca 1964) show that τ computed from σ_ν is at least 5 to 10 times longer than τ obtained from cometary observations. These results imply that the probability of dissociation in a cometary atmosphere is one or more orders of magnitude higher than is determined from the data of photodissociation cross-section and solar radiation flux. The photodissociation process is obviously dominant for the decomposition of the observed radicals, but it seems not to be efficient enough for the dissociation of parent molecules. The determined lifetimes for some components which could be supposed as possible parent molecules are $\tau = 10^{5.5}$ to $10^{6.5}$ seconds (except for NH_3 as a source of NH_2, with $\tau \sim 10^3$ sec). In contrast, previous cometary observations suggest that $\tau \sim 10^4$ seconds.

The differences are so striking that the hypothesis for the production of observed neutral molecules in comets via photodecomposition processes seems to be at stake and other theories are proposed (Wurm 1961; Öpik 1963; Herzberg 1964; Jackson and Donn 1968).

The decomposition of parent molecules was ascribed to the predissociation not considered in σ_ν, or to the chemical reaction in the very inner part of the coma or in the nucleus, or to the presence of free radicals in nuclei. All these proposals have met certain difficulties, which are not discussed here.

Some years ago the author pointed out that profiles of the continuum radiation in a cometary head can be explained by short-living dust particles. Delsemme and Swings (1954) considered that free radicals may be embedded in ice in the form of clathrates. This idea has been modified recently by Delsemme and Miller (1969). They assume that small fragments of ice of submillimetre dimensions expelled from the nucleus into the surrounding halo contain considerable amount of clathrate hydrates formed in the cavities in the water ice lattice, where different molecules, even unsaturated, can be bounded by van der Waals forces. By destruction of the lattice heated by the solar radiation the encaged molecules are liberated into the space and ejected isotropically into the cometary head. If the molecules are free radicals or very short-lived precursors of such radicals the ice particles play the role of parent molecules.

Such an assumption involves explicitly the physical character of small particles in the cometary nucleus. Therefore the behaviour of the continuum and molecular radiation in comets is very significant for any further discussions and can shed light on the formation of the cometary nucleus.

The Structure of a Cometary Atmosphere

The determination of the lifetimes of parent molecules in the solar radiation field is crucial for the further study of chemico-physical processes in comets. Most of the present estimates of molecular lifetimes are based on the observed distribution of molecules in the cometary head and the assumption of a particular kinematical behaviour of the matter in the cometary atmospheres.

The study of kinematics and dynamics of cometary heads and tails has been based upon the analysis of the forms and apparent motions of well defined envelopes, halos, knots in tails and streams. Direct inspection of a large number of photographs (or drawings from the last century) of several bright comets demonstrates that the cometary head is generally a complicated object. The heads consist of nearly circular diffuse patterns with superposition of different features, particularly of curved streams. (This is illustrated by the Atlas of the Cometary Forms compiled by Rahe, Donn and Wurm (1970).)

Comets of small apparent dimensions exhibit few features which could be observed directly and could be used for the interpretation of physical processes. Therefore, for many comets the information available for comparison with theories of the mechanism and tail or head formation was obtained from the distribution of surface density.

It is, however, essential that the observations of this kind should refer, as far as possible, to the radiation emitted or reflected by different kinds of particles (dust, C_2, C_3, CN, CO^+, etc.). It is therefore evident that the interpretation of the structure of comets requires monochromatic observations.

Even if the narrow-band photometry is being used more widely in recent years, the available photometric observations of comets suitable for the study of the different compounds distribution are still very scarce and inadequate. The paucity of accurate photometric observations of comets in monochromatic light is due merely to the fact that it is for instance very difficult to reconcile the needs of cometary photometry with those of stellar photometry.

The results obtained from the wide pass-band photometry must be regarded as tentative only unless it is quite evident that either continuum or emission bands were absent in the spectral region 4 600–5 000 Å. For the CN emission the problem is more complicated because of the possibility of overlapping of CO^+ and CN features.

The acceleration of CN and C_2 molecules due to the light pressure estimated from the oscillator strength for typical bands is 0.3 to 0.5 cm sec^{-2} at 1 AU and leads to some deformation of the CN and C_2 isophotes by shifting them slightly into the tail direction. However, the kinematics of CO^+ ions required obviously larger values of acceleration and the typical "onion-like" form of the isophotes obtained from measurements near the CN emission pass-bands is due to the overlapping of the CN and CO^+ emission (sometimes, of course, also to the scattered light on the dust particles). This effect can easily be demonstrated in many direct photographs or even on the isophotometry charts as is shown in Fig. 6.

The intensity distribution models in the coma which serve for the investigation of the density distribution are based on the following assumptions:

(1) The cometary atmospheres are perfectly optically thin.

(2) The kinematical behaviour of particles can be described by some simple mechanisms, for instance by the Bessel-Bredikhin model where the non-gravitational force is invariably repulsive.

(3) Some simple law of the decay function of particles can be applied.

(4) The power of the particle source remains constant over a sufficiently long time.

The basic equation which must be solved is the equation of continuity

$$\frac{\partial N}{\partial t}=\frac{\partial N}{\partial r_c}\frac{\partial r_c}{\partial t}+\frac{\partial N}{\partial v}+\frac{\partial N}{\partial r_c}\frac{\partial v}{\partial t}=N(v-\gamma); \qquad (2)$$

where $N \equiv N(r_c, v, t)$ is the number of particles, t the time, v the velocity and γ represents the factor of decaying of particles.

The problem was solved during this century by Eddington (1910), Mocknach (1938, 1956), Wallace and Miller (1958), Haser (1957, 1966), Dolginov and Gnedin (1966), Dolginov, Gnedin and Novikov (1971).

Condition (1) is fulfilled in the visual spectral range even if the optical depth very close to the nucleus is not quite negligible. The optical depth is, however, large in the inner part of the H I coma for the Lyman- α radiation and perhaps also the OH atmosphere should not be considered perfectly thin in the vicinity of the nucleus for the wavelength corresponding to the hydroxyl 0–0 emission.

Kinematics of the particles in the tail which fulfil condition (2) are normally described (as far as the time scale and the head dimensions are negligible with respect to the orbital parameters and actual position of the comet in the orbit) by the well-known abbreviated form of the Bessel-Bredikhin expansion of coordinates of moving particles into a Taylor series which leads to the "fountain" model of the coma. Although such a model can be criticized because of its flexibility in regard to the arbitrarily introduced parameters and invalidity for the plasma streams in comets, it seems to be the most realistic aproach to the kinematics of neutral particles.

With respect to condition (3) the time scale τ of the lifetime of parent (as well as dissociated) molecules is defined by the common form

$$\frac{n}{n_0}=1-e^{-\tau/t}, \qquad (3)$$

where n_0 is at $t=0$.

For the *dust particles* in the solar radiation field the lifetime is given by the change of radius a of the particles

$$\frac{da}{dt} = Z_p \frac{M}{\varrho_p}, \tag{4}$$

where M is the mass of produced molecules from the particles of density ϱ_p at production rate Z_p. If a_0 is the original radius of particles then the lifetime

$$\tau \simeq a_0 \frac{\varrho_p}{Z_p M}. \tag{5}$$

Condition (4) is fufilled when the steady state is reached and the power in the source S, i.e. the total number of particles emitted by the source per unit time, is constant over a time interval longer than the characteristic time scale of the studied processes, namely the lifetime of parent or dissociated molecules. Because it holds approximately that $S = S(T, E)$ where T is the nucleus surface temperature and E the latent heat of substituting parameter for the energy required for liberation of the mass from the nucleus. S is not only a function of the heliocentric distance, but also of E which depends on the actual state of the surface temperature of the cometary nucleus. The existence of brightness outbursts, net variations in apparent brightness of comets and expanding halos surrounding the nucleus and other similar phenomena are evidence that S sometimes changes in a very short time interval.[1] For this reason the analysis involving the assumption of a steady state condition, during a longer time interval, must be regarded *as qualitative only*.

In the very simple model where the isotropic ejection with a uniform velocity v_0 is assumed the particle density decreases proportionally to r^2, where r is the distance from the nucleus. The number of particles $n(\varrho)$ along the line of sight only depends on the initial number of particles n_0 and on the projected distance from the nucleus

$$n(\varrho) = \frac{\pi n_0}{v_0 \varrho}. \tag{6}$$

In such case, the surface brightness I is inversely proportional to the radius of the coma ϱ and the isophotes are circles limited by the envelope parabola.

These formulas for the coma models are only valid in the case when the lifetime of particles is very long or infinite as, for instance, the lifetime of non-volatile dust particles at large heliocentric distances.

However, if v_0 is the mean velocity of expansion and τ_0 the mean lifetime and

[1] The accurate simultaneous monochromatic photometry in the continuum region and molecular bands of cometary "flares" can be exploited for determining the upper limit of dissociation lifetime and parent molecules, too.

Table 1. *Ratio of computed lifetimes for the hypothetical precursors of CN and C_2 to adopted τ_p in comets*

τ_p (computed)/τ_p (cometary)

	Product	
Parent	CN	C_2
Adopted τ_p (Comet)	10^4 sec	10^4 sec
C_2N_2	21	
CH_3CN	225	
HC=C–CN	7	
CN–C=C–CN	11	
HC≡CH		27
$H_2C=CH_2$		6

when the equation of continuity for the spherical system is taken into consideration, the $D(r_c)$ density of molecules is given approximately by

$$D(r_c) = \left(\frac{R}{r_c}\right)^2 D(R)\exp[-r\beta]; \quad \beta = (\tau_0 v_0)^{-1} \tag{7}$$

where R is the effective radius of the source or nucleus.

Considering the lifetime of parent molecules τ_p the form of the distribution formula is

$$D(r) = \left(\frac{R}{r_c}\right)^2 D(R)(e^{-r\beta_0} - e^{-r\beta_p}), \tag{8}$$

where $\beta_0 = (\tau_0 v_0)^{-1}$ are reciprocals to the lifetime scale of dissociated molecules, $\beta_p = (\tau_p v_p)^{-1}$ the same of parent molecules, respectively.

The lifetime of parent molecules τ_p and of the produced radicals τ_0 can be estimated only indirectly by determining $\tau_p v_p$ and $\tau_0 v_0$ (where v_0 is the expansion velocity, and is supposed to be constant) from the intensity distribution in the cometary head.

Table 2. *Estimated lifetimes for precursors of CN and C_2*

$v\tau_p$ reduced to 1 AU in units of 10^4 km

	Produced molecule		
Comet	CN	C_2	References
1959k	0.61	0.61	Malaise (1968)
1961f		1.0	O'Dell & Osterbrock (1962)
1961f		5.0	Dewey & Miller (1966)
1963a	1.4 to 5.0	1.28[a]	Malaise (1970)
1963b	1.1		Vanýsek & Tremko (1964)
1964h	0.5 to 1.0	0.3 to 0.8	Vanýsek & Žáček (1967)
1967n	$\lesssim$1.1		Vanýsek (1969)
1970g	1.2		Vanýsek & Žáček (unpublished)

[a] Average value.

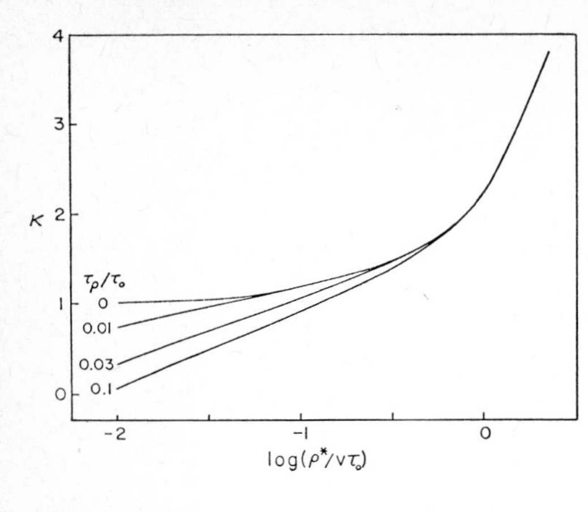

Fig. 1. The behaviour of the gradient $\varkappa$ for the ratio of lifetimes $\tau_p/\tau_0=0$, 0.01, 0.03, 0.1, ϱ^* is linear distance from the nucleus and v is expansion velocity of the parent molecules as well as of radicals. (After Vanýsek, 1969.)

Theoretical expressions for the intensity distribution have been derived by Haser (1957) and fitted to observations of C_2 and CN emission by several authors (see Table 2). The intensity function $I=I(\varrho)$, where ϱ is the distance from the nucleus, has the simple form $\varrho^{-\varkappa}$. If a steady state is reached in the supply of new gaseous matter from the nucleus and $\tau_0=\infty$, $\tau_p=0$, then $n=1$. When $\tau_p/\tau_0>0.01$ then $n<1$ close to the nucleus in the "zone of production". The behaviour of $\varkappa$ is illustrated in Fig. 1 for $\tau_p/\tau_0=0$, 10^{-3}, 3×10^{-2}, and 10^{-1}, respectively.

From the fitting of the observed photometric profiles to the observed ones the following can be estimated:

$$\frac{v_p\,\tau_p}{v_0\,\tau_0}\simeq\frac{\beta_0}{\beta_p}\,. \tag{9}$$

The course of the intensity for several cases of β_0/β_p is shown in Fig. 2.

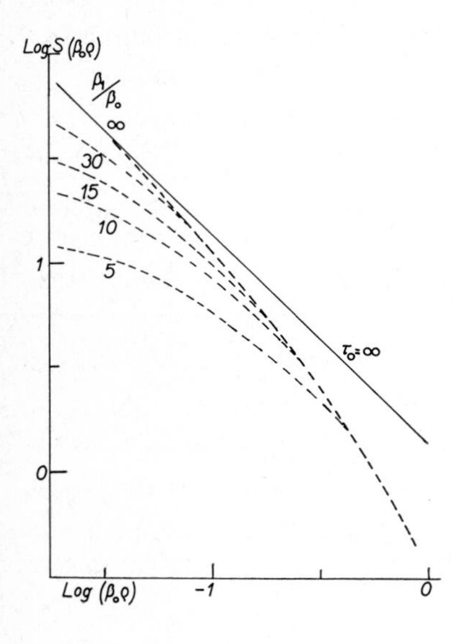

Fig. 2. The family of curves for the surface brightness $S(\beta_0\varrho)\cdot\beta_1/\beta_0=v\tau_p/v_0$. Full line for the ϱ^{-1} law.

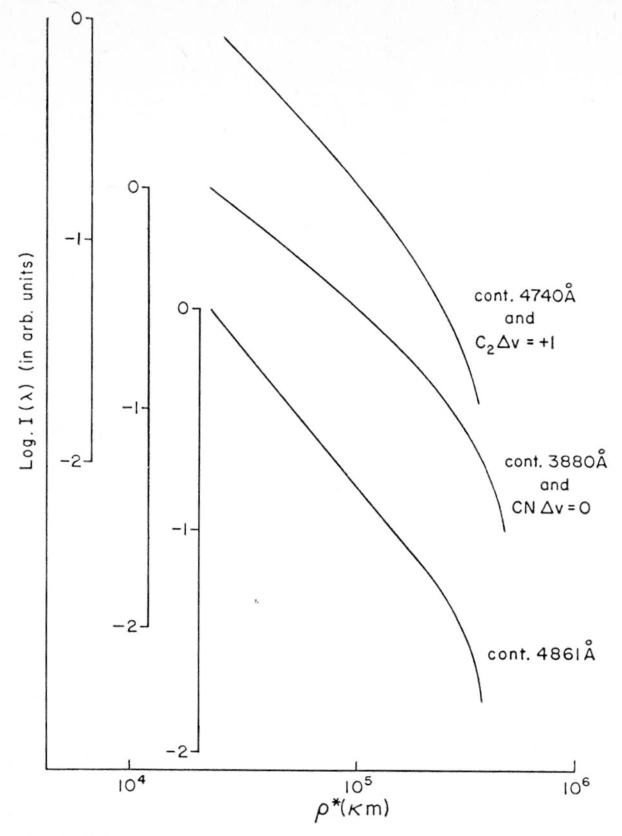

Fig. 3. "Average" profiles of the emission bands and continuum derived from the photoelectric scans over the head of Comet 1967*n* made with 34″ diaphragm. The surface intensity $I(\varrho)$ per area unit is in arbitrary units, and was determined with accuracy 5 % at linear distance $\varrho = 5 \times 10^4$ km and about 15 % at 3×10^5 km from the nucleus.

An example of photometric scans for CN and C_2 in the head of Comet 1969*n* is presented in Fig. 3. (The following values were obtained: $v\tau_0(CN) = 3 \times 10^5$ km and $v\tau_0(C_2) = 2 \times 10^5$ km. Assuming $v = 1$ km sec^{-1} and $\tau_p/\tau_0 = 0.10$ then $\tau_p(CN) = 3 \times 10^4$ and $\tau_p(C_2) = 2 \times 10^4$. Reduced to 1 AU heliocentric distance $\tau_p(CN) = 10^4$ sec and $\tau_p(C_2) = 6 \times 10^3$ sec.)

The linear resolution of photometric scans close to the nucleus is usually too low for an accurate evaluation of $v_p\tau_p/v_0\tau_0$ or $\varkappa$ from the photometric curve.

If the angular resolution of the photometric method is about 6 seconds of arc, then 5×10^3 is the optimal linear resolution for a comet at typical geocentric distances.

Nevertheless the maximum diameter of the "production zone" can easily be estimated. It is evident from Fig. 4 that the continuum decreases rapidly with increasing linear distance from the nucleus. Continuum + C_2 bands at 4 740 Å as well as continuum + CN bands decrease considerably slower at

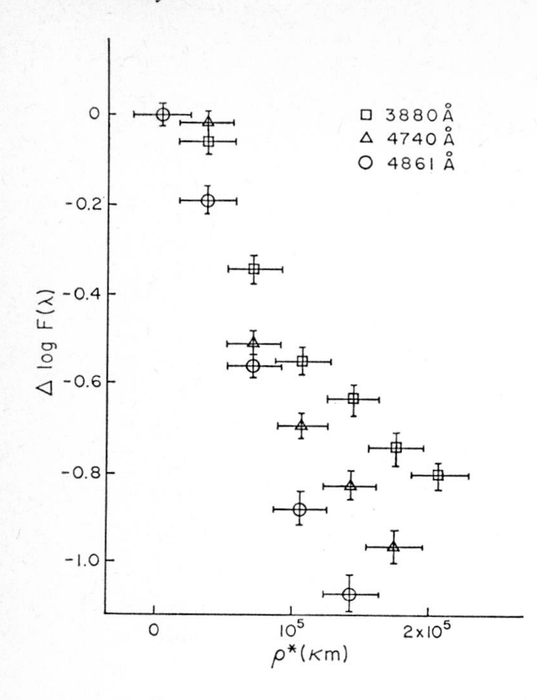

Fig. 4. Photometric measurements in the head of Comet 1967*n* (normalized to flux of central condensation F_0) in 34″ circular diaphragm at different distances from the nucleus in increasing right ascension. Horizontal bars represent the diameter of measured area. (After Vanýsek, 1969.)

least up to 4×10^4 km. If the "zone of production" and $v_p\tau_p$ have the same dimensions then $v_p\tau_p=10^{4.1}$ km.

The situation is quite different when more realistic conditions for the initial velocities are taken into consideration.

The gas as well as dust ejection into a vacuum is a complicated phenomenon which cannot be described by a monovelocity flow.

The velocity distribution of the observed molecules can be characterized by the "radial" temperature T characterizing the distribution of the molecular velocity along the radial direction from the source. If $\bar{v}>2\,kT/m$ (where k is the Boltzmann constant and m the mass of the molecule) the intensity distribution remains nearly the same as in the uniform velocity case.

This may be easily demonstrated on the boundary near the vertex of the paraboloidal envelope. For more details see for instance Arpigny (1965).

The envelope would be slightly "blurred" by a velocity distribution and v_0. However, if v_0 approaches the "thermal" velocity the photometric gradient deviates from the ϱ^{-1} law on the front side of the head.

Wallace and Miller have computed several models in which the initial velocity is not quite uniform but there exists a certain dispersion of velocities. The results are very important because they indicate that a Gaussian distribution (at initial velocities of 1 km/sec$\pm$0.32 km/sec with a limit between 0.05 km/sec and 2.05 km/sec and an acceleration of 0.6 cm sec^{-2}) strongly influences the form of the isophotes: the isophotes are markedly elongated from the Sun.

In such case a variation of intensity with the diameter ϱ^{-1} is in general not valid. On the sunward radius the intensity close to the nucleus varies approximately with $\varkappa=1$ but at greater distances the gradient increases rapidly to $\varkappa=4$. The boundaries of such comae are apparently very sharp on the sunward side. In the direction away from the Sun the intensity varies with ϱ^{-1}.

This effect leads to an exponential decrease of the surface intensity and the intensity profile traced in any direction (except in the prolonged radius vector Sun-comet) *resembles the profile of an optically thin atmosphere with constant velocity and containing particles with limited lifetimes.*

It must be noted that qualitatively the same results are obtained when there is a scatter in the particle acceleration, and such an effect may be very pronounced in the dust component of the cometary atmosphere.

This is confirmed by observations of Dobrovolskij et al. (1966); from their measurements of the position of the tail axis of Comet Arend-Roland 1957 III, they concluded that the tail was not a syndyname or a synchrone, but was produced by a continuous ejection of particles of various dimensions. Finson and Probstein (1968) came to the same conclusion from the analysis of isophotes of the tail of this comet. They definitely showed that this dust tail was produced by a continuous but peaked ejection of dust particles with various values of acceleration.

For dust particles with the effective cross-section σ_{pr} and density ϱ_d in the solar radiation field at a heliocentric distance r the acceleration g due to the light pressure is given by the following relation

$$g = \frac{3\sigma_{pr}\, s}{4\pi a^3 r^2 \varrho_d c}, \tag{10}$$

where a is the radius of particles, s the solar constant.

Because

$$\sigma_{pr} = \frac{1}{\pi a^2}(Q_{ex} - Q_{sca}\overline{\cos\vartheta}) \tag{11}$$

where Q_{ex}, Q_{sca} are the extinction and the scattering efficiency factors, $\overline{\cos\vartheta}$ is the "mean" cosine of the scattering function. Q_{ex} as well as Q_{sca} and $\overline{\cos\vartheta}$ are functions of the size parameter and complex refractive index of the particles.

The distribution of g is determined by the size distribution and the efficiency factor distribution function. Therefore the spreading of the acceleration value is caused not only by the size distribution, but also by the variation in the absorptivity of individual particles.

Moreover, the radii of particles can be functions of time. When a limited lifetime is assumed and the variations of Q_{ex} and Q_{sca} are neglected, then

$$\frac{da}{dt} \sim Z\frac{M}{\varrho_d}. \tag{12}$$

The value of g then increases linearly with time.

Using ice particles, the change of diameter is the following

$$\frac{da}{dt} \simeq 10^{-22,5}\, Z_d. \tag{13}$$

At a distance of 1 AU, assuming the albedo for ice grains 0.3 and Huebner's (1970) data for $Z_d = 10^{17.5}$ molecules cm^{-2} sec^{-1} sterad^{-1}, then for H_2O ice $da/dt = 10^{-5}$ cm sec^{-1}.

Small dust grains with $a = 10^{-4}$ cm covered with an ice mantle 10^{-3} cm thick will be stripped in 10^2 sec. In such case the dust grains become increasingly accelerated by the radiation pressure, because of the diameter decreasing and eventually the albedo decreasing, involving changes of $Q_{sca}\,\overline{\cos\vartheta}$.

The spread in v_0 and g simulates the decay of the particles. This can be demonstrated by photometric analysis of the continuum of Comet Abe 1970g, made recently by Vanýsek and Žáček. They used the photos taken by Börngen

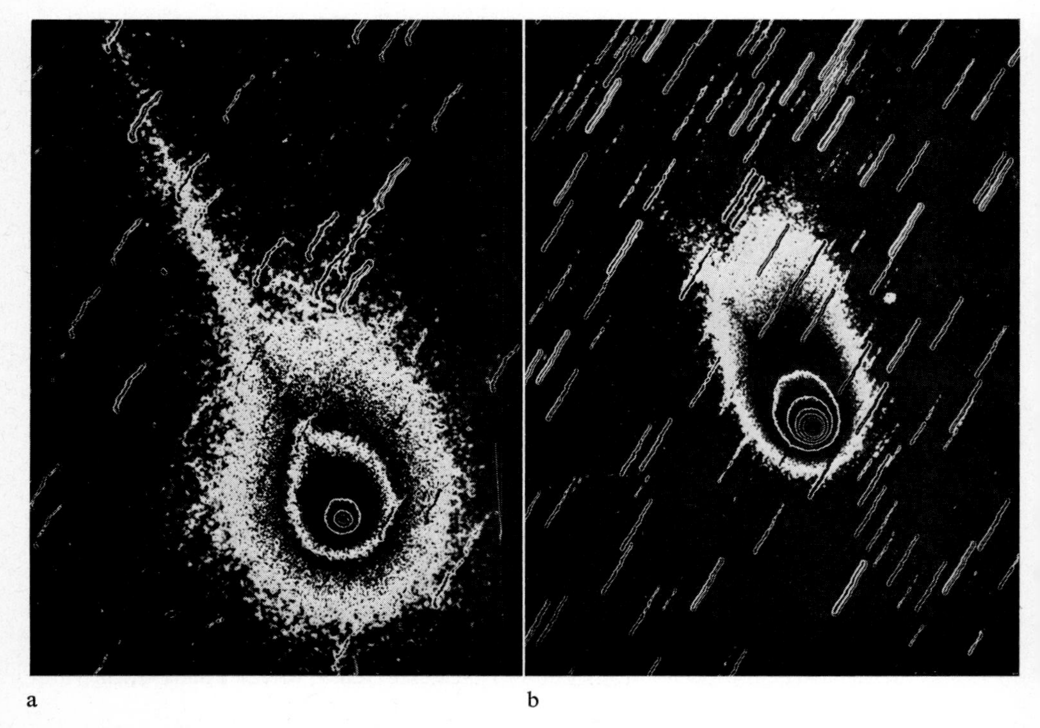

Fig. 5*a*. Equidensities derived by photographic process from red plate ($\lambda_{\mathrm{eff}} = 6\,200$Å) of Comet Abe 1970$g$. The equidensities represent the distribution of the dust.
Fig. 5*b*. Equidensities of CN, CO^+ bands at 3 800–4 000 Å (somewhat contaminated by dust). Comet Abe 1970g, September 10, 1970.

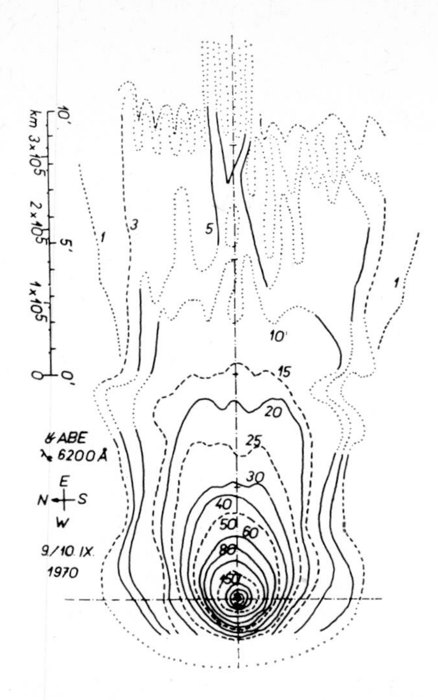

Fig. 6. Charts of isophotes of Comet Abe 1970g in continuum ($\lambda_{eff} = 6\,200$ Å). Intensity units are in arbitrary scale.

and Richter by means of the Tautenburg Observatory Schmidt with the linear resolution 2×10^4 km. (A detailed photometric analysis of Comets Abe 1970g and Bennett 1969i made by Börngen, Richter, Žáček and Vanýsek will be published later.)

The intensity distribution obtained for the continuum and CN, CO^+ coma are presented in Figs. 5a, b. It is evident that the continuum coma on the red plate is considerably smaller than the CN + CO^+ coma, which is very typical of the distribution of continuum in comets. The profiles at different position angles in the continuum coma are more interesting. Figs. 8a, b show the measured

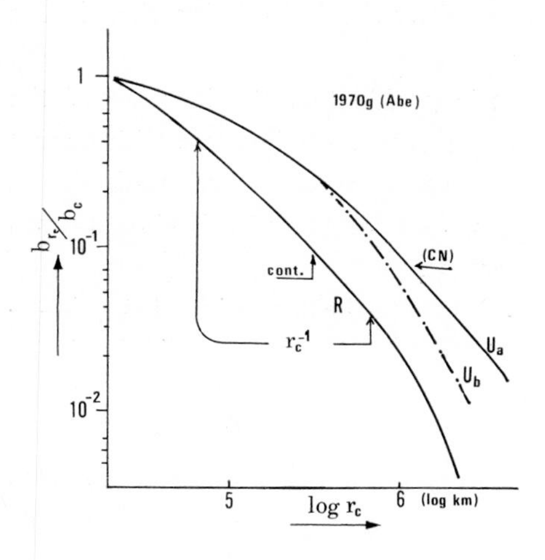

Fig. 7. Profiles of CN + CO^+ (curve U_a) and CN (curve U_b) emission in Comet Abe 1970g compared with the surface brightness distribution of the continuum. (b_r = central brightness, r_c = projected distance from the nucleus.) Curve U_a is derived from the profile along the tail axis where the CO^+ emission obviously deforms the CN profile.

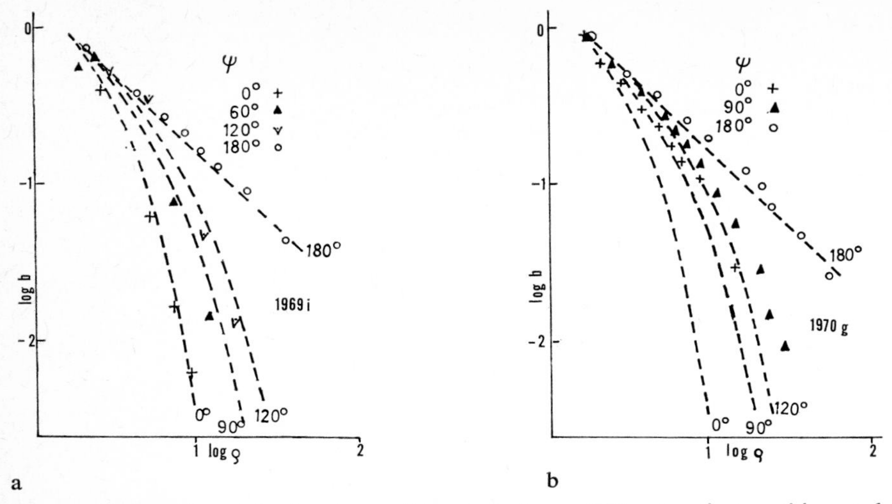

Fig. 8*a*. The profiles of the "continuum" head of Comet 1970*g* at various position angles ($\psi = 180°$ coincides with the tail axis). Dashed curves are predicted profiles for a model of dust coma with the distribution of ejection velocities (see text).
Fig. 8*b*. The profiles of continuum head of Comet Bennett 1969*i* (for meaning of symbols see Fig. 8*a*).

profiles of Comet 1969*i* and 1970*g* compared with the calculated ones for non-decaying particles with a Gaussian distribution of the initial velocities, using the Wallace-Miller model.

Even if the used model is in some sense an arbitrary one the measurements of Comet Bennett 1969*i* agree quite well with the predicted profiles. For Comet Abe 1970*g* there is a considerable discrepancy between the observed and predicted intensity distribution only for the sunward direction. The measurements lead to a lower value of the gradient. The discrepancy is due either to the inadequacy of the chosen model (in Figs. 8*a*, *b* the vertex of the head's parabolic envelope is 5×10^4 km), or to the simple fact that in Comet Abe 1970*g* the continuum was more seriously contaminated by the C_2 emission than in the dust-rich comet 1969*i*.

The increasing slope of the photometric profile (i.e. the increase of the photometric gradient $\varkappa$) for the continuum observed for instance with Comet Burnham (O'Dell 1962) is not surprising because the tracings were made almost perpendicular to the projected radius vector Sun-comet and this phenomenon can be explained by kinematics of dust grains.

The lifetime of ice particles can, of course, lead to the increasing of the photometric gradient $\varkappa$ in the very inner part of the coma. But the available photometric data have resolution of only about 5.0×10^4 to 10^4 km. At heliocentric distances of 1.3 AU to 0.7 AU only large particles with diameters over 10^{-1} cm can reach the distance of 10^4 km, assuming maximum velocities of 0.5 to 1 km sec^{-1}. A size distribution function with the maximum at 10^{-1} cm seems to be unrealistic although the experiments made by Delsemme and Wenger (1969) showed that the flow of evaporated gas stripped relatively large particles from ice. Besides, for such large particles the ratio of acceleration due to the light pressure to the gravitational acceleration is very small and cannot exhibit the observed profiles of the continuum in comets. Moreover, the velocities of dust particles are considerably lower than those of gas. Finson and Probstein (1968) estimated that $v_{0\,\mathrm{dust}} \simeq \frac{1}{2} v_{0\,\mathrm{gas}}$.

The sublimation of ice grains with diameters of 10^{-3} to 10^{-2} cm occurred obviously in a zone of 10^2–10^3 km around the nucleus at heliocentric distances $r \leqslant 1.2$ AU. This zone is observed as the photometric nucleus and the available observational data have inadequate resolution power for any fine analysis of the intensity distribution.

The investigation of the shrinking of the dust coma with heliocentric distance will be more hopeful because at distances of about 2 AU the extent of dust coma may be over 5×10^4 km. The well known effect of shrinking of cometary heads at smaller heliocentric distances and the fact that comets at $r \geqslant 2$ AU mostly exhibit strong continuum spectra seem to be a good basis for the assumption that the observed continuum radiation in the coma and tail is due to the light scattering on small dust grains which may be remnants of original *composed* particles with an ice or clathrate-ice mantle.

If clathrates are a source of the observed radicals then the "zone of production" has the same extent of about 10^3 km as the ice cloud nucleus. However, the observed diameter of the production zone is 10^4 km. Moreover, many comets exhibit no continuum in their spectra, without any significant deficiency in the production of observed radicals.

We cannot exclude the possibility that the photodissociation of parent molecules as well as the destruction of clathrate-ice grains are proceeding simultaneously to produce the observed radicals.

Malaise (1970) deduced from the intensity distribution in the (0–0) band of CN that the minimum value of the total density at $\varrho = 10^4$ km is 10^7 molecules cm^{-3}. Even if some doubt may remain the molecular density at a distance of 10^3 km is relatively high. Because the production rate Z of a nucleus is larger than that of grains the actual production of radicals from the nucleus may be of the same order. However, the bimolecular reactions in a dense region above the nucleus surface can lead to the decreasing of unstable compounds and the effective contribution of the nucleus to the total number of the observed molecules in the inner part of coma will be smaller than the production of free radicals from ice grains in less dense regions. In such case, of course, there would be distinguishable differences in photometric profiles of emission bands of comets with continuum and without it. In particular, the photometric profiles of Comet P/Encke would be extremely interesting because of the absence of continuum.

Dust Component in the Cometary Atmosphere

From the previous paragraphs it is evident that the composition and amount of the dust component in comets may shed light on the structure and composition of these bodies.

The physical character of dust particles is estimated from the selectivity of the scattered light, polarization and infrared emissions.

For several comets colorimetric measurements and spectroscopic results in the continuum resemble the spectral distribution of G8V stars and lead to the conclusions that there exists some reddening of the scattered light and that the scattering on dust grains is selective. However, recent spectrophotometric results for Comets 1967*n*, 1968*b* and 1968*d* by Gebel (1970) show that the reflected or scattered light is grey, in contrast to the reddening previously reported from dust tails.

The continuum energy distributions for these comets follow closely the spectrum of an early G-type star. While this grey scattering conflicts with the reddening deduced from earlier work, much of the light may be directly reflected from a bright nucleus. The "grey" spectra distributions are confirmed for the continuum spectrum of the nucleus of Comet Bennet 1969*i* by Johnson (1971). This agrees with the colour change along the coma radius found in earlier colorimetric measurements. The discrepancy with the older measurements may be solved only by a further study of the variation of the colour of the scattered light through the coma and into the dust tail.

But it seems to be impossible to obtain any reasonable conclusions from the colorimetric observations in the visual spectral range, even if, for instance, the reddening of the cometary continuum implies significant differences between dust particles in comets and in reflection nebulae, where some tendency to blueing on interstellar grains (relatively to the colour of the illuminating star) is observed. Nevertheless, the application to the spectral relative gradient Sun-comet (or colour difference) is de-emphasized by the fact that the size distribution factor for the particles is virtually unknown. Although the integration of the phase function over a large size interval swept out some resonance peaks, the numerical process involving the phase function (particularly the number of integration steps) modified the phase function and the small differences of intensities in two or more wavelengths of scattering light are uncertain. Therefore, the interpretation of direct comparison of computed and observed spectral distribution must be considered as tentative only.

On the other hand, the polarization pattern of a scattering polydispersive medium constituting the cometary dust seems to be more efficient for inspecting actual physical properties of dust particles, especially for a sample of larger particles.

The polarization of cometary continuum has been studied in several comets.

The polarization of cometary light is usually over 10% and increases up to 25% or more as follows from the measurements made in both extended wavelength intervals or in monochromatic light. Rémy-Battiau (1964) shows that the polarization indicated the presence of dielectric particles rather than metallic micrometeorites.

Vanýsek (1970) recently computed many models for a polydisperse optically thin cloud containing various kinds of particles, and from his results it is evident that the change of polarization with phase angle is more important for estimating the physical properties of the scattering medium than the absolute value of polarization degree.

For instance, the fast change of polarization (if detected on real objects) with phase angle in the "backward" direction might be a very significant support for the present hypothesis of the dielectric character of the dust grains in the cometary atmospheres.

The most decisive method for determination of the physical characteristics of cometary grains are the measurements in the infrared. From the interpretation of such data we can estimate the effective albedo and from emission and absorption features in the infrared spectrum (outside possible Ballik-Ramsay C_2 emission wavelengths) derive also the probable chemical composition.

Important infrared measurements of recent bright comets are published by Maas et al. (1970) and Kleinmann et al. (1971). O'Dell (1971) combined the optical wavelength photometry with infrared data. He estimated the particles diameter about 0.1 and found the albedo value 0.3 ± 0.15. This is much lower than the expected albedo for ice-like particles and agrees better with the reflectivity of earthy material. The thermal emission near $\lambda = 10\ \mu m$ may be attributed to the silicate-like meteoric powder. Johnson et al. (1971), however, found no absorption near $\lambda = 1\ \mu m$ which might be expected for ferro-silicates in narrow-band photometric measurements, as made for Comet Bennett 1969*i*.

The comparison of estimated masses of the dust component with the molecular abundance seems to be important for the rôle of dust particles in the production of the molecules.

Unfortunately, this must be limited to CN and C_2, even if it is evident that abundances of neutral hydrogen or OH may be more representative for the estimation of the gas production ability of individual comets. The measurements by OAO-2 and OGO-5 confirmed Biermann's (1969) prediction that $N_H = 10^{36}$ for moderate bright comets; then

$$\frac{N_{CN}}{N_H} < 10^{-5};$$

the same holds for N_{C2}/N_{OH} as was supposed by Arpigny (1965).

From results shown in Table 3 it is evident that the masses of the dust com-

Table 3. *Mass parameter log M of the dust content*[a] *compared with the total number of CN and C_2 molecules*

Comet		log M max.value	log M CN	log M C_2	Note
1954h	Baade	11.91	—	—	Only continuum
1955e	Mrkos	10.8	—	30.8	
1955f	Bakharev-Macfarlane-Krienke	(8.0)	—	29.7	
1956h	Arend-Roland	12.0	32.6	32.3	
1957c	P/Encke	8	30.2	31.0	
1957d	Mrkos	11.2	31.9	31.5	
1958a	Burnham	8.3	—	30.2	
1959b	P/Giacobini-Zinner	9.1	29.8	28.3	
1959k	Burnham	8.2	—	30.3	
1961f	Seki	7.8	29.9	29.9	
1963b	Alcock	9.1	30.3	30.0	
1964h	Everhart	8.3	30.1	29.9	
1967n	Ikeya-Seki	11.5	30.2	29.7	
1969i	Bennett	11.8?	32.0	31.8	

[a] The "mass parameter" is computed from the estimated intensity of the continuum relative to Comet Arend-Roland for which 10^{12} grams is the adopted value for the total mass of dust in the atmosphere.

ponent range from 10^7 to 5×10^{11} grams, where among the comets with maximum of estimated mass of the dust component there are the objects 1957 III and 1957 V and 1967*n*. It may be assumed that these values probably are the highest known ones.

On the contrary the total number of the CN and C_2 molecules varies in a narrower range, with exception of Comet 1967*n*. The maximum number of molecules C_2 is about $10^{32.5}$ for bright comets. In Table 3 only two bright comets, 1957 III and 1957 V, reached this value. For the periodic comet Giacobini-Zinner lower limit for C_2 was found about $10^{28.6}$. However, for the same comet the CN density is ten times higher, i.e. about $10^{29.5}$. Under normal circumstances, a comet with a pure molecular coma and total amount of the C_2 molecules in its atmosphere of about $10^{27.5}$ is a faint object with apparent visual magnitude below 13–14.

Nevertheless, these results show that the relative abundance of the CN and C_2 molecules is independent of the mass of ejected dust grains. (Uncertain results obtained by the author several years ago indicate a slight correlation between the C_3 band and continuum.)

The Secular Changes in the Structure of Cometary Nuclei

The comets, unlike other sizable objects in the solar system, undergo relatively very fast changes. The outflow of gases and dust from cometary nuclei, brightness outbursts and splitting of comets are indisputable evidences of

$MgSiO_3$ as well as SiC can condense at temperatures of about 1 000 K and recondensation may lead to some particles as carbonaceous chondrites. Such particles could produce observed features in infrared spectra of comets (Mass et al. 1970) or in some cool stars like μ Cep. The mass of silicates in such particles is at least 75 per cent while C is a minor contamination.

Composite grains with an ice mantle are probably not common in the interstellar matter as was believed some years ago; however, they might be abundant in some phase of primordial nebulae.

Such grains would undergo considerable changes as was pointed out recently by Donn and Jackson (1970), if they are exposed to the UV and corpuscular radiation. Irradiation of an ice mixture by the energy equivalent to 5 MeV in interstellar space converted the material in less volatile material within 10^8 years.

An even weaker radiation field destroys H_2O molecules in a short time of 10^6 years. It is supposed that less volatile compounds are produced and an icy grain would be either destroyed or partly converted into more complex and less volatile material in a relatively short time, as long as they are not shielded by some dense cloud.

The increasing of stable non-volatile material in grains exposed to the radiation field may be linked directly to the structure of comets. The old non-volatile material is accumulated in the nucleus core and surrounded by a highly volatile young ice conglomerate which is shielded by thin non-volatile material. This thin shell is destroyed by heat at the first approach to the Sun and is a source of a great amount of micrometeoric dust typical of bright comets. If a comet becomes short-periodic the mean source of mass in its atmosphere is the volatile ice layer for the period of less than 10^4 years, until the nucleus shrinks to the less volatile or non-volatile core.

The stratification in a cometary nucleus is suggested also by the decreasing non-gravitational effects. An extensive analysis of secular changes of the orbital elements of many periodic comets by Marsden and Sekanina (from many papers of these authors concerning the non-gravitational effects see for instance Marsden 1969; Sekanina 1969) indicates that non-gravitational effects decrease with time. This agrees with the assumption of secular deactivation of the nucleus or even with the inert asteroidal-like final stage.

The non-volatile core of a comet was formed before the "soft" mantle from loosely accreted silicate grains. The formation process might be similar to one of the three principal processes described by Arrhenius and Alfvén (1971*a*, 1971*b*) for consolidation of the lunar particles which is the bounding by the condensate silicate, sulphate and metal vapor. The process may be somewhat modified in comets. Vaporization due to impacts of partly silicate grains transports some vapour into loosely attached particles. The subsequent condensation would probably lead to a more compact structure, cementing together the remaining particles.

If such a cometary embryo was formed outside the zone where the environment was suitable for ice condensing, then it remains in the form of a "boulder" in the cometary orbit.

It is evident that more information concerning cometary structure is necessary for a more critical discussion with the outlined hypothesis. More information can be obtained by relatively simple observational techniques and such data are—paradoxically—relatively scarce. Beside the very expensive space missions routine ground-based observations can still contribute to our approach to the Solar System history.

Summary

Except for very short-lived cometary phenomena the icy-conglomerate model seems to be the best approach to the real structure of a typical comet. However, the formation of comets and the structure and stratification in a cometary nucleus must be known for their better understanding.

The discrepancies in the lifetimes of the precursors of observed radicals estimated from photometric measurements of comets and laboratory data for supposed parent molecules are so striking that the origin of free radicals cannot be explained only by the photodissociation processes. An alternate hypothesis, namely the role of icy particles, can be taken into consideration.

Recent observations of comets are reviewed and confronted with the model of a cometary atmosphere. It is shown that the available observations are unable to detect the effect due to vaporization of ice particles and the observed continuum must be ascribed to the scattering on less volatile particles.

The differences in physical appearance of comets belonging to different families indicate that the stratification of material in cometary nuclei seems to be very probable and a comet's final stage is an inert asteroid-like body.

A possible process of formation of such a body is shortly outlined.

References

Arpigny, C., 1965*a*, Mém. Acad. R. Belg. *35*, part 5, 83.
Arpigny, C., 1965*b*, Rev. Astr. Astrophys. *3*, 351.
Arrhenius, G. and Alfvén, H., 1971, Paper presented at the 12th Coll. IAU, Tuscon, Arizona, March 1971.
Arrhenius, G. and Alfvén, H., 1971, Earth Planet. Sci. Letters *10*, 253.
Biermann, L., 1969, JILA Astrophys. Report No. 93.
Delsemme, A. H. and Miller, D., 1970, Planet. Space Sci. *18*, 717.
Delsemme, A. H. and Wenger, A., 1970, Planet. Space Sci. *18*, 700.
Delsemme, A. H. and Swings, P., 1954, Ann. Astrophys. *15*, 1.

Dewey, M. E. and Miller, F., 1966, Ap. J. *144*, 1170.
Dobrovolskij, O. V., Faiziev, E. and Ibadinov, K. H., 1966, Dokl. Akad. Nauk Tadjik. SSR *9*, 14
Dolginov, A. Z. and Gnedin, Yu. N., 1966, Icarus *5*, 64.
Dolginov, A. Z., Gnedin, Yu. N. and Novikov, G.G., 1971, Planet. Space Sci. *19*, 143.
Donn, B. and Urey, H. C., 1957, Mém. Soc. R. Sci. Liège, Sér. 4, *18*, 124.
Donn, B. and Jackson, W., 1970, AAS Meeting, Boulder, Abstract.
Eddington, A. S., 1910, M.N.R.A.S. *70*, 442.
Finson, M. L. and Probstein, R. F., 1968, Ap. J. *154*, 327, 353.
Gebel, W. L., 1970, Ap. J. *161*, 765.
Gilman, P. A., 1969, Ap. J. (Letters) *155*, L185.
Gilra, D. P., 1970, AAS Meeting, Boulder, Abstract.
Haser, L., 1957, Bull. Acad. R. Belg. Cl. Sci. *43*, 740.
Haser, L., 1966, Mém. Soc. R. Sci. Liège *37*, 233.
Herzberg, G., 1966, I.A.U. Trans. *12B*, 194.
Huebner, W. F., 1970, Astron. Astrophys. *5*, 286.
Jackson, W. and Donn, B., 1968, Icarus *8*, 270.
Johnson, T. V., Lebofsky, L. A. and McCord, T. B., 1971, Pub. A.S.P. *83*, 93.
Kleinmann, D., Lee, T., Low, F. J. and O'Dell, C. R., 1971, Ap. J. *165*, 633.
Kresák, L., 1965, Bull. Astr. Inst. Czechoslovakia *16*, 355.
Larrimer, W. J. and Anders, E., 1967, Geochim. Cosmoch. Acta *31*, 1215, 1239.
Levin, B. J., Soviet Astr. J. *25*, 246.
Maas, R. W., Ney, E. P. aod Woolf, N. J., 1970, Ap. J. (Letters), *160*, L101.
Malaise, D., 1970, Astr. and Ap. *5*, 209.
Marsden, B. G., 1969, A. J. *74*, 720.
Meisel, D., 1969, Pub. A.S.P. *81*, 65.
Mocknach, D. O., 1938, Leningrad St. Univ. Ann., Astr. Ser. 4.
Mocknach, D. O., 1956, Bull. Inst. Theor. Astr. USSR *6*, 269.
O'Dell, C. R., 1961, Pub. A.S.P. *73*, 35.
O'Dell, C. R., 1971, Ap. J. *166*, 675.
O'Dell, C. R. and Osterbrock, D. E., 1962, Ap. J. *136*, 556.
Oort, J. H. and Schmidt, M., 1951, B.A.N. *11*, 259.
Öpik, E. J., 1963, Irish A. J. *6*, 63.
Potter, A. E. and Del Duca B., 1964, Icarus *3*, 103.
Rahe, J., Donn, B. and Wurm, K., 1970, Atlas of Cometary Forms, NASA SP-198.
Rémy-Battiau, L., 1964, Bull. Acad. R. Belg. Cl. Sci., Sér. 5, *50*, 74.
Sekanina, Z., 1969, A. J. *74*, 720.
Stecher, T. P., 1969, Ap. J. (Letters) *157*, L125.
Urey, H. C., 1952, The Planets, Yale Univ. Press.
Vanýsek, V., 1958, Publ. Astr. Inst. Csl. Acad. No. 37.
Vanýsek, V., 1965, Bull. Astr. Inst. Czechoslovakia *16*, 355.
Vanýsek, V., 1969, Bull. Astr. Inst. Czechoslovakia *20*, 355.
Vanýsek, V., 1970, Publ. Astr. Inst. Charles Univ. No. 58.
Vanýsek, V. and Hruška, A., 1958, Publ. Astr. Inst. Csl. Acad. No. 41.
Vanýsek, V. and Tremko, J., 1964, Bull. Astr. Inst. Czechoslovakia *15*, 233.
Vanýsek, V. and Žáček, P., 1967, Publ. Astr. Inst. Charles Univ. No. 53.
Vsekhsvjatskij, S. K., 1929, M.N.R.A.S. *90*, 712.
Vsekhsvjatskij, S. K., 1958, Physical Characteristics of Comets (in Russian), Moscow.
Wallace, L. V. and Miller, F. D., 1958, A. J. *63*, 213.
Whipple, F. L., 1950, Ap. J. *111*, 375.
Whipple, F. L., 1951, Ap. J. *113*, 464.

Whipple, F. L. and Douglas-Hamilton, D. H., 1966, Mém. Soc. R. Sci. Liège *37*, 469.
Wurm, K., 1961, A. J. *66*, 361.

Discussion

Y. Öhman

In connection with polarization measurements I would like to mention that one of the comets studied by me in 1940 and 1941 (1941 *c*) showed strong polarization in the continuous spectrum and fairly high intensity in the violet. In my opinion the spectropolarimetric and spectrophotometric observations should be combined.

I also want to draw attention to the recent laboratory work of Poulizac, Desesquelles and Dufay 1967 where they produce the CO^+ spectrum by proton irradiation. They conclude that this ionization process may be of some importance in comets in connection with flares.

Poulizac, M. C., Desesquelles, J. and Dufay, M., 1967, Annales d'Astrophysique, *30*, 301.

V. Vanýsek

The cross section is very large for the reaction

$$CO^+ + e \rightarrow C + O.$$

If the concentration of free electrons near the nucleus is high we should not see CO^+. It is therefore puzzling that CO^+ has been observed right up to the nuclei of comets.

M. Wallis

The problem with forming CO^+ from protons is that there are not enough protons in the solar wind. At least a factor 10 more would be required.

T. Gehrels

The Mie theory, of single scattering by small spherical particles, may be used to explain the observations in photometry and polarimetry. However, there are in the literature several wrong results, because of a lack of proper observations I believe. The polarization and colour phenomena depend on, among others, a size parameter, a refractive index (which may be complex), and a fit can be obtained only if at least as many parameters have been observed. Mr L. R. Doose, graduate student at the University of Arizona, has obtained, with three filters avoiding emission features, measurements over a wide range of phase angles on three dust-type comets. So far, it has not been possible to obtain a fit with the Mie theory (contrary to the result of Donn et al. 1967). This might confirm that the particles are the hydrate clathrates of Delsemme and Miller (1970) because these are too large and complex for such fitting.

Delsemme, A. H. and Miller, D. C., 1970, Planet. and Space Sci., *18*, 717.
Donn, B., Powell, R. S., Remy-Battiau, L., 1967, Nature, *213*, 379.

V. Vanýsek
That a fit is not obtained might be due to the non-spherical particles.

A. Elvius
A fit would not be possible if the character of the particles changes with time, as may well be the case because a wide range of phase angles usually corresponds to great variations in the distance of the comet from the Sun.

T. Gehrels (comment added in Proof):
The fitting was tried, without success, at discrete values of phase angle.

Observation and Feature Variations of Comet 1969e before and during the Perihelion Passage

By A. Mrkos

Department of Astronomy and Astrophysics, Charles University, Prague

Periodic comet Honda–Mrkos–Pajdušáková which currently has a period of 5.218 is one of Jupiter's family of comets. The orbital motion of this comet is characterized by west-to-east motion, very low inclination 13°168 and semi-major axis of 3.008 AU from the Sun. As this comet approaches or recedes from the Sun, its brightness changes with unusual steepness. In fact, it has never been observed farther than 1.2 AU from the Sun. This is one of the comets whose motion is markedly affected by non-gravitational forces.

On the basis of the analyses of visual and photographic observations of the head diameter and brightness an attempt is made to estimate the size of its nucleus.

The radius of the nucleus is very uncertain and varies within broad limits 0.4 to 2.9×10^4 cm.

The remarkable periodic comet Honda–Mrkos–Pajdušáková, discovered in 1948, is one of Jupiter's family of comets. Its orbital motion is characterized by west-to-east motion, short period of orbit, very low ecliptical inclination and by unusual behaviour during the perihelion passage. Although it has been observed four times since its discovery, it has never been seen at a greater heliocentric distance than 1.2 AU.

Its orbital elements for the year 1969 are the following (according to Marsden, 1969):

T = 1969 Sept 23.0050 ET	Epoch = 1969 Sept 16.0 ET
ω = 184°1695	e = 0.814291
1950	
Ω = 233°1053	a = 3.008430 AU
i = 13°1684	n^2 = 0.1888836
q = 0.558693 AU	P = 5.218 years

P/Honda–Mrkos–Pajdušáková exhibits some specific properties by which it is distinguished from other comets and which probably are caused by a special structure of its central part. As the comet approaches or recedes from the Sun its brightness changes with unusual steepness. It varies with a very high inverse power of the heliocentric distance r (perhaps even as high as r^{-11}).

During a 65-day interval of observation in the year 1969 the observed brightnesses have differed considerably from those calculated, especially close to the perihelion when the comet often was brighter by 5^m. Strong fluctuations of the total brightness were observed twice during the last return of the comet, namely on August 12th, when its brightness increased by $0\overset{m}{.}5$ within half an hour and on October 15, when it decreased by $0\overset{m}{.}6$ within 20 minutes. These brightness fluctuations were quite real and were not caused by changes of the observational conditions.

However, the curve of the course of the total brightness follows the typical asymmetry of the normal prior and after the perihelion passage.

The appearance of the nucleus of a normal comet usually is starlike. Generally, the nucleus is surrounded by a very dense bright coma. The brightness of the nucleus at a distance of 0.7–1.3 AU usually is by 3–4^m fainter than the total brightness of the coma. When a 25-cm telescope and high magnification were used for the observation of P/Honda–Mrkos–Pajdušáková approaching the perihelion, it was possible to distinguish its nucleus as a very small disk or a group of small dense spots which frequently changed its form. It may be assumed that its nucleus is a conglomerate of small solid bodies, mostly surrounded by gas and dust, which only sometimes resembles a starlike nucleus. The nucleus has never been observed at a heliocentric distance $r > 0.7$ AU even if the comet was closest to the Earth, i.e. 0.309 AU in 1969.

The curve of the brightness variations of the nucleus does not correspond at all to the phase term for an asteroidal body ($r^2\Delta^2$).

The photoelectric measurements of the central part of several comets disclose the brightness fluctuations of $\pm 0\overset{m}{.}2$ within the interval of 20–60 minutes, which corresponds, statistically, to about 10 individual sources of the about 60 minutes lifetime at the approximately same distance as was the distance of the comet at that time, i.e. 1.020 AU. By these local sources the maximum positional residuals of astrometric observations $(O-C)$ can be also explained for most comets at the time of their perigee. (Vanýsek 1968).

At the closest approach to the Earth, when $\Delta = 0.309$ AU and $r = 1.020$ AU, the coma of 1969*e* was abnormally large, quite homogeneous and without any prominent central condensation.

During the approach to the Sun and with increase of the brightness of the comet the diameter of the coma increased. After the perihelion passage, simultaneously with the decrease of the brightness, the diameter of the coma again began to increase (reduced to 1 AU). Such behaviour was found at every observed return of the comet. A proportional dependence of the cometary head diameter on heliocentric distance was studied many years ago by Bobrovnikoff. On about 7 600 single observations of 300 comets he has statistically determined the heliocentric distance 1.4 AU at which the coma reaches its maximum extent (1941). The condensation of the central part of the coma appears

The observations which Mrkos described may be a support for this. The comet Honda–Mrkos–Pajdušáková was first observed as a diffuse cloud, in which later a nucleus was formed. The observations seem to exclude the existence of a nucleus at the early stages, in any case a nucleus large enough to be a reservoir of dust and gas.

Furthermore, the model would fit nicely with what Lal told us about isotropic irradiation of grains in space, and with his picture of a comet as a place where larger bodies are accreted.

Alfvén, H., 1970, "Jet Streams in Space", Astrophys. Space Sci. *6*, 161.
Alvén, H. and Arrhenius, G., 1970, "Origin and Evolution of the Solar System", I and II, Astrophys. Space Sci. *8*, 338, and *9*, 3.
Trulsen, J., 1970, "Formation of Comets in Meteor Streams", presented at IAU Symposium No. 45, "Motion, Orbit, Evolution and Origin of Comets", Leningrad, 1970.

B. A. Lindblad

Mr Chairman. I think one difficulty here is due to the fact that the probability of Jupiter capturing a long-period comet in one single encounter is very, very low. This, I think, is one of the arguments that Danielsson and Professor Alfvén are putting forward here. We cannot explain the present system of short-period comets by assuming that Jupiter has captured all of them in single encounters. We have to put forward some other, more effective capture mechanism, and the question is; what is this mechanism?

H. Alfvén

The alternative is that the new comets we see in short-period orbits have just formed before our eyes, and I think that Mrkos' comet is a case which one should investigate in order to see whether this perhaps is just the formation process we observe.

M. Wallis

I was just going to add that the difficulty of explaining the short-period comets of Jupiter's family as captured is the property that all of them have prograde orbits. Every one of them has an orbital inclination less than thirty degrees, I believe. There are no retrograde orbits at all, which is very hard to understand on the capture-theory.

F. L. Whipple

Should I say something? The man who wrote the Bible wasn't God, I think. First of course you are dealing with the Jupiter family, and Jupiter will disrupt any of these jet streams. Each time each part of the jet stream is affected differently by Jupiter perturbations, it is going to be thrown madly all over the solar system.

H. Alfvén

The Jupiter action will just make the particles in the jet stream coalesce, as shown by Trulsen.

Trulsen, J., 1970. Formation of Comets in meteor streams. IAU Symposium No 45 "Motion, orbit, evolution and origin of Comets". Leningrad 1970.

F. L Whipple

The particles that come by at one time are affected differently than those that come by at another time. We see it in the streams. If we take the famous Leonid shower, in which the stream was directed from the earth and caused the fiasco of 1899, and if you deal with that at all realistically; you find that they just get thrown around too badly. If you did not have Jupiter I do not think there is any chance of the theory working today, but Jupiter is going to wreck it completely.

Unfortunately, Dr Alfvén, a non-icy nucleus made of earthy solids, cannot provide the gas apparently necessary for comet activity or non-gravitational motions as observed. This restriction is even more obvious now than when I first realized it in the late 1940's. Since then the solar wind as measured provides less solar gas than I assumed then, while the huge clouds of neutral hydrogen observed about two comets in Lyman-alpha by the satellites increases markedly the observed amount of gas to be accounted for. Non-icy solid surfaces must replenish the gas lost at each of many cometary perihelion passages, and the solar wind or interplanetary gas is the only source I know to be available. Comet Bennett as reported by Vanýsek produced a cloud of 10^{36} H atoms over an observed maximum diameter of 10^{7} km. At an observed line width of 5 km/sec. this cloud would require replenishing in about 10 days. For a perihelion activity extended over some 100 days we should allow a total loss from the comet nucleus of some 10^{37} H atoms and possible a comparable observed number of OH neutral molecules.

Let us calculate the area needed for a similar comet of period 100 years to accumulate 10^{37} atoms from the solar wind during each period. It would spend most of its time at a solar distance exceeding 20 AU. There the solar wind would surely not contribute more than 10 per cent of the 2×10^{8} protons/sec/cm^{2} that it contributes at 1 AU. Call the average rate 2×10^{7}/sec/cm^{2} for 3×10^{9} seconds or 6×10^{16} p/cm^{2} in all, at 100 per cent collection efficiency. Thus the required cometary surface would be some 10^{21} cm^{2} normal to the solar wind, equivalent to an opaque square of side more than 100 000 km. Such a comet would be quite observable well beyond the distance of Pluto, even if a very poor reflector.

If we should deny the exitence of so much mass loss for fainter periodic comets we still reach a similar impossibility on the basis of only the observed molecules in the normal optical spectrum. A huge reservoir of gas is required, far too much to be held by earthy solids and released near perihelion. Only icy material carried by the comet can supply the demand.

Furthermore, appproximately the observed amount of gas is required to provide the jet action necessary for periodic comets to deviate from Newtonian motion as observed. Some accelerate and some decelerate in their orbital motions. Also a number of long-period comets show the effects of a radial force towards the Sun as expected from an icy discrete nucleus model.

Because the measurement of the amount and velocities of interplanetary dust and solids is now fairly accurate, the system is clearly dissipative for small bodies, as shown in the meteor streams. Reversal of this dissipative action in the present system taxes one's credulity. Also the perturbations of Jupiter, particularly, cause serious and probably unsurmountable difficulties in holding the particles in a coalescing stream, at least for orbits crossing Jupiter's orbit. Could the solid particles survive and accumulate, the resultant object would not appear to be a comet as shown above. Whether the jet-stream concept has merit in the early stages of solar system development, I have no data on which to form an opinion. This ingenious concept certainly deserves a thorough study, applied in more favorable circumstances.

V. Vanýsek

The Mrkos comet shows not only the continuum but also the emission of CN and C_2, very typical of the comets, and then, for instance, only the evaporation of the gas and the exchange of the momentum from one particle to another cause some net acceleration which produces some perturbation. Then, it is also hard to understand some focusing of particles if we assume, for instance, the Poynting-Robertson effect for the small particles.

F. L. Whipple

Not to mention the Sun-grazing comets, for which all the gas is evaporated and thrown away by your plasma.—I mean all the particles are evaporated.

H. Alfvén

I think this is not a very serious objection, because if a comet is formed out of small grains, these have adsorbed gases when they have moved in space, and they have a large surface of adsorption. Hence when they collide in forming a comet, we will have a release of gas. Furthermore, at the time when they collide there will be a grinding of the small particles and a production of more dust. So I think both of these mechanisms—both the production of gas and of dust here are quite reasonable.

F. L. Whipple

The numbers will not bear you out, I'm sorry.

L. Danielsson
I was going to ask Dr Whipple about this huge amount of hydrogen. That has never been observed in a short-period comet, has it?

V. Vanýsek
Not yet.

F. L. Whipple
I do not know, but I believe some was observed around comet Encke.

E. Anders
It would be interesting to put this model on a quantitative basis. If such a jet stream is to be the source of comets, it obviously has to be far enough from the Sun, so that the ice particles are stable over long periods of time, which is probably somewhat beyond the orbit of Saturn. What would the accretion time be for, say, a one-kilometer comet in a jet-stream, 15–20 AU from the Sun? Has this been worked out yet?

H. Alfvén
This is altogether a thing which we are working on, but the quantitative results will have to be checked further on. Dr Trulsen could perhaps say something about the Jupiter perturbations?

J. Trulsen
It has been demonstrated that the perturbation from a planet can produce strong focusing effects in a longitudinally homogeneous jet stream on a short time scale. On the long time scale the same perturbation will act to disperse a stream of non-interacting particles. It is not at all clear that this will happen in a jet stream when collisions between the particles take place. A quantitative theory for this case is needed.

F. L. Whipple
Well, I would make my last comment. Putting Humpty Dumpty together again is easy compared to making a comet in this way. (Humpty Dumpty was an egg, broken in a nursery rhyme.)

B. A. Lindblad
It was mentioned here that all the short-period comets move in low-inclination direct orbits. The same applies to all the short-period meteor streams as well. So the phenomenon is not a property only of the comets—it is a property of the short-period meteor streams as well, and, of course, also of the asteroids. We

Fig. 1. The roving automatic laboratory Lunokhod-1.

slopes with inclinations up to 32 deg when prescribed control system limitations were overruled.

The chassis consists of the locomotion part (4 units of conjugated in-pairs wheels—2 units of left wheels and 2 units of right ones), automatics, motion safety system, a device with a set of sensors for determination of mechanical properties of the soil and for evaluation of the chassis performance.

Each wheel unit is fixed with a bracket to the base of Lunokhod's container. Each of the eight driving wheels has an individual mechanical drive and an independent torsion suspension. Inside the wheel hub the electroengine, reduction gear, brake, mechanism for connection of the power drive and gauges of temperature and of number of revolutions are located.

The turn of Lunokhod is achieved due to the difference in rotation velocities of wheels on the left and on the right boards and also due to reversing of their rotation.

The automatics unit controls the Lunokhod's motion by radiocommands from Earth, provides measurements and control of major parameters of self-propelled chassis and automatic operation of instruments for studying mechanical properties of lunar soil. The chassis is controlled as by command from Earth, so as from the onboard apparatus.

The system of motion control has automatic devices for dosed-in-time direct motion as well as for performance of turns to prescribed angles.

The evaluation of practicability is carried out with the use of a set of sensors which continuously measure the roll and pitch of Lunokhod, currents in driving electroengines, number of revolutions and temperature of wheels. A special

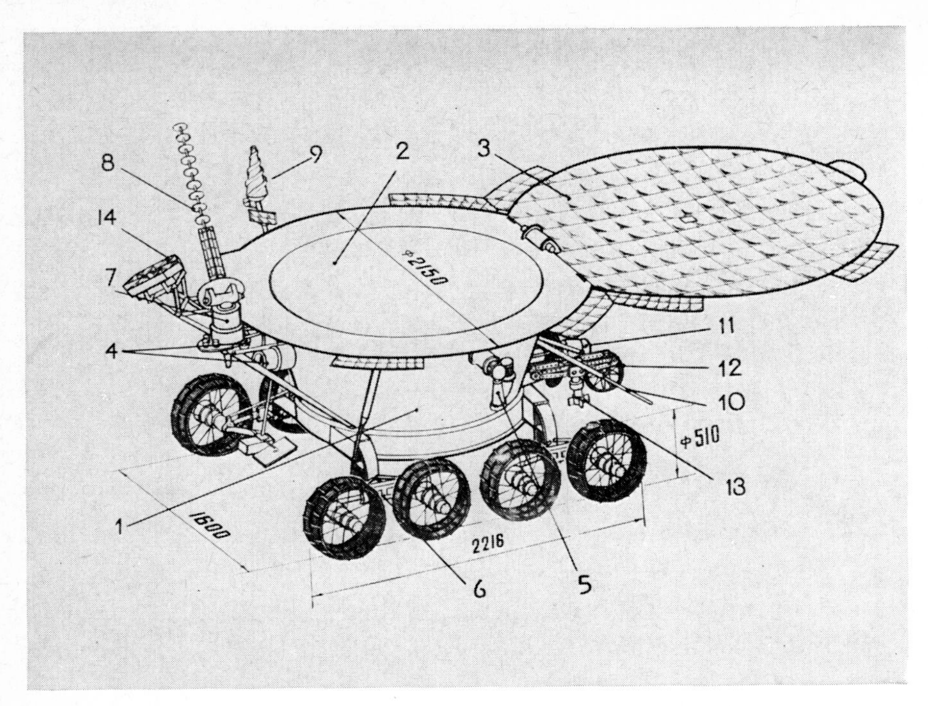

Fig. 2. Equipment of Lunokhod-1. 1) Hermetic equipment container; 2) Cooling radiator; 3) Solar battery; 4) TV porthole; 5) Telephoto cameras; 6) Chassis wheel unit; 7) Driving mechanism for high-gain antenna; 8) High-gain antenna; 9) Low-gain antenna; 10) Spike antenna; 11) Isotope heat generator; 12) 9th wheel; 13) Equipment for determining physico-mechanical soil properties; 14) Optical corner reflector.

instrument gives information about the mechanical properties of lunar soil; it has mechanisms for insertion and turning of a conic-paddle punch in the soil and also a set of sensors which measure the loads affecting the punch, the depth of its insertion into soil and angle of turn.

The length of a path traversed by Lunokhod is measured according to number of revolutions of driving wheels taking into account the correction for their skidding which is determined with the ninth, freely rolling wheel.

Geometrical parameters of the locomotion part of the Lunokhod, specific pressure on the soil, driving characteristics, parameters of elastic suspension and the shape of supporting surfaces of the wheels, —all these factors allowed the Lunokhod to move freely over surface with crumbly quick soil, to overcome steep slope upgrades, to override craters and obstacles like single stones or ridges of stones comparable with the size of the locomotion part.

The electric power system of Lunokhod consists of solar cells connected in parallel with chemical electric power sources which switch-in for recharging.

The remote control system enables to control Lunokhod from the Earth.

Its onboard complex includes: an astronavigation system which determines with the aid of Sun and Earth the location of Lunokhod and its true course in the selenographic coordinates; a course pointing system consisting of a course gyro and coursepointer control unit that inputs the Lunokhod's course into the telemetering system; a system of pointing of the sharply-directional antenna, unit of automatics and radioapparatus (Fig. 2).

Luna-17 station has a TV system consisting of four panoramic telephoto-cameras which are used for obtaining stereopanoramas in the region of Lunokhod motion, for studying Moon's structure and surface, for obtaining images of the rock during Lunokhod's descent from the landing platform and for acquisition of the Sun, Earth and lunar vertical positions required for astro-orientation. In addition, there is a small-frame TV system transmitting from the Lunokhod to Earth images to control its motion.

Six antennas are installed on Lunokhod that ensure the transmission of necessary data and transmission of commands both during the flight and on the Moon.

Apart from the telephotocameras, the following scientific equipment is used on Lunokhod:

(*a*) indicator of distance traversed by Lunokhod; (*b*) system of Lunokhod roll and pitch indicators for construction of a full profile along the path; (*c*) gauges for determination of physico-mechanical properties of the soil; (*d*) X-ray spectrometer RIFMA for determining the chemical composition of the ground surface layer; (*e*) corner reflector (developed and produced in France) for optical location of the Moon; (*f*) patrol dosimeter, controlling the radiational environment in the region of Lunokhod; (*g*) X-ray telescope which enabled the beginning of the study of the far regions of the Universe.

The conducted tests have shown that the vehicle developed is able to

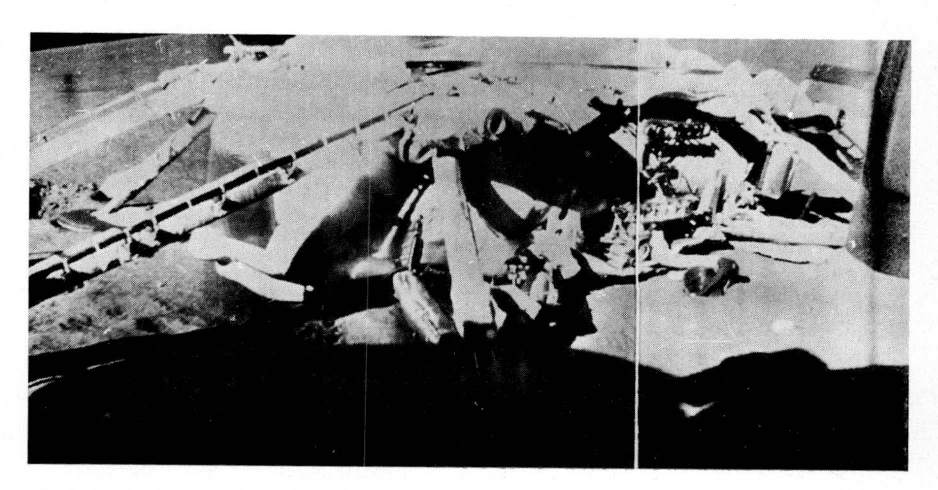

Fig. 3. Test of navigation method.

operate actively during more than eight terrestrial months, to carry sufficient amounts of instruments and has exceptionally high performance.

In the course of experiments the navigation method had been tested (Fig. 3). The success of these tests allow to formulate a number of interesting programs of space investigations, based on possibilities of the tested vehicles, as well as possible improvements and possibilities of combined operation of both roving and unmovable automatic stations on the Moon and of vehicles which can automatically return to Earth.

Among the numerous planned and developed programs of experiments making use of the mentioned above automatic vehicles three groups of experiments directed at studying the origin and evolution of the solar system can be singled out:

1. Delivery to Earth of lunar rock samples from various points of the Moon and, what is of particular importance, from regions related to continents (in the first place for the determination of age of these rocks and time of their exposure on the surface, i.e. the exposition time).

2. Study of the interior structure of the Moon by means of:

a) long-term observations of natural seismicity of the Moon by seismometers installed at different points on the Moon;

b) automatic autonomous determination of a lunar surface point coordinates (latitude and longitude) and their variation in time with accuracy up to 10″ for the purpose of both studying Moon's rotation axis motion in its body and the determination of anomalies of axial rotation velocity for calculating Love numbers which characterize the Moon's body viscosity.

This may be accomplished by installing near the equator a telescope for observation of the stars (the diameter of the lens is of the order of 80 mm with the focal distance of about 1 000 mm and with a special electronic unit). Such a device weighing about 30 kg may be installed on the Moon by an automatic vehicle.

c) Study of the Moon's electric conductivity at depths from 3–5 km up to 400–600 km. For the determination of electric conductivity it is necessary to register electric and magnetic fields on the lunar surface as well as at heights of 50–100 km above it. We mean here the fluctuations of the natural (interplanetary) electromagnetic field within the wide range of frequencies—from tens of kilocycles down to millicycles. The depth of penetration of fluctuations for such frequencies reaches under lunar conditions 3–5 km for high and 400–600 km for low frequencies.

The investigation of the interior electroconductivity of the Moon will make it possible to evaluate the temperature and approximate chemical composition up to the depths of several hundred kilometers.

determine the fluctuations of the meteorite flux, if they were sufficiently large, for the time which we can cover by our measurements.

The success of these studies also depends to a considerable degree on our understanding of the mechanism of impact at very high velocity which must be studied both experimentally and theoretically on Earth.

As I have already mentioned, the basic reason for crater smoothing lies in the influence of impacts of small meteorites that also determines the horizontal displacement of material in the upper layers of lunar surface. This influence (possibly, together with atom bombardment) leads to smoothing of stones protruding on the surface. It is interesting to note that, according to the Lunokhod-1 data, a correlation between the degree of stone roundness and the degree of morphological pronouncement of craters seems to exist.

It would be highly interesting to establish also the correlation between the degree of stone immersion into the regolith and the existence time of a crater to which this stone belongs. It will give also the idea about the rate of the horizontal displacement of substance over the lunar surface that is associated with the meteorite fluxes. The measurement of the immersion degree can be performed with automatic vehicles. But the rate of immersion itself is subjected, apparently, to large fluctuations associated with the local events, i.e. with falling of large meteorites in the crater local.

The relatively cheap and operating for a long time automatic vehicles are capable, when performing measurements over sufficiently large areas, of giving representative statistics.

An important supplement to the described above program of experiments on the Moon must be a program of investigations with automatic Lunar satellites which are necessary for detailed study of the gravitational field and shape of the Moon and for the investigation of magnetic field variations in the nearest surroundings of the Moon which is necessary (along with experiments on the lunar surface) for constructing a model of the Moon's structure.

Also under development is an experiment on observation of the contemporary meteorite flux on the other side of the Moon and a number of other experiments.

I have not described in my report the technical details of conducting the experiments. Many of them are not yet developed in full.

Information obtained by the first automatic space vehicles operated on the Moon is rather extensive and time is required for its processing, understanding and comparison with earlier known data, but, undoubtedly, the possibilities opened by them are extremely promising in various areas of studying the Universe and, as I believe, especially in the field of studying the history of our planetary system evolution.

Potential Contributions of the United States Space Program to Exploration of the Solar System

By Homer E. Newell, Daniel H. Herman and Paul Tarver

US National Aeronautics and Space Administration, Wahington D.C.

Introduction

Long the subject of both scientific and popular interest, the solar system is now getting renewed attention because of a vastly improved ability to observe and measure the planets and their environment. Prior to the 1960's, planetary research had been in a sort of doldrums generated by a long drought of any significant new data and the simultaneous rise of interest in astrophysics where theory and observation could combine to produce some of the most remarkable advances in the history of astronomy. But over the last decade and a half the space rocket and improved observations from the ground have completely altered the complexion of solar system research.

Satellites and space probes arrived on the scene when better detectors and new electronic devices,—including improved photocells, photometers, polarimeters, spectrometers, and infrared detectors,—were giving promise of important new solar system data from ground based measurements. Coupled with these, the capability of spacecraft to provide direct *in situ* measurements, with orders of magnitude better resolution than hitherto, was bound to stimulate a renewed interest in solar system research. The impact of these new and improved techniques was further enhanced by the growing capability of the modern computer which, by strengthening the hand of the theorist, has led to increased activity in thinking about the planets, planetary atmospheres, and Sun-planetary relationships.

Naturally, the earliest results of space exploration pertained to our own planet, the Earth, and derived first from sounding rockets and then from Sputniks and Explorers. The discovery of the Van Allen Radiation Belts from Explorer measurements in the spring of 1958, the continued investigation of which gradually revealed a huge and complex magnetosphere surrounding the Earth (see Fig. 1), marked the first major advance in planetary science contributed by instrumented spacecraft.

There then followed a remarkable period of exploration in which US and Soviet spacecraft reached the nearest planets, Apollo astronauts left their footprints on the moon and brought back samples of the material they found

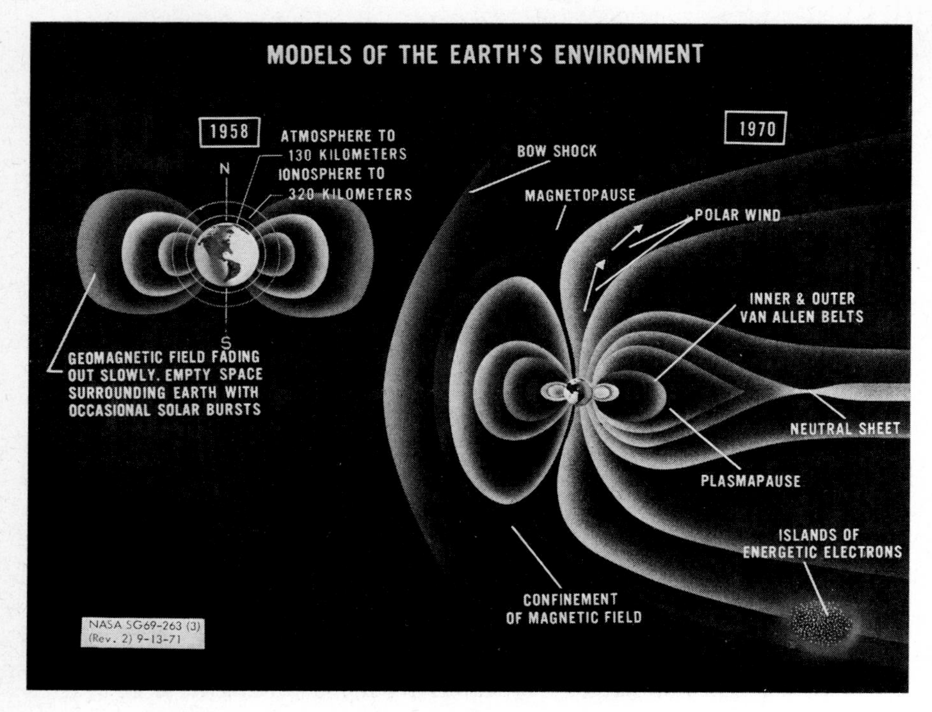

Fig. 1. Models of Earth's Environment—NASA SG69-263.
Compares the 1958 pre-Sputnik concept of the Earth's environment with the 1970 picture as revealed by spacecraft measurements.

there, and both instrumented and manned vehicles ranged for many kilometers over the lunar terrain. Important and fascinating new information has come from these early space explorations of mankind, and hundreds of scientists continue to analyze and study the data that have come back from past space missions. From these early results it is clear that there is much to be gained from continuing and extending the explorations accomplished so far.

We now have the capability to send spacecraft and their instruments throughout the solar system and to send back data from great distances. This capability is most pertinent to the discussions of this Symposium, for it can be applied to obtaining a great deal of the data needed to answer questions raised about the origin and evolution of the solar system.

Because of its pertinence to your discussions, and to your planning for the future, we propose to review briefly in this paper currently approved missions in the NASA program that bear on solar system investigations, and then to outline a number of future possibilities that are in various stages of study and consideration. There is a certain discomfort in devoting a paper to work yet to be done; certainly it is much more satisfying to talk about successes already achieved, results already obtained. But if one is going to plan for the future a program that will benefit from the broadest of participation, then one must make the possibilities known. It is in this spirit that we outline for you our ongoing program in solar system exploration, and our thoughts for the future. We sincerely hope that we will hear from you ideas and suggestions that will help make it the very best program that it can be for all of us.

Objectives and Strategy

President Nixon, in a statement issued in March, 1970, set forth guidelines for planning the U.S. space program of the 1970's. He said that three general purposes should guide the US space program: exploration, the pursuit of scientific knowledge, and practical applications. As specific objectives in the pursuit of scientific knowledge he called for continued exploration of the Moon and a bold exploration of the planets and the universe. The exploration and study of the solar system is a major element in our continuing efforts in space.

A prime objective in exploring and studying the solar system is to further our understanding of its origin and evolution, and in the process to contribute

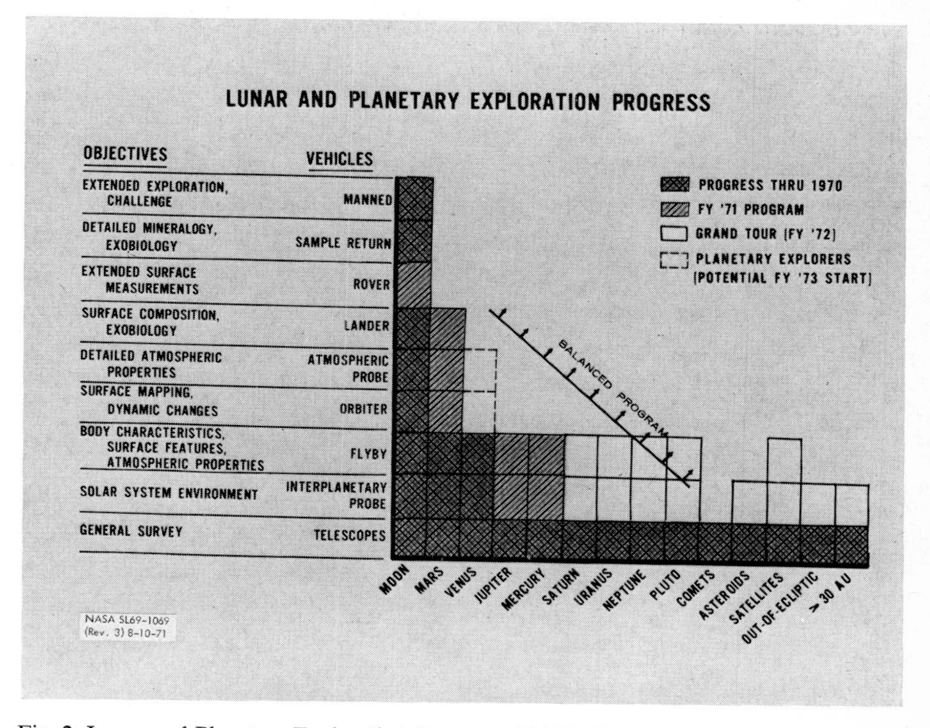

Fig. 2. Lunar and Planetary Exploration Progress—NASA SL69-1069.

Depicts the NASA approach to solar system exploration. Exploration of the Moon and near planets is continuing in depth while exploratory missions are sent to the outer planets, to be followed in turn by in-depth investigations as appropriate.

as much additional insight as possible into the origin and evolution of life in the solar system. An ultimate objective is to understand better our own planet Earth, and in particular to learn more about the dynamic processes that affect and determine the Earth's environment. Clearly these objectives encompass the sorts of problems and questions that have been discussed at this symposium.

The overall approach used in the US program is illustrated schematically in Fig. 2. Along the abscissa are listed various objects and regions of the solar system. One set of ordinates shows various means of investigation while a second set gives in shorthand form some specific objectives that can be furthered by the indicated means with which they are correlated on the figure.

As shown, the earliest available information comes from telescopes. NASA supports a variety of ground based observations,—including optical, radio, and radar,—and has funded the construction and operation of both optical and radio telescopes for lunar and planetary research. As may also be seen in the figure, survey-type missions with interplanetary probes and planetary flybys very naturally tend to come before the more intensive in-depth studies afforded by orbiters, atmospheric entry probes, landers, and rovers. The most difficult missions of all, sample return and manned exploration, come only after a substantial foundation has been laid by earlier less complex missions. Although Apollo astronauts have explored parts of the Moon in person, the United States does not now have any specific plans for manned exploration of the planets.

This overall mission strategy derives from a variety of considerations, —feasibility, technological state of the art, funding, manpower workload, political support, etc.—in addition to the basic consideration of scientific strategy. The approach does, however, seem reasonably well designed to further the scientific aims of the program in an orderly and effective way. The scientific strategy for a specific mission or group of missions is worked out in collaboration with the best scientists available, and is then used to guide the preparation for and conduct of the flight program.

Past Program

To lay the ground work for our discussion of the current and future programs, the briefest of summaries of the past effort will suffice. Tables 1, 2, and 3 indicate the kinds of data that have been acquired from lunar, interplanetary, and planetary missions. Lunar and planetary photography, and reduced instrumental data, are available to interested scientists for at most the cost of reproduction. Those data that are available are catalogued by the NASA Space Science Data Center managed by the Goddard Space Flight Center at Green-

Table 1. *Data obtained from lunar space missions*

MISSION		LAUNCH DATE	OBJECT OR REGION OF STUDY	DATA OBTAINED
RANGER	7	28 JLY 64	Sea of Clouds	4316 TV pictures; up to 2000 times earth-based resolution
	8	17 FEB 65	Sea of Tranquillity	7137 high resolution pictures
	9	21 MAR 65	Alphonsus	5814 high resolution pictures
LUNAR ORBITER	1	10 AUG 66	9 primary, 7 potential Apollo sites	207 frames of medium & high resolution pictures
	2	6 NOV 66	Additional Apollo sites	211 frames of medium & high resolution pictures
	3	5 FEB 67	Additional Apollo sites	211 frames of medium & high resolution pictures
	4	4 MAY 67	Moon frontside mapper	High resolution photos of 99% front face
	5	1 AUG 67	Lunar surface mapper	Detailed coverage; 36 scientific & 5 Apollo sites
SURVEYOR	1	30 MAY 66	Ocean of Storms	Data on morphology & origin, surface bearing strength
	3	17 APR 67	Oceanus Procellarum	6315 pictures, surface data
	5	8 SEP 67	Mare Tranquillitatis	18,000 pictures; surface analysis indicated basaltic character
	6	7 NOV 67	Sinus Medii	30,065 pictures; alpha scattering analysis
	7	7 JAN 68	Tycho ejecta blanket	Combined imaging, alpha scattering, soil sampler
APOLLO	8	21 DEC 68	Lunar surface	Orbital photography; crew and system demonstration
	10	18 MAY 69	Lunar surface	19 color TV transmissions; Lunar Module to 15.2 km altitude
	11	16 JLY 69	1st manned lunar landing Mare Tranquillitatis (Sea of Tranquillity)	Inspection, photography, survey, evaluation, and sampling of lunar soil
	12	14 NOV 69	2nd manned lunar landing East Edge of Oceanus Procellarum (Ocean of Storms)(near Surveyor III)	Surface photos & samples; Lunar Surface Experiment Package which transmits: seismic, heat flow, solar wind, dust, and magnetic data
	14	31 JAN 71	3rd manned lunar landing Fra Mauro	Surface photos & samples
	15	26 JLY 71	4th manned lunar landing Hadley-Apennine Plain	Surface photos & samples

belt, Maryland, USA. Catalogs and information may be obtained by writing directly to the Data Center.

In addition to the material covered by Tables 1 through 3, there is a tremendous quantity of data on our Earth and its environment that has been ac-

Table 2. *Data obtained from interplanetary space missions*

MISSION		LAUNCH DATE	OBJECT OR REGION OF STUDY	DATA OBTAINED
PIONEER	5	11 MAR 60	Heliocentric space, <1.0 A.U.	Forbush decrease in cosmic rays observed in interplanetary space.
	6	16 DEC 65	Heliocentric space, 0.81-0.98 A.U.	Magnetic field lines guide propagation of solar particles. Co-rotation of solar magnetic field (with Explorer 28)
	7	17 AUG 66	Heliocentric space, 1.0-1.125 A.U.	Observation of the earth's magnetospheric wake at 1000 earth radii.
	8	13 DEC 67	Heliocentric space, 0.99-1.09 A.U.	Detection of very low frequency electric fields in interplanetary space. Observation of magnetospheric wake at 500 earth radii.
	9	8 NOV 68	Heliocentric space, 0.75-0.99 A.U.	Anomalous release of solar particles into the solar system (with Pioneer 8)
EXPLORER	10	25 MAR 61	Interplanetary Particles & Fields	Interplanetary plasma from direction of the sun with a velocity of approximately 300 km/sec.
	11	27 APR 71	Gamma Ray Astronomy	Average gamma ray intensity $3 \times 10^{-4} cm^{-2} sec^{-1} ster^{-1}$. Principal cause: cosmic ray interactions with interstellar gas creating π^{o} mesons.
	18	27 NOV 63	Particles & Fields	Existence of detached magnetospheric bow shock wave in solar direction.
	21	4 OCT 64	Cislunar Space	Energy spectra of solar and galactic cosmic rays.
	28	29 MAY 65	Cislunar Space	Differential energy spectra of helium during solar minimum. Co-rotation of the solar magnetic field.
	33	1 JLY 66	Interplanetary Space	Magnetospheric tail beyond the orbit of the moon.
	34	24 MAY 67	Cislunar Space	He/H ratio in the Solar wind. Structure of the magnetospheric boundry.
	35	19 JLY 67	Lunar Orbit	Solar plasma interaction with the moon.
	43	13 MAR 71	Cislunar Space	Onboard computer data processing yields greater detail in cosmic ray composition and structure of bow shock.

Table 3. *Data obtained from planetary space missions*

MISSION	LAUNCH DATE	OBJECT OR REGION OF STUDY	DATA OBTAINED
MARINER 2	27 AUG 62	VENUS	No magnetic field, surface temperature of approximately 800° F
4	28 NOV 64	MARS	First close-up pictures of Mars surface
5	14 JUN 67	VENUS	Strong H_2 corona; 72%–87% CO_2; confirmed high surface temperature
6	25 FEB 69	MARS	TV - three distinct types of terrain: cratered, chaotic, and featureless UV- } atmosphere composed primarily of CO_2, small amounts of O and H. Nitrogen emissions not detected
7	27 MAR 69	MARS	UV- } TV- Polar cap observations
9	30 MAY 71	En route to MARS Orbit	

cumulated from various manned and unmanned missions, that bears directly on any study of the solar system. Certainly any study of planetary origin and evolution must include the Earth, and any study of planetary atmospheres will include a comparison of other atmospheres with that of Earth. Our growing knowledge of the very complex interactions of Earth's magnetosphere with the

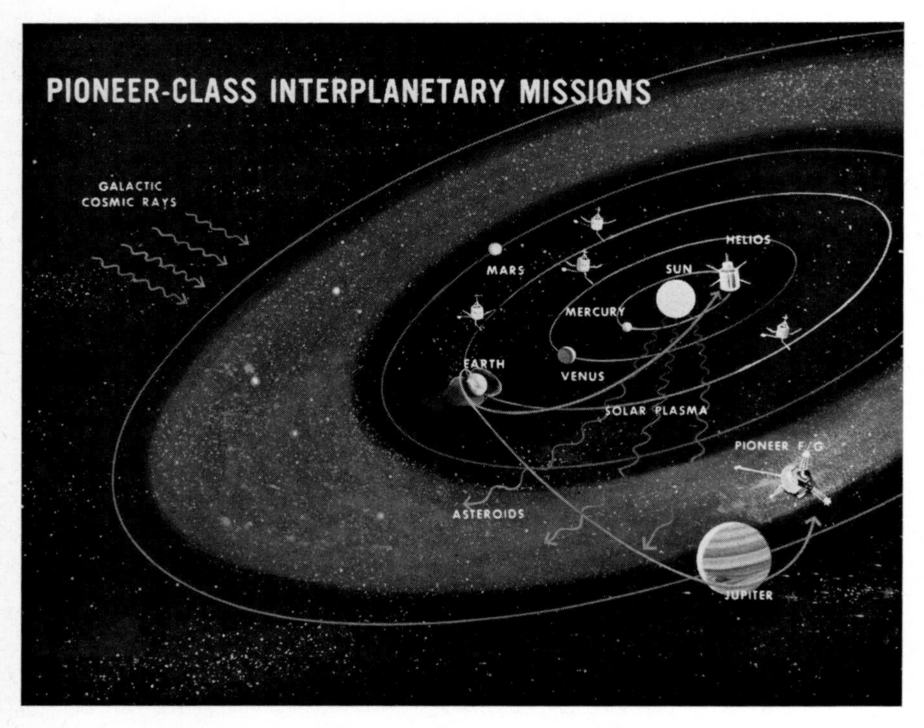

Fig. 3. Pioneer-Class Interplanetary Missions—NASA SG68-480.

The Pioneer family of spacecraft, designed to map the solar wind, magnetic fields, and energetic particles is shown. Pioneers 6–9 are in a heliocentric orbit at approximately 1 AU Helios will investigate the vicinity of the Sun to within 0.25 AU of its surface.

Table 4. *Currently approved lunar and planetary program*

MISSION	71	72	73	74	75	76	77	78	79
APOLLO LUNAR LANDING	15 ▲	16, 17 ▲ ▲							
MARINER MARS ORBITER	▲ ▲								
PIONEER F/G JUPITER FLYYBY		▲	▲						
MARINER VENUS/MERCURY FLYBY			▲						
VIKING MARS ORBITER/LANDER					▲ ▲				
GRAND TOURS						J-S-P △	J-S-P △		J-U-N △ △

▲ CURRENTLY APPROVED
△ PLANNED, FY 72

solar wind must illuminate and be illuminated by what we find out about the interactions of the Moon and planets with the solar wind. But for the purposes of this review it seems adequate to limit the coverage as we have.

Not all the missions tabulated in Tables 1 through 3 are completely in the past. The Apollo 15 astronauts have indeed returned, but the orbital and surface instruments from this and previous Apollo missions continue to radio back data from the Moon. Also, Mariner 9 is still on its way to Mars, and has yet to go into orbit about the planet. Pioneers 6 through 9 are still orbiting the Sun at about 1 Astronomical Unit (AU), and continue to yield valuable interplanetary data (Fig. 3).

Current Program

The current program includes both lunar and planetary missions, as shown in Table 4. At present the only flights to the moon still on the docket are the Apollos through number 17. A variety of planetary missions is scheduled.

Apollo

In shorthand language the instrumentation and experimental objectives of the Apollo program are indicated in Table 5. With the completion of the Apollo 15 flight, only missions 16 and 17 remain. The sample returned from Apollo 15 weighed about 80 kilograms, and contained selected core samples, rocks, rock fragments, and lunar soil.

Using the Lunar Rover, the Apollo 15 astronauts were able to explore three traverses for a total of nearly 30 kilometers within a 6 kilometer radius from the lunar module. The lunar terrain visited included Hadley Rille, a number of craters, the foothills of the Apennines, and the Hadley Plains.

Table 5. *Assignment of approved lunar experiments*

(SURFACE)

	EXPERIMENT OR INSTRUMENT	PRINCIPAL INVESTIGATOR	APOLLO 11	12	13	14	15	16	17
	Lunar Geology Investigation	Shoemaker/US Geol. Survey	A	A					
		Swan/US Geol. Survey			X	A	A		
		Muehlberger/U. of Texas						A	A
	Laser Ranging Retro-Reflector	Alley/U. of Maryland	A						
		Faller/Wesleyan, Mass.				A	A		
	Cosmic Ray Detector	Fleischer/General Elec.	A					A	
		Walker/Washington U. at St. Louis							
		Price/U. of Calif., Los Angeles							
	Lunar Surface Close-up Stereo Photo	Gold/Cornell U.	A	A	X				
	Portable Magnetometer	Dyal/Ames Research Ct.				A		A	
	Lunar Gravity Traverse	Talwani/Columbia U.							A
	Soil Mechanics	Mitchell/U. of Calif., Berkeley				A	A	A	A
	Far UV Camera Spectroscope	Carruthers/Naval Res. Lab.						A	
	Surface Electrical Properties	Simmons/Manned Spacecraft Ctr.							A
ALSEP	Solar Wind Composition	Geiss/Berne Univ.	A	A	X	A			
ALSEP	Lunar Passive Seismology	Latham/Columbia U.	A	A	X	A	A	A	(Note)
ALSEP	Lunar Active Seismology	Kovach/Stanford U.				A		A	
ALSEP	Lunar Tri-Axis Magnetometer	Dyal/Ames Res. Ctr.		A			A	A	
ALSEP	Medium Energy Solar Wind	Snyder/Jet Propulsion Lab.		A			A		
ALSEP	Suprathermal Ion Detector	Freeman/Rice U.		A		A	A		
ALSEP	Lunar Heat Flow (with Drill)	Langseth/Columbia U.			X		A	A	A
ALSEP	Cold Cathode Ionization Gauge	Johnson/U. of Texas (Dallas)		A	X	A	A		
ALSEP	Lunar Ejecta and Meteorites	Berg/Goddard Space Flt. Ctr.							A
ALSEP	Lunar Seismic Profiling	Kovach/Stanford U.							A
ALSEP	Lunar Atmos. Composition	Hoffman/U. of Texas (Dallas)							A
ALSEP	Lunar Surface Gravimeter	Weber/U. of Maryland							A(Note)
ALSEP	Charged Particle-Lunar Environment	O'Brien/U. of Sidney, Australia			X	A			

(ORBITAL, IN-FLIGHT, AND PRE- AND POST-FLIGHT)

EXPERIMENT OR INSTRUMENT	PRINCIPAL INVESTIGATOR	APOLLO 13	14	15	16	17
Gamma-Ray Spectrometer	Arnold U of Calif., San Diego			A	A	
X-Ray Fluorescence	Adler/Goddard Space Flt. Ctr.			A	A	
Alpha Particle Spectrometer	Gorenstein/Amer. Sci & Eng.			A	A	
S-band Transponder (CSM/LM)	Sjogren/Jet Propulsion Lab.	X	A	A	A	A
Mass Spectrometer	Hoffman/U of Texas (Dallas)			A	A	
Far UV Spectrometer	Fastie/Johns Hopkins U.					A
Bistatic Radar	Howard/Stanford U.	X	A	A		
IR Scanning Radiometer	Low/Rice U.					A
Apollo Window Meteoroid	Cour-Palis/Manned Spacecraft Ctr.		A	A		
UV Photography - Earth and Moon	Owen/Illinois Inst. of Tech. Res. Inst.			A	A	
Gegenschein From Lunar Orbit	Dunkelman/Goddard Space Flt. Ctr.		A	A		
Lunar Sounder	Brown/Jet Propul. Lab.					A
	Ward/U of Utah					
Bone Mineral Measurement	Vogel/U.S. Public Health Ser. San Francisco			A	A	A
Total Body Gamma Spectrometry	Benson/Manned Spacecraft Ctr.			A	A	A
Subsatellite:						
S-band Transponder	Sjogren/Jet Propul. Lab.			A		A
Particle Shadows/Boundary Layer	Anderson/U of Calif. Berkeley			A		A
Subsatellite Magnetometer	Coleman/U of Calif. Los Angeles			A		A

A - Assigned
X - Aborted

Note: Lunar Passive Seismology may be substituted for Lunar Surface Gravimeter

A new set of surface experiments[1] was emplaced, including a third laser retroreflector, which now provides ground observers with a network of three widely spaced retroreflectors to use in studying the celestial mechanics of the Earth-Moon system. This latest retroreflector has three times the area (3 400 cm^2) and four times the efficiency of each of the previous two.

Of particular interest is the heat flow experiment now in progress with thermistors inserted into holes to determine the thermal flux through the lunar

[1] Called ALSEP for Apollo Lunar Surface Experiment Package.

surface. In preparing the holes, it was also possible to obtain cores that preserved the stratigraphy to depths of 2.3 meters.

In addition to observations and measurements made by the astronauts themselves while in lunar orbit, an instrumented orbiter was left behind as a satellite of the moon, to measure the lunar magnetic field, solar wind plasma, and the lunar particle and field environment. Also, by tracking the S-band transponder carried on the orbiter, refinements in the lunar gravitational field may be deduced from perturbations in the satellite orbit.

Apollo 16 is targeted for the Descartes region of the Moon. Its crew will be John W. Young, Thomas K. Mattingly, and Charles M. Duke. Like Apollo 15, it too will carry a Rover to extend the traverse capability of the atronauts.

The landing site for Apollo 17 has not been announced yet. The crew is Eugene A. Cernan, Ronald E. Evans, and Harrison H. Schmitt. Apollo 17 also will carry a Lunar Rover. Of especial interest will be the lunar gravimeter, an instrument intended to measure the average elastic properties of the lunar surface and to search for gravitational radiation. It will measure lunar oscillations over the frequency range from 0.05 to 2.0 Hertz. In orbit, a radio sounder operating at frequencies of 5, 15 and 150 MHz, may yield some information about the surface to depths of as much as 1 500 meters.

Mariner 9

Our exploration of Mars is continuing with Mariner 9 presently on the way to the planet. The spacecraft is intended to go into orbit about Mars in November, 1971. Once in orbit, Mariner 9 will seek to obtain topographic maps of the Mars surface like those of the moon obtained by Lunar Orbiter. Since synchronizing the orbit of the spacecraft with the rotation of Earth to facilitate communications also results in a nearly synchronous orbit about Mars, selected areas of the Mars surface will be observed repetitively, permitting a detailed study of temporal variations in surface appearance. Fig. 4 shows the planned orbit of Mariner 9, and the figure legend lists some of the specific objectives of the mission. The instruments, the parameters that these instruments will measure, the objectives of the measurements, and the experimenters are shown in Table 6.

Viking

The Viking mission to Mars is to be launched in 1975. Three sets of objectives are shown in shorthand form on the drawing of Fig. 5. The mission will combine observations from an orbiting spacecraft with surface observations from a survivable lander. During entry of the lander, measurements will be

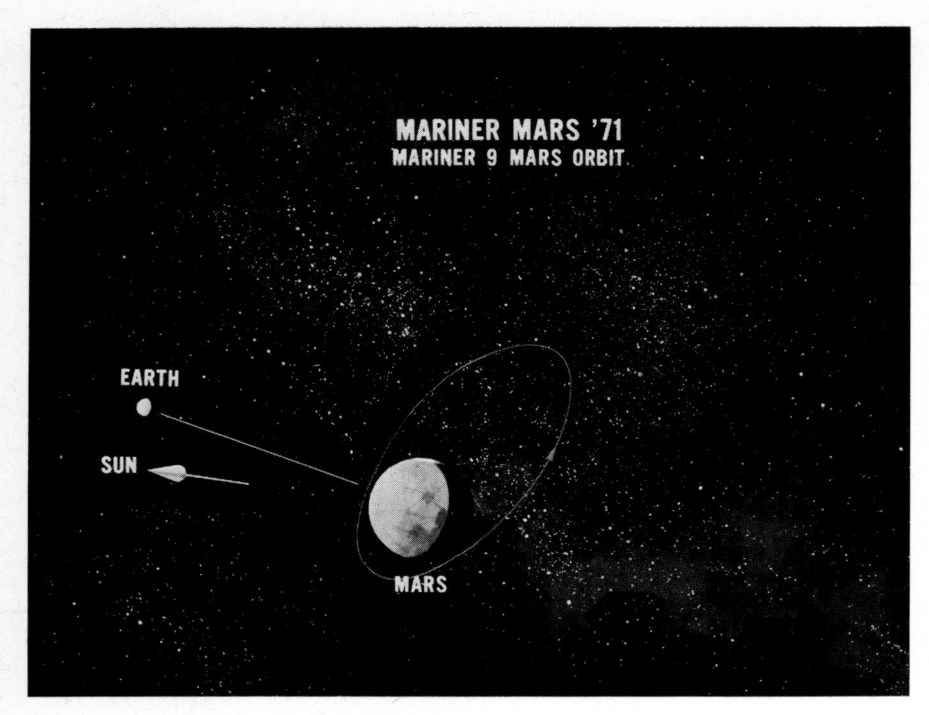

Fig. 4. Mariner Mars 71—NASA SL69-1082.
This Mars orbiter mission will map surface topography and temperatures, search particularly for surface water, obtain atmospheric pressure and density profiles and atmospheric and ionospheric composition, and record temporal changes in surface features.

Table 6. *Mariner Mars 1971 orbiter instruments and experiments*

INSTRUMENT	KEY PARAMETERS	OBJECTIVES	EXPERIMENTERS – PRINCIPAL INVESTIGATORS
1. Television Cameras (2)	1. 50mm focal length 700 x 832 lines 2. 500mm focal length 700 x 832 lines	Investigate visible surface & atmospheric phenomena. Map surface.	Masursky/US Geological Survey
2. Ultraviolet Spectrometer (2 channels)	1. 1100 to 1900 Angstroms .19 x 1.9 degrees 2. 1700 to 3400 Angstroms .19 x .55 degrees	Map surface pressure and atmospheric composition	Barth/ Univ of Colorado
3. Infrared Interferometer Spectrometer	6 to 50 microns 4.5 degrees circular	Determine atmospheric composition and water content, map atmosphere temperature profiles, surface temperature, and types of material.	Hanel/NASA (Goddard Space Flight Center)
4. Infrared Radiometer (2 channels)	1. 8 - 12 microns .53 x .53 degrees 2. 18 - 25 microns .7 x .7 degrees	Map surface temperatures and thermal properties	Neugebauer/California Institute of Technology
5. Radio Subsystem	S-band	By occultation methods obtain atmospheric & ionospheric profiles, radius of Mars.	Kliore/Jet Propulsion Laboratory
6. Radio Subsystem	S-band	Refine Mars ephemeris & gravimetric data for Mars. Test theory of general relativity.	Lorell/Jet Propulsion Laboratory

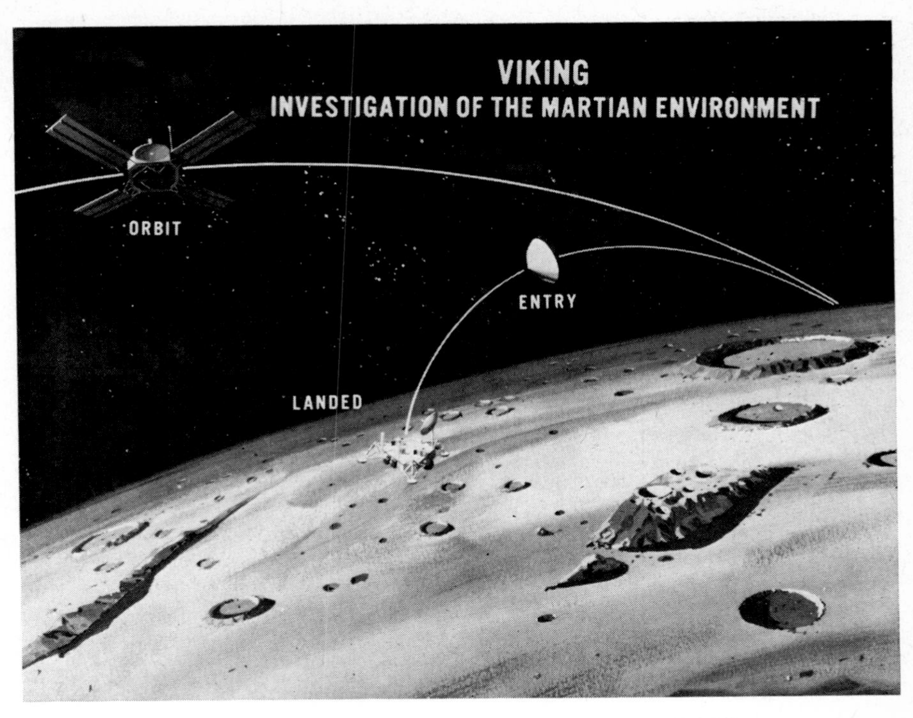

Fig. 5. Viking—Investigation of Martian Environment—NASA SL71-2525.
The Lander will make measurements of the biological, organic, seismic, magnetic, and physical properties of the Martian surface. The Orbiter will map surface and atmospheric temperatures, water vapor, atmospheric pressure, and ionospheric electron densities. The entry capsule will measure ionization in the ionosphere, and atmospheric composition and structure.

taken of ionospheric and atmospheric properties. A list of instruments, parameters to be measured, objectives, and investigators is given by Table 7.

The Viking lander will be instrumented to search for forms of life on the Mars surface, and to study some of the chemistry of the planet. Metabolic activity detectors will look for living micro-organisms, while a gas chromatograph mass spectrometer will search for organic matter in the soil. If life is actually found, and the probability of this is generally conceded to be very low, that will be a most exciting and profound discovery. But even if life is not found on Mars, knowledge of the chemical and physical state of the planet will still be important to the overall investigation of the chemical evolution of life.

Mariner Venus-Mercury 1973

The United States plans to launch a Mariner spacecraft in 1973 to fly by Venus and use the gravitational influence of Venus to deflect the spacecraft towards an encounter with Mercury. The mission is depicted in the drawing of Fig. 6,

Table 7. *Viking 1975 instruments and experiments*

INSTRUMENT	KEY PARAMETERS	OBJECTIVES	EXPERIMENTERS - PRINCIPAL INVESTIGATOR
ORBITER			
Television Cameras	1050mm focal length 1078 x 1210 lines Image motion compensation Color filters, 3300 to 7000Å	Select landing site. Monitor lander vicinity. Map surface.	Carr/US Geological Survey
Infrared Spectrometer	1.4 micron band - Range 1 - 1000 microns of H_2O - Angular resolution 0.4° x 0.1°	Map water vapor distribution	Farmer/Jet Propulsion Laborat
Infrared Radiometer	±1°K accuracy. 140 to 330°K range - Angular resol. 0.3° x 0.3°	Map temperature profiles	Kieffer/University of Californ Los Angeles
ORBITER & LANDER			
Radio Subsystem	Orbiter: S & X-band Lander: S-band, UHF & radar	Study gravitational field, ephemeris, atmosphere, surface, relativity, solar corona.	Michael/NASA (Langley Rese Center)
ENTRY			
Retarding Potential Analyzer	10 - 10^6 ions/cc	Measure ion concentration, ion & electron energies.	Nier/Univ of Minnesota
Mass Spectrometer	1 - 50 AMU	Determine composition & abundance of neutral species.	Nier/Univ of Minnesota
Accelerometers Pressure Gages Temperature Gages	Acceleration: axial 0-150 m/sec^2 lateral ±10 m/sec^2 Pressure: 1 - 200 mb Temperature: 100 - 750°K	Characterize atmospheric structure.	Nier/Univ of Minnesota
LANDER			
Cameras (Facsimile)	.12° color resolution .04°, .12° B/W resolution 100° x 360° field of view 2 camera stereo	Photograph landing site.	Mutch/Brown Univ
Metabolic Activity Detectors	1. Pyrolytic Release 2. Labelled Release (C^{14}) 3. Light Scattering 4. Gas Exchange	Detect microorganisms.	Klein/NASA (Ames Research Center)
Gas Chromatograph/Mass Spectrometer	12-200 AMU	Analyze soil for organic content. Analyze atmosphere.	Biemann/Massachusetts Institu of Technology
Seismometer (3 axis)	0.4 to 4.0 Hertz 76 db dynamic range	Determine seismic background and event activity.	Anderson/California Institute of Technology
Meteorology Package Pressure Gage Temperature Gage Anemometer Adsorption Hygrometer	Pressure: 1-30 millibars Temperature: 130° - 350°K Wind: 2 - 150 meters/sec, direction Water Vapor: 180°-255°K frost point	Measure meteorological environment near surface.	Hess/Florida State Univ
Magnets on Lander	2500 and 500 gauss arrays	Detect ferromagnetic particles in soil.	Hargraves/Princeton Univ
Lander Foot Pads, Soil Sampler, Facsimile Camera	Bearing strength Cohesion Adhesion Grain size distribution	Obtain physical properties of surface.	Shorthill/Boeing

and some of the objectives are listed in abbreviated form. The instruments, parameters to be measured, objectives, and investigators for this mission are set forth in Table 8. If successful, this mission will produce the first close observations of Mercury by a US spacecraft. Additional observations of Venus will, of course, be made during the passage of the spacecraft in the vicinity of that planet on its way to Mercury.

Table 10. *Pioneer F & G instruments and experimenters*

INSTRUMENT	KEY PARAMETERS	OBJECTIVES	EXPERIMENTERS - PRINCIPAL INVESTIGATORS
1. Helium Vector Magnetometer	± 2.5 gamma to ± 1.43 gauss	Map interplanetary field beyond 2 AU; solar wind/Jupiter magnetosphere. Study Jupiter field and solar/galactic boundary.	Smith/Jet Propulsion Laboratory
2. Plasma Analyzer	Ions: 100 ev - 18 kev Electrons: 1-500 ev	Investigate solar plasma variations, solar wind interaction, solar/galactic boundary.	Wolfe/NASA (Ames Research Center)
3. Radio Subsystem	S-band	Determine Jupiter atmospheric & ionospheric profiles, H_2/He ratio	Kliore/Jet Propulsion Laboratory
4. Radio Subsystem	S-band	Refine Jupiter ephemeris & Jupiter system gravimetric data, test theory of general relativity.	Anderson/Jet Propulsion Laboratory
5. Infrared Radiometer	14 - 25 microns 29 - 56 microns	Measure Jupiter radiation balance & temperature distribution, H_2/He ratio	Münch/California Institute of Technology
6. Asteroid/Meteoroid Detector	Mass > 10^{-6} grams	Measure asteroid & meteoroid flux, velocities, mass properties and spectra in asteroid belt.	Soberman/General Electric Co.
7. Meteoroid Detector	10^{-9} - 10^{-13} grams	Map meteoroid flux in asteroid belt & interplanetary space	Kinard/NASA (Lewis Research Center)
8. Charged Particle Instrument	Electrons: 200 kev to 30 Mev Protons: 450 kev to 50 Mev Integral flux: > 1 Bev/nucleon	Investigate solar & galactic cosmic ray flux, interplanetary & Jupiter shock, Jupiter radiation belt	Simpson/Univ of Chicago
9. Geiger Tube Telescope	Electrons: 50 kev to 50 Mev Protons: > 50 Mev	Measure Jupiter radiation belts, radio emissions, magnetic moment, solar & galactic radiation beyond 2 AU.	Van Allen/Univ of Iowa
10. Cosmic Ray Telescope	Differential energy spectra H_2 & He: 50 - 800 Mev/nucleon Stopping particles: 22 - 50 Mev/nucleon	Observe patterns, spectra & distribution of galactic & solar radiation, extent of solar cavity.	McDonald/NASA (Goddard Space Flight Center)
11. Trapped Radiation Detector	Electrons: 2, 5, and 11 Mev Total electron flux: 1.5 Mev Protons > 450 Mev	Measure protons & electrons in Jupiter magnetic field; radio emission correlation.	Fillius/Univ of Cal San Diego
12. Ultraviolet Photometer	200 - 800Å	Measure H_2/He ratio and temperature of the Jupiter atmosphere; interplanetary hydrogen density and interaction with solar wind.	Judge/Univ of S. Cal.
13. Imaging Photopolarimeter	Polarization phase angles 0° to 90° 45° to 135° Spectral band 3900 to 5000Å 5900 to 7200Å Imaging Resolution 120-400 km	Study photometric and photopolarization of Jupiter and satellites, quantity and distribution of particles in space.	Gehrels/Univ of Arizona

a series of missions. Both US and European scientists have expressed interest, and we expect the Science Steering Group to have an international composition.

NASA has several studies underway to help determine the next steps in the exploration of Mars. Data from the Viking 1975 mission should aid in deciding what observations and measurements should be attempted in any follow-on missions. One possibility for 1979 would be to land on Mars an adaptive science laboratory which, with appropriate commands from Earth, could accommodate in realtime or nearly realtime to results being obtained during the course of the investigation. A crude block diagram for such a laboratory is shown in Fig. 10.

Studies are being made to determine how best to use such a laboratory, for example, how to ensure that the samples analyzed by the laboratory are adequately distributed geographically over the Mars surface. A roving vehicle

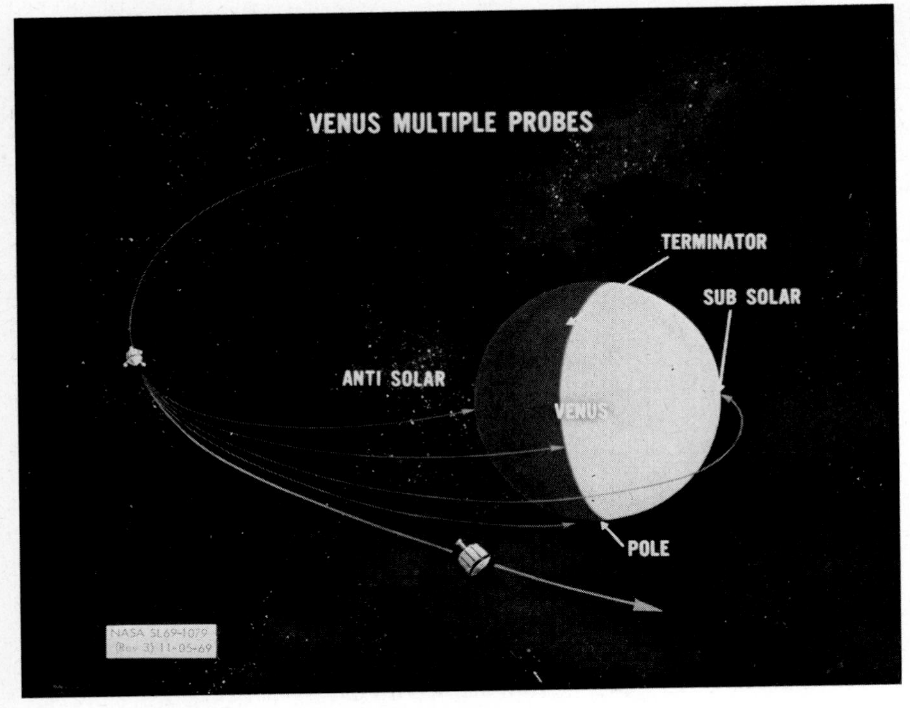

Fig. 9. Venus Multiple Probes—NASA SL69-1079.
Shown are multiple trajectories which impact on antisolar, terminator, and sub-polar points of the planet Venus. The purpose of this mission is to obtain data on the nature and dynamics of the Venus atmosphere.

could bring samples collected from widely separated sites to a fixed laboratory. Alternatively, the laboratory itself could be made mobile.

We are also studying the feasibility of landing a spacecraft on Phobos or Deimos and returning a sample from one of these satellites of Mars. Other studies are looking into the feasibility, cost, and scientific value of returning to Earth samples of Mars itself.

A second attractive line of investigation will be to continue making a general survey of the solar system. Here the outer planets are clearly of great interest. Direct flights to Jupiter pose no great problem as far as propulsion and flight time are concerned, but to go beyond Jupiter by direct flight becomes increasingly difficult or impossible with current capabilities. Without substantial increases in propulsion, using nuclear, nuclear-electric, or solar-electric power plants, the outermost planets will either remain unreachable by direct flight or the required flight times of decades will remain unacceptably long.

Fortunately the huge gravitational influence of Jupiter on a closely passing spacecraft can be used to slingshot the craft out to any of the outer planets. Moreover, the transit times from Jupiter to the other planets would be at most

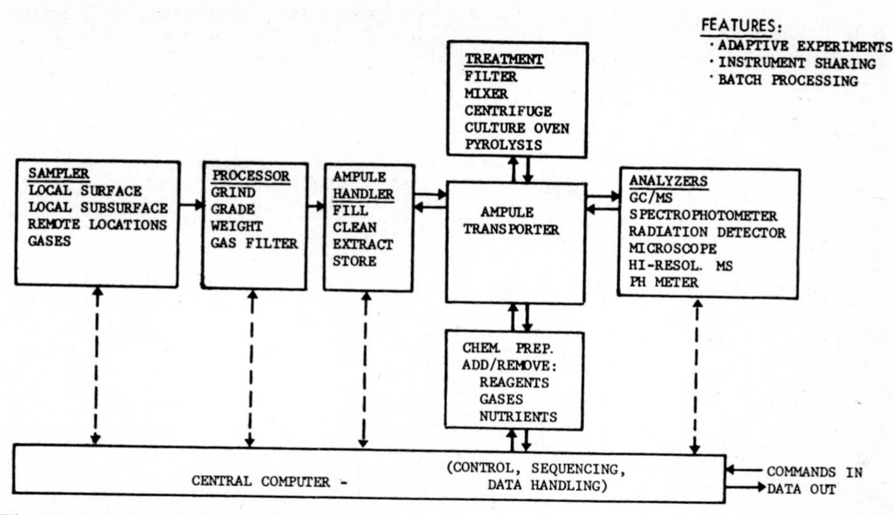

Fig. 10. Adaptive Science Laboratory.

This functional diagram portrays automated features of adaptive experiments, instrument sharing, and batch processing which would be desirable in an automated laboratory to be landed on a planet like Mars. Control, sequencing, and data handling are shown as being responsive to a central, on-board computer, with override by commands from Earth.

a decade in many of the most attractive cases. For example, transit times from the Jupiter encounter to Saturn would be 2 years and to Neptune would be $7\frac{1}{2}$ years. Because of the obvious importance of Jupiter swingby missions to a continuing exploration of the outer solar system, such missions have been under study for many years. In the course of such studies the Jet Propulsion Laboratory discovered an exciting set of multiple planet missions that have since become widely known as Grand Tour missions.

Grand Tours

A unique alignment of the planets which occurs in the late 1970's and does not recur for approximately 180 years makes it possible with current launch vehicles to send spacecraft to all of the outer planets of the solar system in a series of multiple planet missions. Two sets of three-planet Grand Tours, which have been proposed in NASA's budget request for Fiscal Year 1972, are illustrated schematically in Fig. 11.

In the budget request it is proposed that a single spacecraft be launched in 1976 and another in 1977 to fly by Jupiter, be deflected towards Saturn with an increased velocity, pass under the South Pole of Saturn (thus avoiding the rings), and be deflected out of the ecliptic plane to encounter Pluto. These flights would be known as JSP missions, for Jupiter, Saturn, Pluto. In 1979 two spacecraft would be launched to pass by Jupiter, be deflected toward

Fig. 11. Outer Planet Grand Tours—3-Planet Swingbys—NASA SL71-2504.
The positions of the planets during this decade permit using the gravitational field of Jupiter to effect a slingshot maneuver and reach the outer planets in much reduced transit times and with the expenditure of greatly reduced propulsion energy. Portrayed are trajectories for Jupiter-Uranus-Neptune (JUN) and Jupiter-Saturn-Pluto (JSP) missions.

Uranus with an increased velocity, and then in flying by Uranus be deflected towards Neptune.

Thus, with four launches, we can arrange that two spacecraft fly by each of the outer planets, with all four of the spacecraft passing close to Jupiter. In each encounter with a planet, the altitude of encounter is constrained by the navigation required to ensure encountering the succeeding planet. Targeting flexibility will exist only with respect to the last planet on the tour. But, although the encounter altitudes will be constrained, the times of arrival can be varied appreciably. By appropriately adjusting these times of arrival, close encounters with a number of satellites of the outer planets will be achievable. Possible encounters with Io, Ganymede, and Callisto,—satellites of Jupiter,—are illustrated in the drawing of Fig. 12.

As tabulated in Fig. 13, it is planned to adjust the timing of encounters in the four Grand Tours included in the current NASA budget proposal so as to achieve 27 encounters, flying by a total of 16 different bodies in all. In this light, the Grand Tour missions are clearly a powerful means for providing a broad survey of the outer solar system.

Fig. 12. Outer Planets Grand Tours—Possible Jupiter Flyby—NASA SL71-2505.
A properly chosen trajectory in passing the planet Jupiter permits flybys of several of Jupiter's satellites as well. Such missions would greatly expand the extent of our survey of the solar system.

In these Grand Tour missions, the transit times from Earth to the outermost planets will be up to about 10 years. Communication times with the spacecraft will range up to 4 hours one way, and must be provided for in the mission operations. Adequate power must be supplied, and in view of the constantly increasing distance of the spacecraft from the Sun, solar energy cannot be used. Long life must be built into the spacecraft and its equipment. Because of the long communication times involved, it will also be necessary to provide an adaptive electronics control system within the spacecraft to protect against damage to the spacecraft systems from malfunctioning elements that must be quickly turned off or isolated from the rest of the system long before any commands could be gotten to the spacecraft from the ground.

These severe requirements have led to the design of a special spacecraft, called TOPS for Thermoelectric Outer Planets Spacecraft, for use in the Grand Tour missions. An artist's concept of TOPS is shown in Fig. 14, where, to give some idea of size, the dish antenna is about 4.3 meters across. The TOPS spacecraft will be powered by radioisotope thermoelectric generators, will have adaptive data mode control, and will achieve the reliability required for a ten-

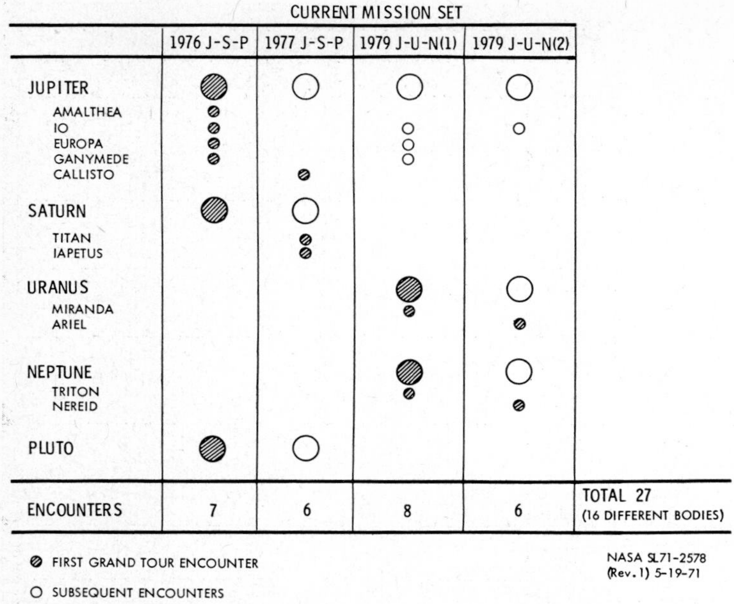

CURRENT MISSION SET

	1976 J-S-P	1977 J-S-P	1979 J-U-N(1)	1979 J-U-N(2)	
JUPITER	⊘	○	○	○	
AMALTHEA	⊘				
IO	⊘		○	○	
EUROPA	⊘		○		
GANYMEDE	⊘		○		
CALLISTO		⊘			
SATURN	⊘	○			
TITAN		⊘			
IAPETUS		⊘			
URANUS			⊘	○	
MIRANDA			⊘		
ARIEL				⊘	
NEPTUNE			⊘	○	
TRITON			⊘		
NEREID				⊘	
PLUTO	⊘	○			
ENCOUNTERS	7	6	8	6	TOTAL 27 (16 DIFFERENT BODIES)

⊘ FIRST GRAND TOUR ENCOUNTER

○ SUBSEQUENT ENCOUNTERS

NASA SL71-2578 (Rev. 1) 5-19-71

Fig. 13. Outer Planets Grand Tours—Typical Encounter Plan—NASA SL71-2578.
By selecting a Jupiter-Saturn-Pluto mission in 1976 and again in 1977 and a double Jupiter-Uranus-Neptune mission in 1979, sixteen different bodies of the solar system may be investigated during a total of 27 different encounters.

year mission by the efficient use of redundancy. This redundant control will be accomplished by a Self Test and Repair computer, called STAR, which will

Table 11. *Team leaders and number of scientists on outer planets Grand Tours mission definition phase*

TEAM	LEADER	NUMBER OF MEMBERS	NUMBER OF NON-U. S. MEMBERS
1 IMAGING	M. J. S. BELTON/University of Hawaii	10	-
2 RADIO SCIENCE	V. R. ESHLEMAN/Stanford University	8	1
3 PHOTO POLARIMETRY	A. M. J. GEHRELS/University of Arizona	6	1
4 ENERGETIC PARTICLES	R. E. VOGT/California Institute of Technology	8	-
5 PLASMA	J. H. WOLFE/Ames Research Center, NASA	16	2
6 UV SPECTROSCOPY	T. M. DONAHUE/University of Pittsburgh	8	-
7 PLANETARY RADIO ASTRONOMY	J. WARWICK/University of Colorado	12	4
8 METEOROID	R. K. SOBERMAN/Drexel University	8	2
9 IR SPECTROSCOPY	C. B. FARMER/Jet Propulsion Laboratory	16	1
10 MAGNETIC FIELDS	P. J. COLEMAN/University of California at Los Angeles	13	3
11 HYDROGEN LYMAN ALPHA	J. E. BLAMONT/University of Paris	3	2
12 PLASMA WAVE	F. L. SCARF/TRW, Inc.	7	1
13 PLANETARY X-RAY	K. A. ANDERSON/University of California at Berkeley	5	-

Fig. 14. Outer Planets Spacecraft Concept—NASA SL71-2539.
Since transit times for Grand Tour missions are about a decade and communication times will range up to four hours one way, a Thermoelectric Outer Planets Spacecraft (TOPS) has been conceived, an artist's concept of which is shown. The spacecraft would weigh about 700 kilograms, provide about 440 watts of power from radioisotope thermoelectric generators, use a 4.3 meter dish antenna for communications with Earth, and employ adaptive data mode control and high redundancy.

detect anomalies and will have the capability to switch in standby redundant elements.

Although designed especially for use on Grand Tour missions, the versatility of the TOPS spacecraft will make it very useful for a wide variety of deep space missions. If built, TOPS will also be available for conducting in-depth investigations of individual planets, like Jupiter, for missions to comets, and to carry measuring instruments out of the ecliptic to make observations of a region about which we can only speculate at the present time

A Science Steering Group has been formed to develop an overall science strategy for the Grand Tours, and to define the specific objectives for the individual flights. The composition of the Steering Group, listed in Table 11, is also international. Although the scientific payloads of the Grand Tour missions have not yet been selected, a typical complement of instruments, and the objectives served, might be as listed in Table 12.

Table 12. *Outher Planets Grand Tours instruments and experiments (Illustrative)*

Instrumentation	Key Parameters	Objectives	Experimenters Principal Investigators
2 cameras: Narrow angle / Wide angle { 2.54 cm vidicons	3500-9000Å range, with band-pass & polarizing filters; optics undefined, depends on trajectory	Measure cloud circulations, velocities, and internal patterns. Observe surface configurations and internal structure of the planets and satellites. Determine planetary and satellite dynamics.	Imaging M. Belton, Kitt Peak Nat'l. Observatory
Photopolarimeter with telescope & prism to provide dual optical paths.	2700-9200Å range, polarization measurements to an accuracy of 0.12	Determine atmospheric structure and dynamics. Measure the brightness of the galactic background.	Photopolarimetry, T. Gehrels, Univ. of Ariz.
Spacecraft radio system	Dual Frequency, S-X band	Detect atmospheric densities. Define ephemerides and test the general theory of relativity.	Radio Science (Celest. Mech.) V. Eshleman Stanford U.
Trapped radiation detectors: cosmic ray telescope: low energy particle detector	Energy range from about 10^4 ev to $>10^9$ ev: high & low energy nucleons & electrons, high intensity fluxes.	Obtain interstellar cosmic ray spectra. Measure energies of particles in the interplanetary medium and around outer planets.	Energetic Particles R. Vogt Calif. Instit. of Tech.
Two detectors: 1) Meas. solar wind ions 2) Meas. solar wind electrons and non-solar wind ions	Electrostatic deflection. range .5 ev to about 40 kev.	Measure solar wind plasma, electron distribution function, and superthermal particles. Measure interaction of solar wind with planets, satellites, and interstellar gas.	Plasma J. Wolfe Ames Research Center
Silicon solid state detectors	X-ray energies 1.5 kev to 30 kev at an accuracy of 0.3 - 0.6 kev	Detect and measure energetic plasmas around planets and satellites.	X-Ray R. Anderson Univ. of Calif. Berkeley
Deep mounted electric dipole for electric fields; search coil & loop antenna for magnetic fields.	Frequency range of 1 Hz to 200 kHz. For electric & magnetic wave amplitudes	Measure interaction of solar wind with planets. Locate orgin of satellite radio noise modulation. Measure radial gradients of solar wind, and heliosphere-interstellar field region. Detect planetary emissions, lightning, and whistlers.	Plasma Wave F. Scarf TRW Systems

Instrumentation	Key Parameters	Objectives	Experiment Principal Investigator
2 dipole antenna pairs 10 m. long	Receiver response: 1 0 kHz to 100 M Hz., dynamic range > 100 db.	Measure planetary radio emissions, plasma phenomana in magnetospheres, and ionospheres.	Radio Astronomy J. Warwick Univ. of Colorado
Boom mounted, redundant vector helium and vector fluxgate magnetometers	Freq. range; 0-6 Hz (max) 0-0.1 Hz (min) Dynamic ranges: 0±8 gamma 0±256 gamma 0±8000 gamma 0±10 gauss	Define magnetic properties of planets and satellites, their interaction with plasma environment, solar wind, and interstellar medium.	Magnetic Fields P. Coleman Univ. of Calif. Los Angeles
Optical particle detector photometer, impact compositional analyzer, ionization detector	Particle size range of .1μ to several kilometers	Measure size of particles in or transiting solar system, and interaction of this material with solar and planetary forces: zodiacal light and interstellar grains.	Meteoroid R. Seberman Drexel Univ.
Multichannel radiometer and spectrometer	Wavelength range of .3μ to 1000μ	Measure radiation balance, atmospheric structure dynamics, composition, and cloud structure of the outer planets and their satellites. Measure surface composition and thermal properties of satellites and Pluto.	Infrared C. B. Farmer Jet Propulsion Lab.
Objective grating spectrometer	Airglow measurements at HLα 584Å & 1607Å; occultation data at 6 wavelengths, one on either side of 918Å, 803Å and 504Å.	Determine atmospheric constituents of outer planets and their satellites, and the associated hydrogen-helium abundance.	Ultra-violet Spectroscopy I. Donahue Univ. of Pittsburgh
High resolution Lα photometer	H, He and OH resonance lines H 1216Å OH 1025Å He 584Å He+ 304Å	Measure deuterium/hydrogen ratios, interstellar medium and heliosphere interactions.	Lyman-Alpha J. Blamont Univ. of Paris

Other Jupiter Missions

There is considerable debate about whether the Grand Tours as I have described them, or some Grand Tours plus other types of missions, provide the best next step in exploring the outer solar system. If it is true that Jupiter and Saturn are alike, while Uranus and Neptune (perhaps being made up of comet-like material) are also alike but different from Jupiter and Saturn, then one might simplify the early survey of the outer planets by concentrating on Jupiter and either Uranus or Neptune.

To keep this possibility open, as we have gone forward with our planning for the Grand Tours described above we have also been studying the possibility of using the TOPS and other spacecraft for a concentrated study of Jupiter, perhaps eliminating one or both of the JSP Grand Tours in favor of purely Jupiter missions. The National Academy of Sciences' Space Science Board has been studying these questions this summer, and will be giving us their recommendations shortly. Inasmuch as we must prepare now for the JSP missions if we are going to carry them out, a decision must be made this fall.

Especially interesting would be missions to send probes into the atmosphere

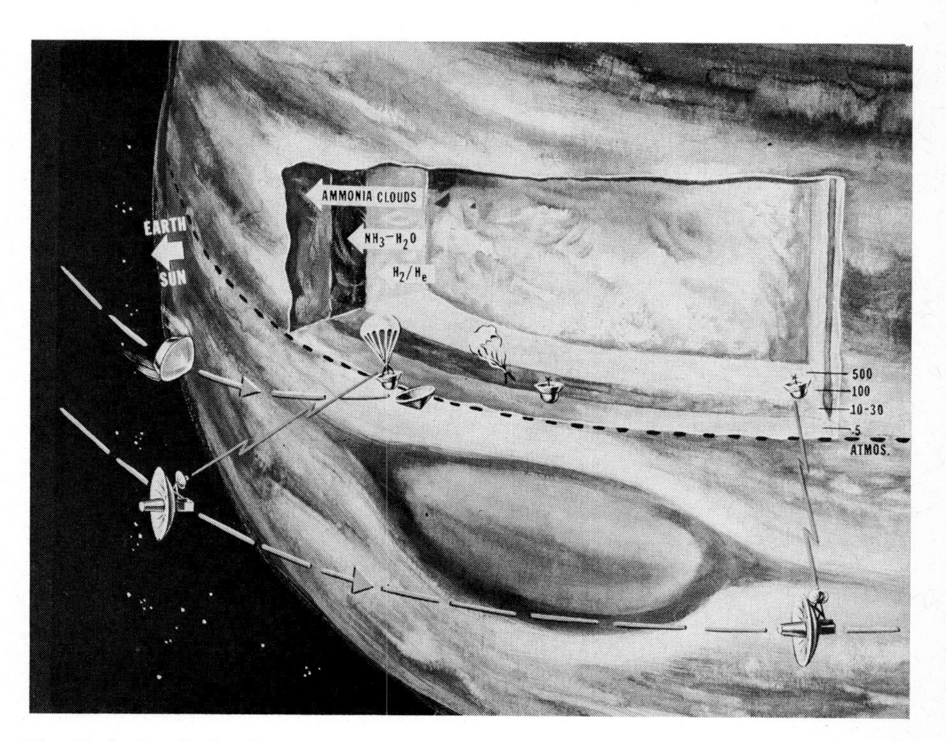

Fig. 15. Jupiter Probe Concept.

A mission of especial interest would be to send probes into the dense atmosphere of Jupiter. The high entry velocity of about 55 kilometers per second would require special design using high temperature materials still to be developed.

of Jupiter. One concept which is under study is illustrated in the drawing of Fig. 15. The probe would be released from a passing spacecraft which would then serve as a communications relay to Earth. There are formidable technological hurdles to overcome before this kind of mission can become a reality. The entry velocity into the dense Jupiter atmosphere would be on the order of 50 kilometers per second. Materials must be developed to withstand the high radiative and convective heat loads that would be imposed. NASA is proceeding to develop the necessary technology.

The best design approach for a Jupiter orbiter is also being sought in NASA studies. An initial orbiter would probably be devoted to the detailed mapping of the Jupiter radiation belts, and to an analysis of the planet's magnetic field. A follow-on orbiter might then begin an intensive investigation of the Gallilean satellites which could be encountered in each orbital period.

Comets and Asteroids

Alfvén, Arrhenius, Whipple, and others have pointed to the potential importance of small bodies in the solar system for providing clues to the state of affairs in the earliest period of the solar system. Bodies the size of the Moon and planets will necessarily have undergone substantial evolution since the time of their formation, and evolutionary processes will have altered much of the initial record of their formation. Smaller bodies may still consist of the primordial material (like the comets perhaps) or still remain pretty much as they were when they formed (as in the case of the asteroids perhaps). These possibilities make the comets and asteroids especially interesting in our exploration of the solar system.

NASA has supported a number of studies of comets, to determine those which are likely to be most interesting scientifically and at the same time reasonably accessible with space probes. Table 13 lists a number of comets, giving parameters such as the minimum distance from Earth, the time of possible recovery of the comet by Earth-based telescopes, and some astronomical parameters pertinent to observing a comet during the time the comet is under observation from a deep space probe. Table 14 lists a number of the more attractive comets from the point of view of scientific interest and accessibility by space probes, together with some of the more important mission parameters.

Of those listed in Table 14, Halley's Comet is probably the best known, and might well be the most spectacular. Unfortunately, the retrograde orbit of Halley presents formidable problems for achieving a low relative velocity between the spacecraft and the comet. With presently available launch vehicles, encounter velocities would be on the order of 55 kilometers per second. In contrast, Encke's Comet is fairly easily accessible for either a slow flyby or rendezvous, and is of considerable scientific interest. The most efficient transit

Table 13. *Comet classification based on Earth observation parameters*

CLASS	COMET YEAR	COMET NAME	MINIMUM EARTH DISTANCE, AU	WEIGHTED OBSERVATION INDEX IN HOURS FROM APPARITION TO 120 DAYS PRIOR TO PERIHELION**	MAXIMUM BRIGHTNESS, TOTAL MAGNITUDE	OBSERVABLE PERIOD* WHEN TOTAL MAG. < 12 FROM	TO
GOOD	1980	ENCKE	0.325	27	4.5	-60d	-10d***
	82	D'ARREST	0.942	284	11.2	-60	40
	82	GRIGG-SKJELLERUP	0.273	462	10.6	-20	40
	83	KOPFF	0.754	403	9.7	-130	90
	84	ENCKE	0.647	512	3.6	-40	-10
	85	GIACOBINI-ZINNER	0.479	72	8.6	-70	70
	86	HALLEY	0.412	738	1.5	-140	-10
						20	140
	87	BORRELLY	0.442	20	9.2	-80	70
	88	TEMPLE-2	0.721	427	11.8	-60	10
	91	FAYE	0.605	94	11.0	-60	40
	93	FORBES	0.629	211	10.8	-90	40
	93	SCHAUMASSE	0.553	364	9.0	-100	80
	94	TUTTLE	0.380	97	7.5	60	0
	95	PERRINE-MRKOS	0.322	20	11.3	-20	30
	96	KOPFF	0.593	210	8.8	-110	100
	98	GIACOBINI-ZINNER	0.315	285	7.7	-70	70
FAIR	1980	BROOKS-2	0.960	219	13.2	-	-
	80	FORBES	1.052	574	12.4	-	-
	81	SCHWASSMANN-WACHMANN-2	1.197	416	13.0	-	-
	87	ENCKE	0.886	91	4.9	-	-
	87	GRIGG-SKJELLERUP	0.805	708	13.1	-	-
	89	PONS-WINNECKE	1.002	630	13.6	-	-
	90	ENCKE	0.703	5	4.8	-50	-20
	90	KOPFF	1.760	166	12.0	-	-
	90	TUTTLE-GIACOBINI-KRESAK	0.185	12	9.7	-60	50
	92	GIACOBINI-ZINNER	2.028	75	11.7	-	-
	95	D'ARREST	1.711	76	12.7	-	-
	97	ENCKE	0.467	208	4.7	50	60
POOR	1978	ASHBROOK-JACKSON	1.302	0	11.0	-50	100
	80	TUTTLE	1.936	3	11.1	-	-
	81	BORRELLY	1.307	12	11.9	-30	10
	84	AREND-RIGAUX	0.571	0	11.4	-20	60
	84	SCHAUMASSE	1.177	4	10.7	-50	60
	85	HONDA-MRKOS-PADJUSKOVA	1.543	0	9.2	-	-
	86	WHIPPLE	2.176	207	15.6	-	-
	89	D'ARREST	1.448	1	12.4	-	-
	90	HONDA-MRKOS-PADJUSKOVA	0.179	0	7.8	-50	20
	92	DANIEL	1.442	0	16.4	-	-

* FOR A 35° NORTH LATITUDE OBSERVATORY SITE

** INDEX IS WEIGHTED FUNCTION OF BRIGHTNESS, DAILY OBSERVATION PERIOD, AND ELEVATION. USED BY ASTRONOMERS AS COMPARATIVE MEASURE OF RECOVERY. HIGH NUMBERS INDICATE GOOD RECOVERY CONDITIONS AND ARE ASSOCIATED WITH A MORE EFFECTIVE INTERCEPT BY A SPACE PROBE.

***DAYS BEFORE (-) AND AFTER (+) PERIHELION

Table 14. *Comet missions summary*

COMET	APPARITION ENCOUNTER MODE	SCIENCE PAYLOAD (KG)	LAUNCH DATE	PROPULSION	ENCOUNTER PARAMETERS DATE	Days From Per Passage	Relative Velocity (KmPS)	Duration
d'ARREST	1976/FFB*	35	3 JUN 76	**TAT(6C)/DELTA/BURNER II	17 AUG 76	+4	12.4	0.9h***
ENCKE	1980/SFB*	50	1 MAR 79	TITAN 3C/SEP‡(10KW)	20 NOV 80	-10	1.5	7.4h
d'ARREST	1982/R*	70	13 AUG 80	TITAN 3D/CENTAUR/SEP (15KW)	23 AUG 82	-25	0	100d
KOPFF	1983/R	70	14 JLY 81	TITAN 3D/CENT/SEP (15KW)	24 JLY 83	-25	0	100d
ENCKE	1984/R	70	18 MAR 82	TITAN 3D/CENT/SEP (15KW)	16 FEB 84	-40	0	100d
HALLEY	1986/FFB	50	4 JLY 85	TITAN 3B/CENT/BURNER II	11 DEC 85	-60	55	0.2h
		50	7 SEP 85	TITAN 3B/CENT/BURNER II	5 MAR 86	+25	75	0.15h
HALLEY	1986/SFB	50	29 SEP 78	TITAN 3D/CENT/SEP (15KW)	12 SEP 85	-150	6.4	1.7h
HALLEY	1986/R	70	16 MAY 83	TITAN 3D(7)/CENT/NEP‡(60KW)	21 DEC 85	-50	0	100d

*FFB = Fast Flyby. SFB = Slow Flyby. R = Rendezvous.
+, - denote after, before perihelion passage respectively.
‡SEP = Solar Electric Power. NEP = Nuclear Electric Power.

** THRUST AUGMENTED THOR

*** h = HOURS
d = DAYS

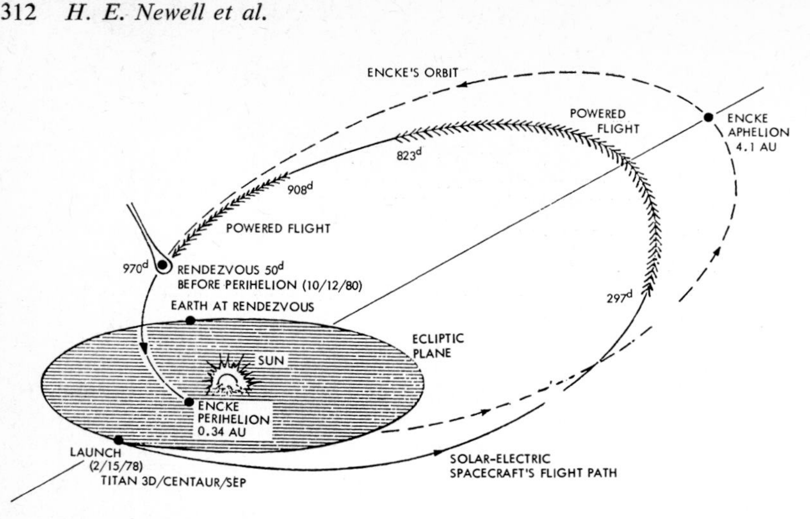

Fig. 16. Comet Encke Solar-Electric Rendezvous Mission.
The drawing shows a spacecraft rendezvous with the Comet Encke 50 days before perihelion. Encke is one of the more easily accessible comets, and could be reached with either a slow flyby or a rendezvous.

to Encke can be accomplished by using a spacecraft with a low-thrust solar-electric propulsion unit. Such a spacecraft would have to be developed for the purpose. The mission profile for a flight to Encke during its 1980 apparition is shown schematically in Fig. 16. With a launch in February of 1978, rendezvous with the comet would occur in 1980, 50 days before perihelion, permitting exploration of the comet and its environment during perihelion passage. Details of such a mission and the design of the required spacecraft are presently under study.

A working group of interested scientists convened at Yerkes Observatory in June, 1971, to review the possibilities for exploring comets with deep-space probes. The group has recommended the development of a solar-electric spacecraft to be used for a slow flyby of, or a rendezvous with, Encke in both the 1980 and 1984 opportunities. It was further recommended that a flyby of Comet d'Arrest be attempted at the 1976 apparition to evaluate the instruments that would be used for the more sustained investigation of Encke later, and that instruments be developed to function at the high encounter velocities of a flyby of Halley's Comet.

The solar electric propulsion spacecraft developed for comet missions would also be useful for rendezvous and sample return missions to some of the asteroids. Such missions are presently being studied by both the Jet Propulsion Laboratory and NASA's Marshall Space Flight Center. A science advisory group co-chaired by Ernst Stuhlinger of the Marshall Space Flight Center and Hannes Alfvén has been formed to define scientific objectives and instrument requirements for asteroid missions.

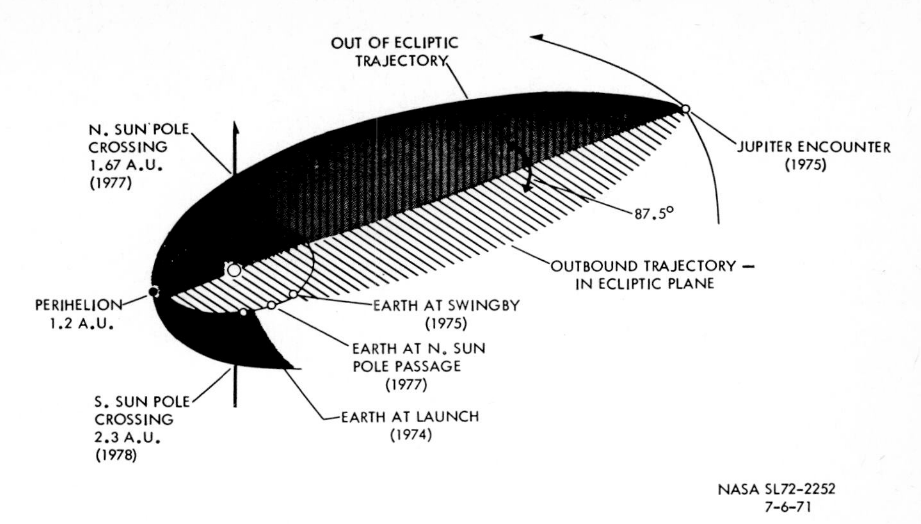

Fig. 17. Jupiter Out of the Ecliptic Mission—NASA SL72-2252.
The gravitational field of Jupiter may be used to throw a spacecraft slingshot fashion out of the ecliptic plane and thus permit investigation of regions of solar influence previously inaccessible to direct measurement.

Miscellaneous Missions

We can think of many other kinds of missions that will be important to the scientific study of the solar system, and indeed many others are also under study by NASA. For example, there are numerous flybys of Jupiter, that are of potential interest. Between now and the end of the century there are numerous opportunities to fly by Jupiter in such a way as to encounter any chosen one of the other outer planets; these are also under study. Such a mission to Saturn would enable us to explore Saturn's rings, a possibility that is being looked at for the late 1970's.

Some of the more intriguing Jupiter flybys permit sending spacecraft very close to the Sun. One use of the Jupiter swingby technique is to throw a spacecraft as far as possible out of the ecliptic plane to investigate regions of solar influence previously inaccessible to direct measurement. NASA is studying the possibility of sending a Pioneer spacecraft on such a mission, as depicted schematically in the drawing of Fig. 17. On such a mission, instruments would attempt to measure the dependence of the solar wind and the production, storage, and propagation of solar cosmic rays on heliographic latitude, and would seek to determine whether magnetic field lines issuing from near the poles of the sun connect directly to the interstellar medium. Missions out of the ecliptic plane would alter our exploration of the solar system from its previously primarily two-dimensional character to a truly three-dimensional exploration.

Conclusion

The power of instrumented space probes and ground-based telescopes outfitted with modern instrumentation for investigating the solar system bids fair to amass for us in the years ahead a tremendous quantity of new data. The new data should enable the theorist to delve more perceptively into solar system and planetary evolution than has been possible in the past, and one may anticipate something of a revolution in solar system studies over the decades ahead. To make the most of the opportunities that lie before us, it is essential to develop the best possible scientific strategy and to fly the most illuminating space missions. To this end, NASA seeks and welcomes the advice of those experts who may be interested. In concluding, then, let me repeat the hope expressed earlier that we will hear from participants in this symposium ideas and suggestions that will help us put together the very best program that we can for the exploration of the solar system.

The Critical Velocity of Gas-Plasma Interaction and Its Possible Hetegonic Relevance

By J. C. Sherman

Department of Electron and Plasma Physics, Royal Institute of Technology, Stockholm, Sweden[1]

I. *Introduction*

Many years ago Alfvén (1954, 1960) proposed that when the relative velocity between a neutral gas and a magnetized plasma increases to a value of $(2eV_i/M)^{1/2}$, then ionization of the gas will increase abruptly. As a consequence of this it was also proposed that the strong coupling due to ionization would limit the relative velocity to the critical value of $(2eV_i/M)^{1/2}$ or less. Here V_i and M are the ionization potential and mass of the neutral atom, and the relative velocity is usually perpendicular to the magnetic field. This proposal is referred to as Alfvén's critical velocity hypothesis and is the subject of this contribution.

Alfvén's original proposal was made in connection with his theory of the hetegonic process. (Hetegony is the term proposed by Alfvén and Arrhenius (1970) to describe the formation of secondary bodies around primary bodies.) The rôle of the critical velocity in Alfvén's theory is to ionize and, because of the magnetic field, to stop an infalling cloud of gas when it has reached a certain distance from the central body. The resulting cloud of plasma is then taken to be the starting point for the processes of condensation and accretion which form the secondary body and which have been among the subjects discussed at this symposium. However, apart from this, the interaction of a magnetized plasma and a neutral gas in relative motion is a topic of general interest, which seems to have attracted relatively little attention.

Before speculating on the processes which might have occurred during the formation of the solar system, it is worthwhile to discuss some of the laboratory experiments and theories which relate to the critical velocity concept. The experiments which can be interpreted in terms of the critical velocity are discussed in Part II, this being partly based on the review of Danielsson (1970*a*), and some of the theories which have been proposed are discussed in Part III. The relevance of these experiments and theories to Alfvén's, hetegonic model is then considered in Part IV, and some conclusions are drawn in Part V.

[1] Present address: Department of Electrical Engineering, University of Liverpool, Brownlow Hill, Liverpool, England.

Although the critical velocity is usually discussed in terms of a relative plasma-gas velocity, this can lead to confusion in those cases where the ion and electron components of the plasma are also in relative motion. For this reason attention will often be focussed on the value of $\mathbf{E} \times \mathbf{B}/B^2 = \mathbf{U}$, where $\mathbf{E}$ and $\mathbf{B}$ are the electric and magnetic fields in a frame of reference in which the neutral gas is on average at rest. For simplicity, and since it is also appropriate to the laboratory experiments, $\mathbf{E}$ is taken to be perpendicular to $\mathbf{B}$. Furthermore, whenever critical velocity effects have been experimentally observed the electron Hall parameter $(\omega\tau)_e$ has been larger than unity. In these circumstances the average electron velocity relative to the gas is equal to $\mathbf{U}$, but, since the ion Hall parameter is not necessarily greater than unity, the average ion velocity through the gas can have a smaller magnitude than, and direction different from, $\mathbf{U}$.

At first sight it might seem reasonable that a plasma-gas ionizing interaction should become appreciable when the relative velocity (for the moment assumed to be the same for both ions and electrons) reaches a value of $(2eV_i/M)^{1/2}$; however a serious difficulty is soon apparent. Ionizing collisions between the ions and the neutrals will not occur unless the ion kinetic energy in the gas frame of reference exceeds $2eV_i$ (assuming equal ion and neutral masses and negligible neutral random kinetic energy). This follows from the fact that the maximum inelastic energy transfer equals the kinetic energy in the centre of mass system of the colliding particles. It is then obvious that any theoretical justification of the critical velocity must include considerations of the electron and/or ion random energies or temperature.

II. *Experimental Observations of the Critical Velocity*

A voltage limitation effect which can be interpreted in terms of the critical velocity is now a well known observational fact which has been reported many times (Danielsson 1970*a*). The effect is observed in discharges where the voltage and current are applied across a magnetic field. It appears as an upper limit to the burning voltage which varies linearly with the magnetic field strength, but is independent of the plasma pressure, the gas pressure, and the current. There must, however, be some neutral gas present within the discharge volume; once ionization is complete throughout the volume the critical velocity concept is no longer valid.

The connection between the voltage limitation and the critical velocity arises as follows. With the same notation as above, and assuming the gas is at rest, the burning voltage can be written

$$V = V_0 + \int (E/B)\, B dx$$

where V_0 is attributed to electrode effects and the integration goes from one electrode to the other. The experiments show that $(E/B) = U$ takes on a constant value which depends almost entirely on the particular gas used and is near the value given by $U_c = (2eV_i/M)^{1/2}$. This is the same as the critical velocity proposed by Alfvén but now U_c has a somewhat different meaning.

Most observations of the critical velocity effect have been accidental and the effect has appeared as an unwanted limitation of the energy storage in various plasma devices, for example, thermonuclear machines like the Ixion, the early Homopolars and Lehnert's F machines. Special arrangements have sometimes been made to avoid the limitation and to date these have concentrated on keeping the plasma and the gas apart.

The experiments which exhibit the critical velocity effect have been reviewed by Danielsson (1970*a*). Among the various types of experiment included in Danielsson's review are rotating plasma devices, plasma guns, shock tubes, and a plasma-gas impact experiment. The earliest reports were those of Simon (1959), using a P.I.G. discharge, and Alfvén (1960) and Fahleson (1961), using a homopolar rotating plasma. Only four experiments have been aimed at a study of the voltage limitation in detail: these are the experiments of Angerth, Block, Fahleson and Soop (1962), Eninger (1965), Lehnert, Bergström and Holmberg (1966) and Danielsson (1970*b*). Several theoretical investigations have been based on the experiment of Angerth et al. and Danielsson's experiment demonstrates the critical velocity effect in a particularly striking way. Both these experiments are discussed in more detail below. Eninger's (1965) experiment is important as regards possible hetegonic applications and is also briefly described below. One remarkable aspect of Danielsson's review is the very wide range of configurations, currents and gas pressures over which the effect is observed.

Table 1. *Examples of the critical velocity*

Atom or molecule	Created ion	Ionized mass	Ionization potential (V)	Critical velocity (km/s)
H	H^+	1	13.6	51.0
H_2	H_2^+	2	15.4	38.6
H_2	H^+	1	18.0	59.0
He		4	24.5	34.4
C		12	11.3	13.5
N	N^+	14	14.6	14.2
N_2	N^+	14	24.5	18.4
N_2	N_2^+	28	15.6	10.4
O		16	13.6	12.8
Ne		20.2	21.6	14.4
Ar		39.9	15.6	8.7
Fe		55.8	7.8	5.2

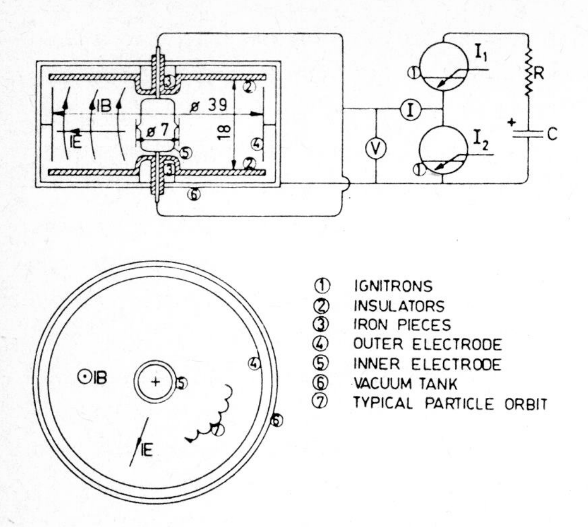

Fig. 1. Experimental arrangement (Angerth et al. 1962).

When considering atomic gases the expression $(2eV_i/M)^{1/2}$ has an unambiguous value. However in the case of molecular gases there are several possible values of V_i and M, and the value of critical velocity that occurs depends on the experimental conditions. Table I gives some examples. A similar ambiguity arises in the hetegonically important case of gas mixtures.

1. **The Rotating Plasma Experiments of Angerth, Block, Fahleson and Soop**

A. *Experimental device and results*

These experiments were carried out in a co-axial cylindrical geometry with a magnetic field aligned along the axis. The apparatus is shown in Fig. 1. The magnetic field on the axis varied from 0.1 to 1 Wb/m² and decreased to half the central value at the outer cylindrical electrode (cathode). Currents of 30 to 10^4 A from a condenser bank were passed through seven different gases, the initial gas pressure ranging from 5 to 200 microns, and the current pulses lasting up to 70 μsec.

The main experimental results concerned the discharge voltage during the burning phase, and the associated radial electric field E_r. It turned out that E_r/B was almost independent of radius, pressure and current, for currents less than around 2kA. For higher currents the experiment was less reproducible but an increase in current or a decrease in initial pressure tended to increase the discharge voltage. For all the gases used the constant values of E_r/B were near to $(2eV_i/M)^{1/2}$, based on the atomic mass and ionization potential.

B. *Discharge processes*

As a background to the theoretical discussion in Part III it is helpful to note here some of the more important experimental parameters which occur under typical conditions. This will give some idea of the type of processes occurring in the discharge. These parameters have been studied in some detail by Sherman (1970*a*), but only a brief outline of the more important points can be given here.

Photomultiplier and magnetic probe measurements (Fahleson 1961) indicated that there were no large azimuthal variations in plasma density during the roughly steady state burning phase, so that it is reasonable to assume that there was no azimuthal electric field. It is also assumed that any axial electric field (parallel to B) was much smaller than E_r. The molecular gases are taken to be fully dissociated and the ions to be singly ionized and atomic.

The experimental current pulses lasted around 70 μs, but the constant values of E_r/B were observed only a few μs after breakdown. Consideration of the neutral gas acceleration (Sherman 1970*a*) shows that although gas motion and nonuniformity could well build up towards the end of a current pulse, when the constant values of E_r/B are first observed the gas can be considered as stationary and uniform. Similar results hold for the gas temperature, so that it can be concluded that gas motion and heating are not essential for the occurrence of constant values of E_r/B. It should also be noted that some Doppler shift measurements did not indicate neutral gas rotation. On the other hand, given sufficient time, even very small currents will produce thin rotating layers of gas on the end plates and outer electrode, as outlined by Angerth et al. (1962).

Estimates of the electron Hall parameter $(\omega\tau)_e$, including collisions with both neutrals and ions, show that throughout the experimental range $(\omega\tau)_e \gg 1$. The high values of $(\omega\tau)_e$ mean that the average electron velocity is mainly azimuthal and equals E_r/B.

Turning now to the ion motion, symmetric resonant charge exchange collisions are the most important type of ion collision at the energies and electron densities of interest here. The motion of ions in perpendicular and uniform strong electric and variable magnetic fields with this type of collision has been studied by Sherman (1970*b*). The strong electric field condition refers to conditions such that the neutral temperature can be ignored. It turns out that the charge exchange collisions lead to several unusual effects, some of which are briefly mentioned here.

The special character of symmetric resonant charge exchange collisions, i.e. the very low ion velocity after the collision, means that between collisions the ions move along a cycloidal path and that between collisions the ion energy is given by $W = M(E/B)^2 (1 - \cos \omega t)$. Here ω is the ion cyclotron frequency and t is the time measured from the previous collision. Under free orbiting conditions the average ion energy is $M(E/B)^2$ and the peak energy is $2M(E/B)^2$, i.e. $2eV_i$ and $4eV_i$ respectively if $E/B = (2eV_i/M)^{1/2}$. Any departure from free orbiting conditions caused by an increase in the collision frequency reduces the average ion energy, since the effective value of t is reduced. Expressions for the average ion energy and for various other ion velocity moments have been given by Sherman (1970*b*) and under typical experimental conditions the average ion energy is reduced from $2eV_i$ but is still in the region of 10 eV or more.

The average velocity formulas show that typically the ion velocity is radial rather than azimuthal, in direct contrast to the electrons. The average ion velocity is substantially less than E_r/B in most cases; only hydrogen and helium at the lowest pressures and highest magnetic fields used have azimuthal velocities near to E_r/B. In fact this experiment is one case where the distinction between E_r/B and plasma velocity must be kept in mind. Typically most of the radial current is carried by the radial ion motion and the electron density is in the range 10^{18} to 10^{19} m^{-3} which corresponds to degrees of ionization of less than 1 percent.

The Doppler shift measurements made by Fahleson (1961) and Angerth et al. (1962) indicated a rotational ion velocity in the region of E_r/B. The discrepancy between these measurements and the above description of the ion motion can be resolved when it is noted that the measurements were made using doubly ionized carbon and silicon ions in a hydrogen discharge. Although these ions are heavier, the Hall parameter for them will be much larger than for protons for three reasons. Firstly these ions are not affected by the large symmetric resonant charge exchange cross-section. Secondly, they are doubly ionized, and thirdly, heavy ions moving in a light gas show a large persistence of velocity effect (Chapman and Cowling 1960). Because of their large Hall parameter these impurity ions will indeed move azimuthally with a velocity near E_r/B as observed.

If only symmetric resonant charge exchange collisions are allowed for, the ion velocity is only in a plane perpendicular to **B**. However, Coulomb collisions with other ions, elastic collisions with molecules in the case of incomplete dissociation, and the small fraction of charge exchange collisions that result in an appreciable ion velocity after the collision, all serve to scatter ion energy into a direction parallel to **B** and so reduce the very strong anisotropy in the ion velocity distribution. The collisional effects have been treated by Sherman (1970*b*) as small perturbations. It turns out that in the experiments of Angerth et al. the anisotropy is still strong under typical conditions, and an important consequence of this is mentioned in Part III Section 1 C.

When considering whether ion energy and momentum are transferred to the gas or directly to the chamber walls, the neutral-neutral mean free path is important. Under the conditions of interest here the neutral formed in a symmetric resonant charge exchange collision has, typically, an energy of 10 eV or more. At these high energies the pronounced forward scattering leads to a longer mean free path for momentum or energy loss than would otherwise occur. It turns out that under some conditions the power and momentum input that is

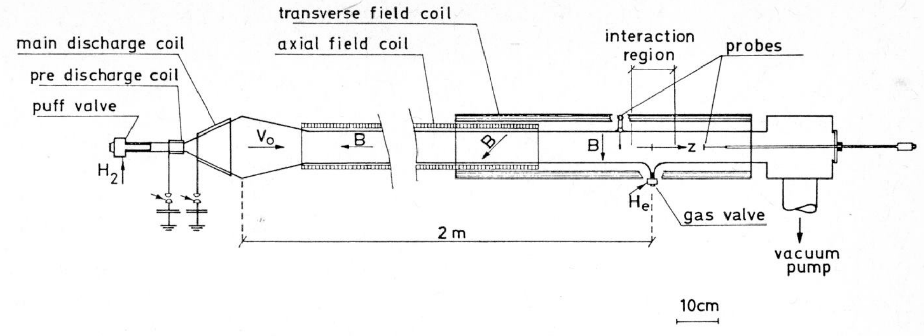

Fig. 2. Experimental arrangement (Danielsson 1970*b*).

transferred from the ions to the fast neutrals is carried by the latter directly to the outer electrode.

2. The Plasma-Gas Impact Experiment of Danielsson

This recent experiment (Danielsson 1970*b*) demonstrates a critical velocity effect in a more direct way than any of the voltage limitation experiments.

A. *Experimental device*

The apparatus used is fundamentally different from that used in the voltage limitation experiments and is shown in Fig. 2. A hydrogen plasma is generated in an electrodeless plasma gun (a conical theta pinch) and is accelerated along a magnetic field into a drift tube. The magnetic field gradually changes direction until it is transverse to the drift tube axis, but the field magnitude remains constant and can be varied up to 0.5 V sec/m^2. As the plasma proceeds along the drift tube much of it is lost but the plasma sets up a polarization electric field, and, by virtue of this field, some plasma ends up drifting across the transverse magnetic field. Initially the apparatus is under high vacuum (3×10^{-7} torr). By means of fast electromagnetic valves hydrogen is injected into the gun and another gas, usually helium, is injected into the transverse field region. The whole process is synchronized so that the gas cloud has an axial spread of 5 cm and an average density of 10^{20} m^{-3} by the time the drifting plasma reaches the interaction region. The plasma drift velocity can be varied between 2×10^4 and 4×10^5 m/sec while its density is between 10^{17} and 10^{18} m^{-3}. The length of the drifting plasma column is around 1 m; the duration of the interaction varies from 2 μsec at the highest velocity to 60 μsec at the lowest velocity.

B. *Plasma velocity and ion production*

Double probes were used to measure the polarization electric field **E**. When the upstream plasma velocity v_0 was high enough (see below) it was found that **E** decreased from its upstream value to a relatively low downstream value over

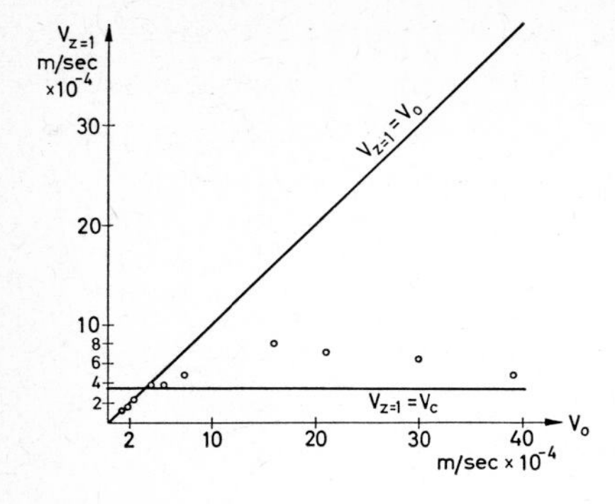

Fig. 3. Plasma velocity 1 cm behind the centre of the gas cloud as a function of the original velocity. $B=0.44$ Vsec/m².

distances of one or two centimetres. Although the connection between **E** and the plasma velocity **v** is not straightforward in the centre of the interaction region, where the scale length for variations in **E** and the proton Larmor radius are of the same order, upstream and downstream **E** is relatively uniform so that the plasma velocity can be deduced from $\mathbf{v}=\mathbf{E}\times\mathbf{B}/B^2$. In other words an interaction with the neutral gas cloud strongly decelerates the plasma beam over distances of a few centimetres.

The relationship between the upstream velocity v_0 and the downstream velocity 1 cm behind the centre of the gas cloud $v_{z=1}$ is shown in Fig. 3. It is found that for very low v_0 the plasma is not retarded as it penetrates the gas cloud, whereas for $v_0>4\times10^4$ m/sec the relative retardation increases with v_0. The critical velocity $v_c=3.5\ 10^4$ m/sec, based on the ionization potential of helium and the mass of *helium*, is also shown in Fig. 3. It has been found that the plasma velocity is never reduced below v_c. Measurements with capacitatively coupled probes placed outside the drift tube indicate that substantial numbers of helium ions are formed in the interaction. Absolute emission intensity measurements and 4 mm microwave measurements also indicate a substantial production of helium ions.

C. *Radiation emission and the electron energy*

Before the interaction no radiation was detected from either the gas or the plasma (fully ionized protons) and measurements in the plasma gun region indicated an electron temperature <5 eV. However, when the plasma penetrates the gas cloud, emission is observed with a fast rise time ($<0.5\ \mu$sec), even from energetic levels of He^+. The low electron density of interest here then implies that step-wise processes cannot be responsible for the rapid onset of emission, and so the line intensities must be interpreted in terms of the cross-section for single step electron excitation of a particular line. Measurements of the relative intensity at 4 686 Å in He^+ and 4 713 Å in He then indicate a group of electrons with an energy of 85 eV (assuming a monoenergetic group to be responsible for the emission). Alternatively a Maxwellian distribution with a temperature corresponding to 70 eV leads to the same observed intensity ratios. These intensity measurements were made with $v_0 = 4 \times 10^5$ m/sec and a corresponding proton energy of 840 eV. Also it should be noted that several tests were made to check that these large and rapid changes in electron energy were in fact caused by the interaction and not by some other extraneous effect.

D. *Effect of neutral gas density and magnetic field strength*
It was found that, provided the gas density was above roughly 5×10^{19} m^{-3}, then gas density variations had little effect on the plasma retardation. On the other hand the retardation quickly disappeared when the density was reduced below 2×10^{19} m^{-3}. Similarly it was found that although a magnetic field was necessary for the retardation to occur, doubling its strength from 0.18 to 0.36 V sec/m^2 had relatively little effect on the retardation.

E. *Interaction mechanisms*
Consideration of the typical gas density and *undisturbed* plasma parameters (Danielsson 1970*b*) leads to the conclusion that simple binary collisions between the undisturbed plasma and gas cannot account for the observed retardation. On the other hand estimates of the production rate of helium ions produced by the observed *high energy* electrons are high enough to account for the retardation. A theoretical description of the interaction must then be along the lines of a collective interaction between the plasma and gas which leads to the production of high energy electrons. We should note here how well the retardation observed in this experiment supports Alfvén's critical velocity hypothesis.

3. The Co-Axial Gun Experiments of Eninger

These experiments (Wilcox, Pugh, Dattner and Eninger 1964; Eninger 1965) must be distinguished from normal plasma gun experiments. In this case the current and energy input are relatively low so that the plasma is only partly ionized and the ionization process plays an important part in the plasma motion.

A. *Experimental arrangement*
The annular region between the electrodes of a co-axial plasma gun with inner electrode (cathode) radius 1.25 cm and outer electrode radius 3.60 cm was filled with gas, and a constant azimuthal magnetic field was generated by a current in a central conductor. A constant current pulse from an artificial transmission line then generated a current sheet which propagated down the gun. The applied magnetic field (~ 10 kgauss) was much stronger than the field generated by the plasma current.

B. *Experimental results*
The experiments covered a large range of current (5–5 500A) and pressure (0.01–100 torr) and the results obtained varied in a complicated way with

these parameters and with the magnetic field. In particular it was found that with the two molecular gases used (hydrogen and nitrogen), the discharge could operate in one of two distinct modes. This behaviour was not observed with the atomic gases (helium and argon). However, as regards possible hetegonic applications, the important results were as follows. With atomic gases, high magnetic fields and low pressures it was found that the current sheet propagation velocity, the ion velocity, E/B, and $(2eV_i/M)^{1/2}$ were all equal. Here E is the radial electric field which was observed to follow the $1/r$ dependence of the applied azimuthal magnetic field B. The three velocities were equal, independent of the current as long as the current was not so large that the plasma was no longer partially ionized.

III. *Critical Velocity Theories*

The theories can be divided into two classes. In the first class the theory concentrates on explaining the results of a particular experiment without directly invoking the critical velocity. In the second class the gas-plasma interaction is discussed in more general terms, and a particular experiment is seen as one example of the more general process.

1. **Theories Based on a Particular Experiment**

In fact all the theories of this type have been aimed at explaining the results of Fahleson (1961) and of Angerth et al. (1962). These theories have been discussed by Sherman (1970*a*) and only the more important points are mentioned here.

A. *Sockol's theory*

This work (Sockol 1968) considers a steady state continuum model of the discharge. In this model both the ionized and neutral components of the plasma move with almost the same velocity in the azimuthal direction, and the $j_r B$ volume force is balanced by viscous drag on the end plates. Away from the end plates the rotational velocity approaches E_r/B and most of the current flows in thin Hartmann boundary layers on the end plates.

The radial electric field or overall voltage required by this model has been calculated by Sockol for hydrogen, nitrogen and argon, assuming the molecular gases to be completely dissociated. For these three gases the dependence of the electric field or voltage on current, pressure and magnetic field is in fairly good agreement with experiment, but the magnitudes are too small by a factor of two.

The major weakness of this theory in explaining the observations has already been pointed out by Sockol (1968). This is that the constant values of E_r/B are observed long before the neutral gas has accelerated to the velocity re-

A. *The non-thermal ion space charge theory of Lehnert*

This theory (Lehnert 1967) has been extended and discussed by Sherman (1970*c*), and only the main points are included here.

Lehnert's theory concerns the space charge, and associated electric potential, produced by non-thermal ions in a magnetic field. It is an analysis of the following idealized situation. With a steady uniform magnetic field aligned along the z axis of a right-handed Cartesian co-ordinate system, non-thermal ions are formed moving in the x direction only, all with the same initial velocity v_0. The ion source density is taken to be a function of x alone. It is assumed that there is no initial electric field, but it turns out that under certain conditions a space-carrying (roughly sinusoidal) electric field is set up in the x direction. This field is caused by the ion space charge and in turn the field affects the ion orbits and space charge. The field is constant in time if the number of non-thermal ions is also constant.

The self-consistent solution of the complicated orbits and space charges produced can only be computed numerically after considerable effort, so that the number of solutions computed to date is rather limited. Hence the parameter range which results in a fluctuating potential is not well understood, although in general the fluctuations occur when the ion density and density gradient are sufficiently large. The physical basis for this effect is that, owing to their different Larmor radii, the electron and ion formed at a certain point move over different regions of space. A gradient in the source density then leads to a space charge.

As a possible explanation of Alfvén's critical velocity hypothesis, an important aspect of this space charge mechanism is that potential differences can occur which are in the region of $Mv_0^2/2e$, where M is the non-thermal ion mass. If some electrons can gain this energy and if the non-thermal ions are formed from a beam of atoms moving along the x axis with velocity v_0, then Lehnert's theory predicts an abrupt increase in ionization when $v_0 = (2eV_i/M)^{1/2}$. This is just Alfvén's hypothesis. It should be noted that the relative velocity between the plasma and the gas beam enters the theory in a natural way.

From the point of view of Alfvén's hypothesis the mechanism by which the electrons gain energy from the space charge fields is vitally important, but this has not been discussed in any detail. The energy gain of an electron moving in an undisturbed orbit is not large (Sherman 1970*a*), so that collisions must occur if the electrons are to gain appreciable amounts of energy. Thus for electron heating the electric fields produced are of as much importance as the potential differences. Also it should be noted that those electrons which gain energy by moving from low to high potentials tend to reduce the space charge which causes the potential differences.

We turn now to the formation of the non-thermal ion space charge potentials.

The potential formation is a necessary but perhaps not sufficient condition for an explanation of the critical velocity in terms of Lehnert's theory. Lehnert's calculations are based on the following assumptions:

(i) There is no thermal background plasma
(ii) The situation is steady and collisionless, with a uniform magnetic field
(iii) Every ion starts with the same velocity in the x direction only, but at random times
(iv) Variations are in the x direction only, and the only electric field is that caused by the space charge, i.e. there are no initial electric fields in the frame of reference chosen
(v) The non-thermal ion source distribution is stationary.

An extension of Lehnert's theory to include a thermal (low temperature) background plasma has been given by Sherman (1970*c*). In general the effect of a background is to increase greatly the non-thermal ion density required to give appreciable effects. This is basically due to the dielectric behaviour of the background plasma. In fact appreciable potentials and electric fields are not produced unless (Sherman 1970*c*; 1969)

$$n_{nt}/n_b \gg 2M_b/M_{nt}.$$

Here n_{nt} and n_b are the non-thermal and background ion densities and M_{nt} and M_b are their masses.

We consider now the application of Lehnert's theory to two of the experiments discussed in Part II. Consider first the experiments of Fahleson (1961) and Angerth et al. (1962). Lehnert's theory is worked out for collisionless conditions, and in fact the theory relies on the organized orbital motion of the non-thermal ions. Without further numerical calculations it is difficult to see exactly how important non-thermal ion collisions are, but too high a collision frequency will prohibit the effect. However, as was mentioned in Part II Section 1 B, in these experiments the ions typically complete only a small fraction of an orbital cycle before colliding. Thus, except perhaps for hydrogen and helium at the highest magnetic fields and lowest pressures, it seems unlikely that Lehnert's mechanism can be important in these experiments.

Secondly consider the plasma/gas impact experiment of Danielsson (1970*b*). Several difficulties arise when an attempt is made to apply Lehnert's theory to this experiment, and these are discussed in more detail by Sherman (1970*c*). Non-uniformities in the laboratory electric field make it difficult to choose the moving frame of reference required by Lehnert's theory in which these electric fields are zero (assumption iv). Other transformation difficulties affect assumption (v), and also, when allowance is made for the background (proton beam) plasma, make it unlikely that a sufficiently high non-thermal ion density can be built up in the transit time available. Also the boundary conditions in this experiment will have a strongly inhibiting effect on the potential formation.

B. *An electrostatic two-stream instability*

This instability and its relevance to the critical velocity are currently being investigated by the author, and so what is mentioned here is rather preliminary in nature. The well known electrostatic two-stream instability (Buneman 1959) has also been investigated by Buneman (1961) but now including the effect of a magnetic field. The latter work has been extended by Sherman (1969) who concentrates on wave propagation almost perpendicular to the magnetic field.

Sherman (1969) considers a situation in which cold ions drift through cold electrons in a frame of reference in which there is a magnetic field but no electric field. Conditions are taken to be uniform, steady and collisionless. The assumption of a steady ion drift velocity is strictly inconsistent with the other assumptions, but it turns out that this can be justified by the shortness of the relevant time scale. Similar reasoning applies to the collisionless assumption. A linear perturbation analysis then leads to complex frequencies for a range of real wave numbers. The instability is convective with a phase velocity between the electron and ion drift velocities.

Compared to the normal two-stream instability, the anisotropy introduced by the magnetic field, in conjunction with the almost perpendicular propagation, produces important new effects. There are indications that the growth stops when the electron energy is of the order $Mv_d^2/2$, where M and v_d are the mass and velocity of the drifting ions. In contrast the order of the electron energy when the normal two-stream growth stops is $mv_d^2/2$, where m is the electron mass. This large change arises from the fact that in the magnetic field case the important comparison is between the random electron velocity parallel to the magnetic field and the component of the phase velocity in the same direction. Since the propagation is almost perpendicular to the field, the phase velocity component along the field is comparatively very large and this leads to a large increase in the electron energy limit.

The ion motion is little affected by the magnetic field, and so is similar to that found in longitudinal electrostatic instabilities. In contrast the electrons are tightly bound to the magnetic field and their important motion is along the magnetic field under the influence of the relatively small electric field component in this direction. The important electron motion is thus almost transverse to the wave propagation direction.

Because the electron motion parallel to the magnetic field is fundamental to this instability, the wavelength in this direction must not be much longer than the scale length for non-uniformities in the same direction. Since the propagation is almost perpendicular to the field the wavelength parallel to the field is

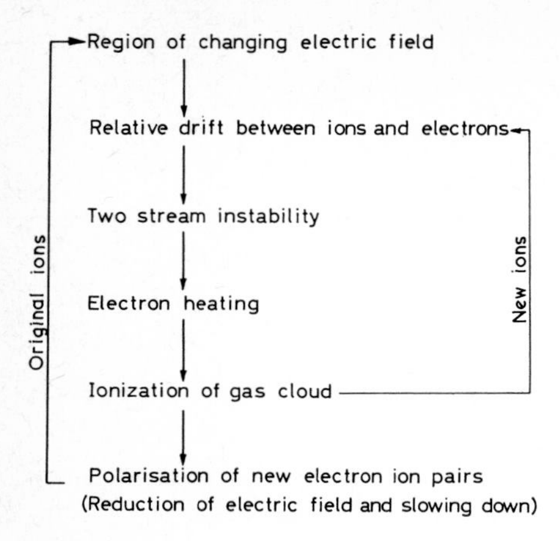

Fig. 4. Circular process.

relatively large and non-uniformities in the field direction can be important (Sherman 1969).

We turn now to the different mechanisms by which a relative drift between electrons and ions can be produced, each mechanism being appropriate to a particular situation. Consider a situation in which the ion Hall parameter $(\omega\tau)_i$ is much less than unity but the electron Hall parameter $(\omega\tau)_e$ is much greater than unity. This is a quite common experimental situation, and in particular is appropriate to the experiments of Fahleson (1961) and Angerth et al. (1962). In this situation ion collisions with the neutral gas are frequent enough to hold the ions relatively stationary in the gas frame of reference, but in the same frame the electron velocity is E/B. We then have $v_d = E/B$. Since the instability growth time can be very much less than an ion cyclotron period it is still possible for the instability to occur even in the presence of ion-neutral collisions. If the instability does occur it could lead to an electron energy in this case of order $M(E/B)^2/2$, and so we have a direct explanation of the results of Fahleson and Angerth et al. However this application has not been worked out in detail yet, an important consideration being non-uniformities parallel to the magnetic field.

Another mechanism which produces a relative drift has been proposed by Sherman (1969) in connection with Danielsson's (1970*b*) experiment. In this experiment the scale length of the interaction region in which the proton beam is decelerated is comparable to the proton Larmor radius. The proton inertia then leads to an electron-ion relative drift, but in this case the relative drift velocity is not simply connected to E/B. Alternatively Wallis (1971) has pointed out that since the helium ion Larmor radius is comparable to the interaction scale length and the tube radius, these ions are relatively stationary in the gas frame through which the electrons move with the E/B velocity. The instability can then be modified to include stationary electrons and ions (protons) with a

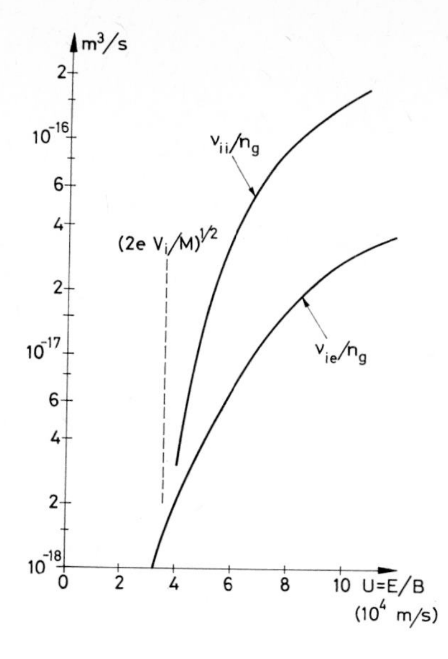

Fig. 5. Ionization frequencies in helium for large $(\omega\tau)_i$.

second ion species (helium) having a relative drift. In this case the drift velocity is just E/B, but again the modified theory has not been worked out in detail.

One difficulty with both the above interpretations is that they rely on the circular argument indicated in Fig. 4. In this figure 'original ions' refers to the proton beam (Sherman 1969), and 'new ions' refers to the modification suggested by Wallis. The problem posed by a circular argument is: how does the self-consistent process begin? A similar difficulty occurs when applying Lehnerts space charge theory to Danielsson's experiment. In the case of Lehnert's theory a sizeable helium ion density is required before electron heating in the space charge fields can occur, and it is difficult to produce a sufficient density unless electron heating occurs.

C. *Non-collective ionization processes*

In this section we consider some simple ionization processes in a cold and stationary gas with perpendicular electric and magnetic fields. When $(\omega\tau)_e \gg 1$ and electron collisions with ions and excited atoms can be ignored, then the zeroth order electron velocity distribution for electrons moving through a cold gas under the influence of perpendicular electric and magnetic fields is a function of E/B alone (Allis 1956). Consequently the normalized electron ionization frequency ν_{ie}/n_g is also a function of E/B alone. This has been calculated by Lehnert (1966) for helium on the assumption of a Maxwellian distribution and his results are shown in Fig. 5. The curve of ν_{ie}/n_g is almost parallel to the ordinate axis when $E/B \simeq 2 \times 10^4$ m/s.

It is well known that a single electron has a much higher ionization frequency than an equal energy ion, unless the energy is very large. However, for the

Maxwellian distribution which leads to the results of Fig. 5, only a small fraction of electrons have energies in excess of the ionization threshold. On the other hand, for singly ionized ions moving through their parent atomic gas (symmetric resonant charge exchange collisions), if the Hall parameter is large, i.e. $(\omega\tau)_i \gg 1$, then every ion can have energies in excess of the ionization threshold over a fraction of its orbit. The possibility then occurs that, compared to the electrons, the relatively large number of ions above ionization threshold can counteract their relatively small ionization frequency.

The measurement of the small ion-neutral ionization cross-sections at the energies of interest here is difficult, and serious discrepancies exist in the literature. Gilbody and Hasted's (1957) measurements for He^+ in He are around ten times larger than the same measurements of Rostagni (1936). On the other hand Hayden and Utterback's (1964) results for He^+ in He are 5 to 100 times smaller than Rostagni's for He^+ in He, and in general the ionization cross-section for ions and neutrals are not widely different (Rostagni 1936; Massey and Burhop 1952).

With symmetric resonant charge exchange collisions, the ion velocity v is given by $v = 2U|\sin\alpha|$ where $\alpha = \omega t/2$, and $U = E/B$. Under free orbiting conditions the ion's ionization frequency ν_{ii} is then given by

$$\nu_{ii}/n_g = (2U/\pi)\int_0^{\pi} q_{ii}(2U\sin\alpha)\sin\alpha\, d\alpha.$$

Here $q_{ii}(v)$ is the ion-neutral ionization cross-section, and, using the data of Rostagni (1936) these integrals have been carried out graphically for helium, the results being shown in Fig. 5. It can be seen that in this case the ions contribute more to the total ionization than the electrons, but three important points must be kept in mind. Firstly the calculation of ν_{ie} ignores ionization from excited states and assumes a Maxwellian distribution. Secondly the calculation of ν_{ii} is restricted to the ion velocity distribution produced by symmetric resonant charge exchange conditions when $(\omega\tau)_i \gg 1$. Thirdly large discrepancies exist in the values of the ion-neutral ionization cross-section.

As well as the straightforward impact ionization considered above, there are two other mechanisms whereby the ions can contribute to the ionization. Firstly, it has been observed (Utterback 1964; Hangsjaa, Amme and Utterback 1969) that a small fraction of symmetric resonant charge exchange collisions result in the production of a metastable atom. This process can occur at relatively low impact energies. Since a metastable atom can then be rapidly ionized by low energy electrons this process might also give a significant contribution to the total ionization. Secondly the normal ground state atoms formed in symmetric resonant charge exchange collisions have roughly the same average energy as the ions. Since the ionization cross-sections for ion-neutral and neutral-

neutral collisions are of the same order of magnitude, the fast neutrals can also make a significant contribution. When comparing the ionization rates of the ions and fast neutrals the major factor is their relative number densities. The relative number densities vary from case to case, but, for example, in the experiments of Fahleson (1961) and Angerth et al. (1962) the fast neutral density is typically larger than the ion density (Sherman 1970*a*). However, because of the restriction to large $(\omega\tau)_i$, the heavy particle ionization processes mentioned above are probably only important in these experiments for helium at high magnetic fields and low pressure.

The non-collective ionization processes discussed here might be important in some critical velocity situations. The observed critical velocity effects can then be interpreted as simply a consequence of the fact that the total ionization frequency is a rapidly varying function of E/B which quickly becomes extremely small when E/B is reduced much below $(2eV_i/M)^{1/2}$. This can be seen from Fig. 5 and other atomic gases have a similar variation of ionization frequency. However for molecular gases the ν_{ie}/n_g curve lies further to the right of $(2eV_i/M)^{1/2}$ owing to the energy losses to nuclear vibration and rotation. On the other hand some critical velocity experiments, notably Danielsson's, require a faster ionization frequency than the simple non-collective processes mentioned here can provide.

IV. *Relevance of Critical Velocity Experiment and Theory to Alfvén's Hetegonic Theory*

The rôle of the critical velocity in Alfvén's hetegonic theory (Alfvén 1954) is that it determines the region of space in which the matter which later forms the secondary body accumulates. Alfvén proposed that the magnetic field which surrounds the central body supports a plasma, and when an in-falling gas cloud reaches a velocity $(2eV_i/M)^{1/2}$ it is ionized. The ionized gas is then stopped by the magnetic field. The gas falls from infinity under the influence of the central body's gravitation, so that the gas is then stopped when its distance from the central body is near to $\varkappa M_c M/eV_i$. Here $\varkappa$ and M_c are the gravitational constant and mass of the central body respectively.

It is not necessary that in-falling gas be completely ionized by the critical velocity process. Firstly, if the mean free paths are short enough, even a partially ionized plasma can be supported by the magnetic field. Secondly, large amounts of energy are released when any partially ionized plasma is accelerated during the partial co-rotation process (Alfvén 1954; Alfvén 1967), and this energy can increase the degree of ionization. The critical velocity process can then be regarded as only the first stage of a more complicated process, the net result of which is to accumulate matter in a region determined by the critical velocity.

Although the conditions prevailing during the formation of satellites according to Alfvén's theory are rather speculative, some features are reasonably certain. It is likely that the Hall parameters $(\omega\tau)_i$ and $(\omega\tau)_e$ are both very much greater than unity, and similarly the ion and electron Larmor radii are likely to be much smaller than the other dimensions of interest. The gas and ions are mostly atomic, and the ions are mostly singly ionized. However the gas and plasma will not in general consist of only one atomic species.

1. Relevance of the Laboratory Experiments

Of the three experiments discussed in Part II, only Eninger's (1965) comes close to Alfvén's hetegonic situation. As pointed out in Part II Section 1 the ion Hall parameter in Angerth et al.'s (1962) experiment is typically less than unity, and so these experiments are not directly applicable to the envisaged hetegonic situation.

Danielsson's (1970*b*) experiment is also not directly relevant, the major difficulties arising from the change in the frame of reference that is necessary. In the envisaged hetegonic situation a plasma is held relatively stationary in a magnetic field, and a falling neutral gas is ionized by an interaction with the plasma. In other words, in a frame of reference in which the interaction region is stationary the gas is moving (with the critical velocity) and the plasma is not. In contrast, in Danielsson's experiment the interaction region is relatively stationary in the laboratory frame and in this frame the gas is stationary and the plasma is rapidly moving. Furthermore the plasma is rapidly decelerated in Danielsson's experiment, i.e. the plasma motion is strongly non-uniform. Also in this experiment the Hall parameter for the He^+ ions is approximately three (owing to their large charge exchange cross-section), and the Larmor radius of these ions can be, depending on their position of formation, comparable to the tube diameter.

In one part of the parameter range used, i.e. low pressures, high magnetic fields, and not too high currents, Eninger's (1965) experiment using helium conforms with several of the important features of the hetegonic situation. A straightforward change in the frame of reference to one moving with the current sheet gives a situation similar to that envisaged. The maximum value of $(\omega\tau)_i$ is around 10 and the ion Larmor radius is small compared to the gun radius. It should also be noted that in this experiment the azimuthal magnetic field does not intersect any walls. On the other hand there are radial currents which correspond to azimuthal currents in the hetegonic case (see below). Under the above conditions the plasma velocity through the neutral gas is limited to

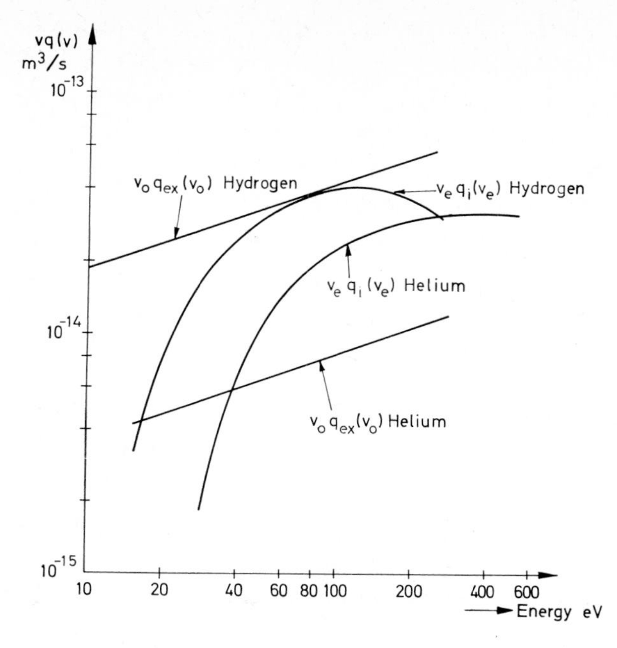

Fig. 6. Ionization and charge exchange cross-sections for atomic hydrogen and helium.

$(2eV_i/M)^{1/2}$. This is the most direct experimental evidence available in support of Alfvén's use of the critical velocity concept in his hetegonic theory.

2. Relevance of the Theories

The theories which are based on a particular experiment (Part III Section 1) cannot be of importance here. Of the three more general theories discussed above, the two-stream instability approach does not seem to be applicable in the hetegonic case since the large Hall parameter and small Larmor radii make it unlikely that a relative electron-ion drift occurs.

Turning to Lehnert's non-thermal ion space charge theory (Lehnert 1967), an important point arises from the uniformity assumed by Lehnert. In the situation analysed by Lehnert currents flow in the y direction (see Part III Section 2) which corresponds to the azimuthal direction in the hetegonic case. Lehnert assumes that conditions are infinite and uniform in this direction so that in his situation these currents have no effect. However, azimuthal non-uniformities are likely in the hetegonic case, and if Lehnert's mechanism is to be relevant in this case azimuthal currents must still flow. Currents can exist in the accumulating plasma even in the presence of very strong plasma density changes if the circuit can be completed by currents along the magnetic field lines to the central body, and a linking current near the central body.

Another point is that Lehnert's process cannot on its own lead to the accumulation of a large fraction of infalling clouds of hydrogen or helium, the two elements with the largest cosmic abundances. The relative probabilities of scattering by a charge exchange collision or ionization by an electron for a neutral atom moving through the plasma in the situation envisaged by Lehnert are approximately $v_0 q_{ex}(v_0) : \langle v_e q_i(v_e) \rangle$. Here v_0 is the atom velocity, q_{ex} is the charge exchange cross-section, v_e is the electron thermal velocity, and q_i is the electron ionization cross-section. Fig. 6 shows curves of $v_0 q_{ex}(v_0)$ and $v_e q_i(v_e)$ versus the neutral atom energy $Mv_0^2/2e$. All electrons are taken to have the maximum energy predicted by Lehnert's process; i.e. $Mv_0^2/2e$. It can be seen from Fig. 6 that near the critical energy an atom is more likely to be scattered by charge exchange than ionized for these two gases. Thus if most of the neutral gas is accumulated the ions that are eventually formed are thermal in character and Lehnert's process cannot apply. However, non-thermal ions can still be formed if most of the gas passes straight through the tenuous plasma. Lehnert's process can then act as the first stage of the more complicated process mentioned above.

Turning to the non-collective ionization process, a change in the frame of reference in the hetegonic case to one moving with the neutral gas then gives a situation which is similar to that assumed in Part III Section 3. The simple ionizing processes discussed in that section could then be important in the hetegonic case. Some simple calculations based on the situation envisaged by Alfvén (1954) do in fact show that these simple ionizing processes could well provide a sufficiently high ionization frequency.

V. *Conclusions*

It can be seen from the discussion in Part II that a wide range of laboratory experiments have exhibited critical velocity effects. These experiments are of several completely different types and cover a wide range of pressure and current. Similarly the theories discussed in Part III also support the critical velocity idea. However most of the theories are not directly applicable to the experiments. To date none of the experiments has been satisfactorily explained in detail, although several theoretical approaches seem to be promising. Furthermore the wide variety of experimental conditions under which the critical velocity effects occur indicates that several different theories are necessary.

With regard to possible hetegonic applications, Eninger's (1965) experiment comes closest to providing evidence that critical velocity effects can occur in the assumed hetegonic situation. Most of the theories run into difficulties when an attempt is made to apply them to the assumed hetegonic situation but simple non-collective ionization processes seem to be adequate in this case. In addition the wide variety of experiments and theoretical processes which exhibit critical velocity effects strongly suggest that the critical velocity could play a part in the hetegonic process.

References

Alfvén, H., 1954, On the Origin of the Solar System. Oxford University Press.
Alfvén, H., 1960, Rev. Mod. Phys., *32*, 710.
Alfvén, H., 1967, Icarus, *7*, 387.
Alfvén, H. and Arrhenius, G., 1970, Astrophys. Space Sci., *8*, 338.
Allis, W. P., 1956, Handbuch der Physik, *21*, 414. Springer-Verlag.
Angerth, B., Block, L., Fahleson, U. and Soop, K., 1962, Nucl. Fusion Suppl., Pt. 1, 39.
Beard, D. B., 1965, J. Geophys. Res., *70*, 4181.
Buneman, O., 1959, Phys. Rev., *115*, 503.
Buneman, O., 1961, Stanford Electronics Lab., Tech. Report No. 251-1, Stanford University, Stanford.
Chapman, S. and Cowling, T. G., 1960, The Mathematical Theory of Non-Uniform Gases, Chapter 5. Cambridge University Press.
Colgate, S.A., 1961, Univ. Cal. Lawrence Radn. Lab. Report, UCRL-6176.
Danielsson, L., 1970*a*, Dept. of Electron and Plasma Physics Report 70-05, Royal Institute of Technology, Stockholm.
Danielsson, L., 1970*b*, Phys. Fluids, *13*, 2288.
Drobyshevskii, E. M., 1964, Soviet Phys., Tech. Phys., *8*, 903.
Drobyshevskii, E. M., 1967, Soviet Phys., Tech. Phys., *11*, 870.
Drobyshevskii, E. M. and Rozov, S. I., 1967, Soviet Phys., Tech. Phys., *11*, 878.
Eninger, J., 1965, Proc. VII Int. Conf. on Phenomena in Ionized Gases, Belgrade, *1*, 520.
Fahleson, U. V., 1961, Phys. Fluids, *4*, 123.
Gilbody, H. B. and Hasted, J. B., 1957, Proc. Roy. Soc., A *240*, 382.
Hayden, H. C. and Utterback, N. G., 1964, Phys. Rev., *135*, A1575.
Lehnert, B., 1966, Phys. Fluids, *9*, 774.
Lehnert, B., 1967, Phys. Fluids, *10*, 2216.
Lehnert, B., Bergström, J. and Holmberg, S., 1966, Nucl. Fusion, *6*, 231.
Lin, S. C., 1961, Phys. Fluids, *4*, 1277.
Massey, H. S. W. and Burhop, E. H. S., 1952, Electronic and Ionic Impact Phenomena, Chap. 8. Oxford University Press.
Rosenbluth, M., 1957, Magnetohydrodynamics, p. 57 (ed. Landshoff), Stanford University Press.
Rostagni, A., 1936, Il Nuovo Cim., *13*, 389.
Sherman, J. C., 1967, D.Phil. thesis, Oxford University.
Sherman, J. C., 1969, Dept. of Electron and Plasma Physics Report 69-29, Royal Institute of Technology, Stockholm.
Sherman, J. C., 1970*a*, Dept. of Electron and Plasma Physics Report 70-30, Royal Institute of Technology, Stockholm.
Sherman, J. C., 1970*b*, Dept. of Electron and Plasma Physics Report 70-15, Royal Institute of Technology, Stockholm.
Sherman, J. C., 1970*c*, Dept. of Electron and Plasma Physics Report 70-14, Royal Institute of Technology, Stockholm.
Simon, A., 1959, General Electric Research Lab. Report, No. 59-RL-2322E.
Sockol, P. M., 1968, Phys. Fluids, *11*, 637.
Utterback, N. G., 1964, Phys. Rev. Lett., *12*, 295.
Hangsjaa, P. O., Amme, R. C. and Utterback, N. G., 1969, Phys. Rev. Lett., *22*, 322.
Wallis, M., 1971, Private communication.
Wilcox, J. M., 1959, Rev. Mod. Phys., *31*, 1045.
Wilcox, J. M., Pugh, E. R., Dattner, A. and Eninger, J., 1964, Phys. Fluids Suppl., *7*, S51.

Discussion

H. Alfvén

May I just point out that if M in Sherman's formula $V_{crit} = (2eV/M)^{1/2}$ is the electron mass, it is nothing but Frank and Hertz' experiment. The interesting thing is that it is the ionic or atomic mass. This may motivate that we call this the "plasma version of the Frank and Hertz' experiment". It seems to be almost as fundamental.

B. Lehnert

First I wish to congratulate you to this interesting review. In addition I should like to make two comments:

1. In our experiments at the Royal Inst. of Technology on fully ionized rotating plasmas, there is strong evidence that the critical velocity phenomenon is very sharply defined and localized within a thin neutral gas layer, less than 1 mm in thickness at the end walls of our magnetic bottles. Since the dimension of the interaction region in the magnetic field direction is very small in this particular case, there may arise some difficulty in explaining the plasma behaviour by the two-stream instability theory.

2. You mentioned that my theory on the space-charge effects may have difficulties in the cosmical case, on account of the boundary conditions involved. However, the special calculations carried out on this theory by means of a computer could be generalized to situations which better correspond to the cosmical case, without losing the general physical features of the mechanism involved. Also the time scale of development of these effects is important in this connection. It is therefore not certain that one could rule out this as a possible way of explaining the critical velocity phenomenon in cosmos.

J. Sherman

With regard to your first point, your experiments demonstrate very well the importance of neutral gas in the critical velocity phenomena. I have not been able to discuss them in detail here due to lack of space. You are right that the dimension in the magnetic field direction is an important factor in the two-stream instability approach. However no detailed calculations on the application of this theory to your experiment have yet been carried out. It should be noted that the fastest growing wavelength can be small for high electron densities, and also that smaller wavelengths than the optimum also grow rapidly.

With regard to your second point, I agree that a less specialized theory might avoid the difficulties I outlined.

F. L. Whipple

For space application of the critical velocity effect I wonder what are the limitations as to the space density of volume of the atoms involved?

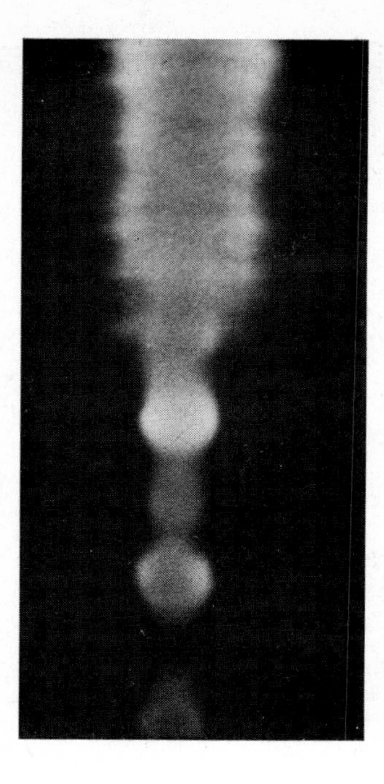

Fig. 2. Hα spectrogram of a prominence filament (Feb. 19, 1968) showing circular Doppler features indicating rotation.

been shown by me before (Öhman 1955) that in order to explain some absorption phenomena of prominences it seems necessary to assume that the prominences contain sometimes regions of rather cold gas embedded in a hot gas of about the same pressure. In fact some observations indicate that this hot gas may be sometimes more or less invisible due to very high ionization (corona infiltration). The effect we observe in Fig. 1 may be of a related type. The appearance of the neutral gas is particularly interesting with respect to Lehnert's recent results (Lehnert 1970) that such a gas, when present in a plasma, may produce abrupt disappearances, as well as rotation.

Fig. 1 seems in my opinion to present some new examples of solar plasma phenomena similar to those studied before by Alfvén and his collaborators in the laboratory. I have shown before (Öhman 1968*b* and 1969) that the plasma ring experiment (Lindberg et al. 1960) may be related to some ringshaped Doppler structures observed by me in prominences. Fig. 2 shows such structures observed by me on February 19, 1968 at the Swedish Solar Observatory in Anacapri. If a rotating smoke ring is observed edgewise and with the edge on the slit such circular images may result. A rapid contraction or expansion of gas may produce similar images (Öhman 1970). But a third possibility seems to be offered by the *pinch effect*. According to H. Persson (Hopfgarten et al. 1971) the angular momentum is conserved if a rotating plasma is subject to a pinch effect. Suppose that the prominence in Fig. 2 (lined up with the solar limb) is a rotating filament imaged on the slit, then the circular features may well be places where pinch effects appear, resulting in increased Doppler rotational displacements if the angular momentum is conserved.

A more detailed discussion will be presented at the Prominence Colloquium on Sept. 29, 1971 at Anacapri.

References

Alfvén, H. and Carlqvist, P, 1967, Solar Physics, *1*, 220.
Carlqvist, P., 1968, Nobel Symposium 9, 193.
Hopfgarten, N., Johansson, R. B., Nilsson, B. H. and Persson, H., 1971, IV Europ. Conference, Contr. Fusion and Plasma Physics, Rome 1970.
Lehnert, B., 1970, Cosmical Electrodynamics, *1*, 397.
Lindberg, L., Witalis, E. and Jacobsen, C., 1960, Nature, *185*, 452.
Öhman, Y., 1955, Stockholms Obs. Meddel., No. 71.
Öhman, Y., 1968*a*, Symposium No. 35 of the IAU, 240.
Öhman, Y., Hosinsky, G. and Kusoffsky, U., 1968*b*, Nobel Symposium 9, 95.
Öhman, Y., 1969, Solar Physics, *9*, 427.
Öhman, Y., 1970, Scientia, *105*, (Ser. 7), 3.

Discussion

B. Lehnert

You have mentioned that in some cases there is a relative velocity between the neutral gas and the rotating solar plasma of the order of 20 km/sec., and that the neutral gas then disappears suddenly. This could possibly be connected with the critical velocity phenomenon and its associated ionization process.

Y. Öhman

This is a very interesting suggestion. I take this opportunity to thank Professor Lehnert for many valuable discussions of theoretical aspects on rotational phenomena on the Sun.

B. A. Lindblad

You mentioned this neutral gas having velocities of the order of 12 or 13 km/sec. What is the velocity measured relative to and what is the direction of the velocity vector?

Y. Öhman

What we observe in the Al I line is probably the motion of the area of ejection on a tube-shaped rotating flare object.

S. K. Runcorn

Could you comment on the sense of the rotation? It is of interest whether the Coriolis force is of importance in these motions. And is there any association with sunspots?

Y. Öhman

The first spectroscopic observation of rotational motion in prominences was made by Ellison (1947). In an eruptive arch he found scalar velocities of the order of 200 km/s. My observations of quiescent prominences show that rotational motion is sometimes present in such objects too but with scalar velocities of the order of 10 km/s only.

Ellison, M. A., 1947, The Journal of the British Astronomical Association, *57*, 229.

A "Cometary" Suggestion

By D. Lal

Tata Institute of Fundamental Research, Bombay, India and
Scripps Institution of Oceanography, La Jolla, California

The purpose of this discussion is to consider one of the processes which takes place when a comet comes close to the Sun, $\leqslant 1$ A U, namely a possible thermalisation of the impacting polar plasma. In order to obtain an idea of the scale of the process, let us calculate how much solar material, M, impinges on a coma of cross-sectional area A in time t:

$$M_r = A \cdot F_r \cdot t \tag{1}$$

where F denotes the solar-wind ion flux and the subscript r denotes the distance of coma from the Sun. With $F_{1\ \mathrm{AU}} = 2 \times 10^8$ protons/cm^2 sec., and assuming the diameter of the coma to be 10^5 km, one obtains:

$$M_{1\ \mathrm{AU}} \simeq 2 \times 10^{28}\ \text{protons} \cdot \text{sec}^{-1}$$

$$M_{0.3\ \mathrm{AU}} \simeq 2 \times 10^{29}\ \text{protons} \cdot \text{sec}^{-1}$$

In spite of the approximations involved, the value of M is clearly of an order of magnitude that it seems necessary to consider the importance of thermalisation of impinging solar wind ions in a coma. The magnetohydrodynamical processes which will be operative here are not clearly understood but some educated guesses can be made in the framework of theoretical (Alfvén 1964) and experimental work (Danielsson 1970; Sherman 1971) on the behaviour of plasma in comparable situations. Collective interaction between plasma and groups of particles, similar to the space charge produced by potential fluctuation outlined by Lehnert (1967) may lead to appreciable energy losses. Thus, thermalisation of plasma due to electrical fields may be quite important but this point has to be examined in greater detail.

If thermalisation of solar plasma occurs efficiently in a coma, the solar wind —comet interaction may constitute an important phenomenon at distances $\leqslant 1$ AU and it seems likely that a part of the gas observed streaming out from comets near perihelion is not intrinsic to these objects.

This note is a culmination of some remarks made by me during discussions at the conference. I am grateful to Profs. Alfvén and Arrhenius for encouragement to write up this suggestion. I would also like to acknowledge my appreciation to Profs. Anders, Mrkos, and Whipple for stimulating discussions and to Drs Lars Danielsson and Lennart Lindberg for helpful remarks.

References

Alfvén, H., 1964, "On the Origin of Solar System", Chapter III. Oxford University Press, Oxford.
Danielsson, L., 1970, Phys. Fluids, *13*, 2288.
Lenhert, B., 1967, Phys. Fluids, *10*, 2216.
Sherman, J. C., 1971, proceedings (this conference); also see Report No. 70-30, 1970 (Royal Inst. Tech. Stockholm, Sweden).

Note Added After the Conference

Recent observations of comets by satellite borne instruments have led to the discovery of existence of huge atmospheres of atomic hydrogen (visible in the resonance line Lyman α 1216 Å) around the heads of the two bright comets of 1970, Bennett and Tago-Sato-Kosaka (L. Biermann, 1971, Nature, *230*, 156; See also L. Biermann, "Comets in the Solar Wind": Preprint (1971)). In the case of comet Bennett, whose head is estimated to have the usual diameter of several 100,000 km, Biermann estimates the production rate of hydrogen to be of the order of 10^{30} atoms/sec. Thus, for most of the observed hydrogen to be of a solar-wind origin, the efficiency for thermalisation in the coma would have to be close to 1.

Discussion

T. Gehrels

Could you observe the effect on the Moon?

D. Lal

The cross-sectional area is about 300 times smaller and I do not know whether it would be possible to see the effect.

V. Vanýsek

Maybe I misunderstood something in Dr Lal's paper. If we take into consideration the cross-section of the particles and the density of the particles, then we get a much lower number of captured protons. I made some rough calculations. Say that the mean density of particles is 10^{-6} cm^{-3} and the cross-section of such particles is 10^{-8} cm^{2}, then the effective cross section of the coma is much smaller than given by Dr Lal and we get a smaller number of protons. I should like to describe the H I coma of two comets. Unfortunately I have not the data concerning 19 observations of Comet Bennett made by Professor Blamont, so I must recollect it from my memory.

For Comet Bennett, the visible coma, CN, C_2 and dust, is about 10^6 km or so, say 5×10^5 km. The general feature of a H I cloud is shown in Fig. 1. The H I atmosphere around the comet looks like such a pear. The dimensions are 10^7 km

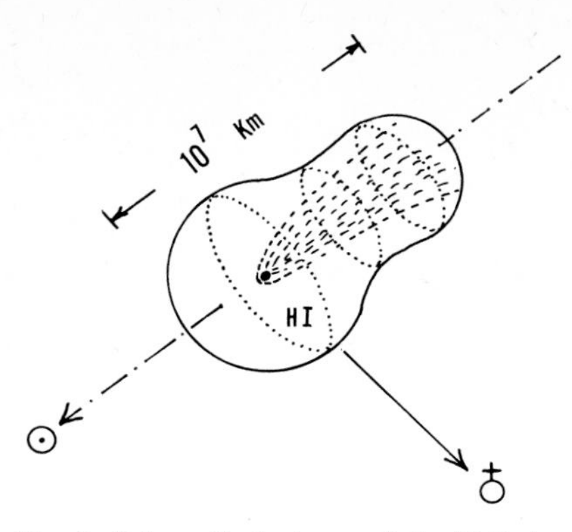

Fig. 1. Schematic features of the H I cloud around the Comet Bennett 1969*i* (first week in April 1970). The boundary of "pear" shape of the hydrogen cloud correspond to the intensity of about 1 kRay in Ly-α. The typical visible appearance of the comet is inside the H I cloud.

(length). The boundary isophote corresponds to one kilorayleigh. The maximum intensity near the nucleus was between 10–25 kRay. The dimension of the dense H I coma is about several million kilometers. (I am not sure of the figure but I think it may be some 4×10^6 km.) This means that here it may be some anisotropic expansion of H I from the nucleus. The hydrogen very far from the nucleus is hard to explain by the acceleration of the hydrogen atoms due to the light pressure even if the value of the acceleration is 0.3 cm sec^{-2}. (Nevertheless, through the dissociation process from H_2O the hydrogen atom gains the energy of at least 1 eV which means the velocity of 13 km sec^{-1}.) We cannot explain the H I tail. One speculation is that the dust carried hydrogen out there. But it may be some effect of the kind Dr Lal described here. However, it is also interesting that the second comet (Tago-Sato-Kosaka 1969*g*) 2 months before this one, was observed last year also in Lyman-alpha by Code with the OAO-II satellite. It showed that the intensity of the H I radiation was about the same (maybe by a factor of 2 lower). *But the dust content was at least 10 times lower or perhaps 100 times lower.* Perhaps Dr Lal's mechanism can work in some part of the comet. Maybe we have 2 components, 2 sources of the H I radiation? More calculations should be made taking into account the *effective cross-section* of the coma.

M. Wallis

We should compare the figure of 10^{36} hydrogen atoms per year with the figure 10^{36} per day or few days required for comet Bennett's hydrogen coma. For the time scale for existence for H atoms in the coma cannot be greater than a few days. Simple radiation pressure gives an acceleration out of a region 10^6 km across in 10^6 sec and charge exchange interaction with the solar wind protons destroys the atoms in 2×10^5 secs. Even apart from the difficulties involved in maintaining ice particles over as large a scale as 10^5 km (for neutralization of the protons), the proton supply would be too small by at least three factors of ten (two factors of ten with the actual production rate of H in comet Bennett, namely 10^{35} per day).

F. L. Whipple

Dr Lal's discussion covers one of the most exciting results of plasma physics. Without plasma physics we can explain neither the ionization nor the motions of Type I (or ion) comet tails. The solar wind, moving at supercritical velocity, produces a collisionless shock wave in the neutral gases in the outer coma of a comet by ionizing some of the gas escaping from the nucleus. The momentum of the solar wind carries the ions away from the Sun and coma to produce the classical ion tail of a comet. The neutral atoms and molecules appear not to be very much affected. I am very pleased at this triumph of the plasma concept.

On the Existence of a Resonance-Captured "Quasi-Satellite" of the Earth

By L. Danielsson and W.-H. Ip[1]

Department of Plasma Physics, Royal Institute of Technology, Stockholm 70

The present orbital elements of the asteroid 1685 Toro shows that its period is almost exactly 1.6 years. Further the relative positions of Toro and the Earth exhibit a certain symmetry including close approaches. This quasi-resonance—if we may call it so—has been investigated more in detail by integrating the perturbed Toro-orbit for 100 years backwards and forwards, i.e. totally 200 years. The perturbation by Venus, Earth, Mars, Jupiter and Saturn are included in the orbit integration according to Cowell's method. The accuracy of the computation is extremely high; after 100 years which requires 10^4 integration steps the relative error is about 10^{-8} (absolute error in position 10^{-5} AU).

Naturally a time-span of 200 years is insufficient for determining whether Toro is caught in a real resonance. Since, however, we know that Toro comes at least as close as 0.1 AU to the Earth an investigation based on the secular variation of the elements is considered rather unreliable.

The present orbits of Toro and the terrestrial planets are oriented as shown in Fig. 1. The possible positions of the Earth and Venus when Toro is at perihelium are marked. These positions are quasi-stable. The orbit of Toro relative to the Earth, i.e. in a coordinate system rotating with the Earth, is shown in Fig. 2. This particular loop is produced during eight years around year 2020. The loop is not exactly closed, instead it can be said to oscillate around its center (the Sun) with an amplitude of 9° and a period of 144 years. In the figure this is best illustrated by keeping the 8-year Toro loop fixed and letting the Earth oscillate along an 18°-arc. At the turning points the distance between Earth and Toro is 0.13 AU (August 12, 1964 and January 17, 1896 & 2040). At the August (every 8th year) close encounter, in the vicinity of its ascending node, Toro is always at least 2°.5 ahead of the Earth and 7° above the ecliptic plane. Thus Toro is retarded, i.e. the angular momentum and the period are decreased, which tends to move it away from the Earth and the encounter will be more and more distant. Instead Toro will approach the Earth in the vicinity of its descending node (January every 8th year). At the January encounter Toro is always at least 6°.9 behind the Earth and 3°.2 below the ecliptic plane. Toro

[1] Permanent address: Department APIS, University of California, San Diego, La Jolla, California.

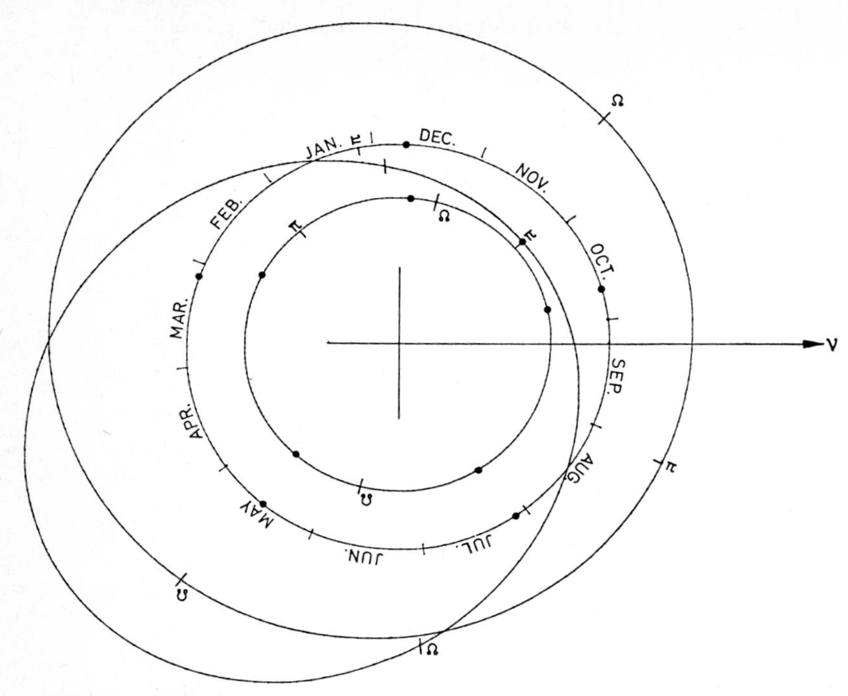

Fig. 1. The orbit of the terrestrial planets and Toro. The positions of Venus and Earth when Toro passes its perihelium are marked. The data used are from the perihelium epoch of December 1967 and four following perihelia.

is accelerated, its period increased and again it tends to move away from the Earth at this point to approach it in the vicinity of the ascending node (August). This process can be seen in the variation of the period, Fig. 3. The decaying slope is the result of upward steps every 8th year. Fig. 4, showing the variation of the longitude, is also illustrative.

The described configuration would be favourable for a stable resonance but two circumstances make us foresee that Toro might get out of the coupling to the Earth. The first is that Toro's orbit is at a rather large distance from the ecliptic plane at the close encounters. At the August encounter the lateral distance is three times the longitudinal. Secular variations or other perturbations might bring Toro to the "wrong side" of the Earth here. Secondly Toro has close encounters also to Venus. During our integration the closest approach is about 0.15 AU and at present the effect of Venus is to help keeping Toro in resonance. As Venus moves out of this configuration (see Fig. 1) it might very well either "capture" Toro from the Earth or perturb it out of resonance altogether. This will, however, take a very long time due to the near commensurability of the periods.

In summarizing we state that the 200-year accurate orbit calculation has given

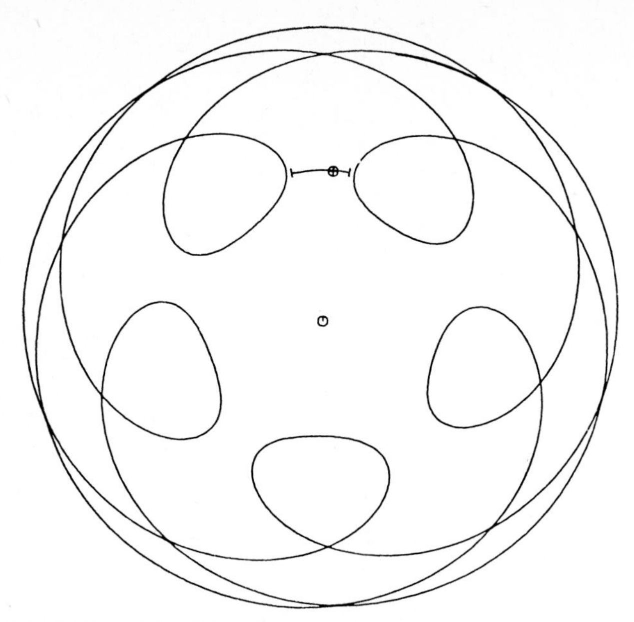

Fig. 2. The orbit of Toro in a coordinate system rotating with the Earth. The fact that Toro's period oscillates around 1.6 years is shown by superimposing an oscillation on the rotation of the coordinate system. This makes the Earth oscillate along an 18° arc with a period of 144 years.

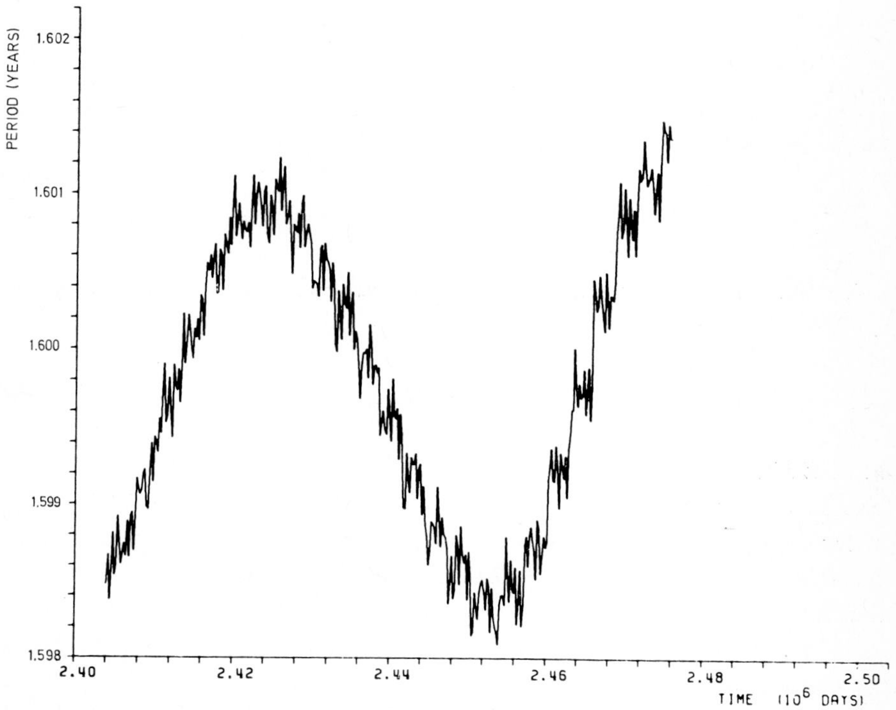

Fig. 3. Variation of Toro's period. The time is in Julian Date.

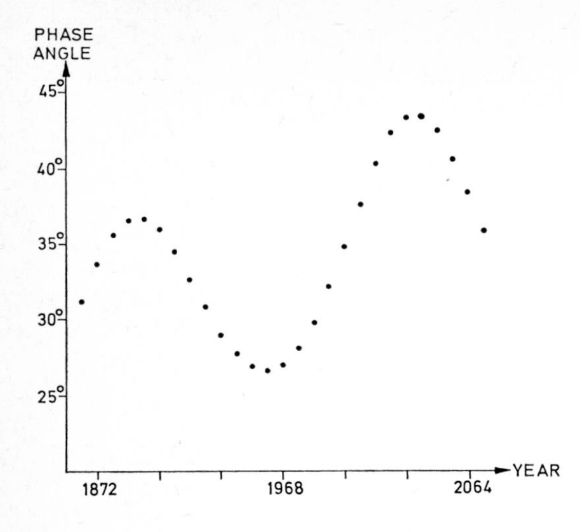

Fig. 4. The difference in longitude between Toro and the Earth at the perihelium in October every 8th year.

good evidence that the asteroid 1685 Toro is in, at least, a temporary 5/8 resonance with the Earth. The amplitude of the libration around the equilibrium is about 9° and the period of libration is 144 years. Toro is also very close to a 5/13 resonance with Venus.

One of the objectives for studying orbits like Toro's in detail is that this is one of the possibilities for a pre-capture orbit for the moon (Alfvén and Arrhenius 1969). Another is that the existence of orbits like Toro's may be important for determining the lifetimes of Apollo asteroids.

The authors are indebted to Mr N. Carlborg of Stockholm Observatory at Saltsjöbaden for many helpful discussions and kindly making his Cowell program available for the orbit integration and to Mr R. Mehra for assistance in the computer work.

Reference

Alfvén, H. and Arrhenius, G., 1969, "Two Alternatives for the History of the Moon"; Science *165*, 11.

Discussion

F. L. Whipple

The connection between the motion of Toro and the origin of comets may be obscure to some of you. The question is whether the Apollo, or Earth-orbit crossing asteroids, are true asteroids or old comet nuclei. Öpik has shown that the average lifetime of Earth-crossing objects is the order of 10^8 years. There seem not to be enough such asteroids near Mars to maintain the supply of Apollo asteroids by perturbations. Hence many or most may be old comet nuclei. If the Apollo asteroids tend to be in stable orbits with lifetimes much longer than 10^8 years, then they may consist principally or entirely of true asteroids. For many years I have considered the evidence inadequate to settle the question. Perhaps Danielsson can do so.

H. Alfvén

I think that what Danielsson says here is correct, namely that one cannot be sure of how this will develop in a very long time. But as he is very careful he stressed a little too much that it may very well be that it is not permanent. As far as I can see there is nothing in his calculations which does not say that it may also be permanent.

L. Danielsson

Yes, but unless Venus and Earth also are in resonance I know that Venus will come even closer to Toro than the Earth does at these points.

T. Gehrels

I have three comments or questions. Firstly, perhaps Dr Anders could comment on the large number of Hungaria asteroids. Secondly I wonder if Dr Danielsson could comment on how this relates to the Moon, because the Moon has a similar possibility, which I think you pointed out yourself, of perhaps having been captured. The third question is: What can we do with the beautiful opportunity in 1972 to make physical observations and study the question whether Toro is a cometary or asteroidal object. I can myself give at least a partial answer. At least we can get a light-curve, spin rates, and an approximate determination of the shape of Toro.

L. Danielsson

Answering the last question first I can say that next year is a good opportunity to study Toro. It comes close and it does not move nearly as fast as Icarus did. So you have a much better possibility with this object. The other question Hannes Alfvén may answer better.

H. Alfvén

The problem of the capture of the Moon is a long story.

D. Lal

Are there other objects than Toro which could be similarly studied?

L. Danielsson

We know only very few Apollo asteroids with well established orbits. Of these Toro is the only one with a period which is commensurable with the Earth's.

One more has similar characteristics, namely 1948EA, which is a very uncertain object according to a table which Tom Gehrels showed earlier in the symposium.

E. Anders

Yes, the discovery of so many Hungaria asteroids shows once again how incomplete the count of Mars-crossing asteroids is. Most Mars asteroids have high inclinations and are therefore often missed in asteroid searches centered on the ecliptic. This is borne out by the fact that most Mars asteroids have high numbers; evidently their discovery has lagged behind that of the other asteroids. Among the last few hundred numbered asteroids, Mars and Apollo asteroids comprise about 5 %, if I remember correctly.

In spite of my great admiration for Öpik, I disagree strongly with his conclusion that Mars asteroids are not an adequate source of Apollo asteroids. I hope to discuss the matter in print some day.

B. A. Lindblad

Which planets were included in Dr Danielsson's calculations?

L. Danielsson

Venus through Saturn.

has steadily increased in strength. I have tried to pick a term which will be best understood by the greatest number of people, and it does depend on the physical observation of the dynamics of the situation. I agree in general with what Professor Alfvén says, but at the moment I think that the term "cometary meteoroid" is a suitable one.

H. Alfvén
Yes, I said "meteorites" and I am not quite sure whether this was correct. In your case you refer to something which is related to observations, namely that you have certain classes of objects which move in the same orbits. And in that case I have no objection.

P. M. Millman
I am simply following our IAU terminology which defines a meteorite as something that has fallen to Earth, and which can be studied in the laboratory. A meteoroid is the object in space.

H. Alfvén
Yes, and in the case the orbits really are similar, I have no objection to that.

Z. Kopal
I should like to comment on Professor Alfvén's discussion of some terms. I agree with most of his views, but should like to point out that the word "old" often has little meaning. It may possess definite meaning for evolving systems, but usually we talk about "old" systems when we mean "time-independent" systems which have already attained the state of equilibrium.

I have a more important comment on Professor Alfvén's diagram. I think your insistence on experiments as intermediate link between theory and observations should be qualified. This link is very useful, and in fact indispensable, whenever phenomena are essentially linear—so that you can scale up laboratory results to the scale you observe in the universe. You are very fortunate in having chosen a field which is more linear than others.

H. Alfvén
Is plasma physics linear? Definitely not.

Z. Kopal
More linear (Maxwell's equations!) than some others (like celestial mechanics, for instance), which are non-linear from the very beginning. This is why we should underline the importance of space experiments, because in space alone we are in a position to bridge the gap between laboratory experiments and the actual scale on which the phenomena occur.

My last comment on theory and observation being separated is that each one must offer inspiration for the other. Observations without inquiry into their meaning would be sterile, just as a theory which does not seek confrontation with the observations. In many cases the loop which is indicated on the blackboard just cannot be completed.

T. Gehrels

I like to make the following remarks to Professor Alfvén's paper:

1) Space research is in the "observation" category, rather than in that of a laboratory. Once a telescope is on a mission it becomes merely (with due respect to the effort) an extension of ground-based telescopes.

2) You give astronomical observers too much credit. In fact, we have been working with only little consultation with theoreticians, resulting in waste, and should associate with them more.

3) The "comet" versus "asteroid" semantics is purely observational: When a coma is observed, the object is called a comet, when not, an asteroid.

H. E. Newell

With respect to Gehrels' suggestion that space research is entirely observation, I would comment that whereas such has been largely the case in the early years of space research, with the passage of time more and more "experimenting" is being done—like the use of particle accelerators to experiment with particle lifetimes in the radiation belts or the use of barium clouds to experiment with magnetic fields in space. I believe that in the future with continuing improvement in space capability and flexibility the amount of controlled experimenting as compared with passive observation will increase and should serve as a powerful connecting link between theory and observation as Professor Alfvén suggests.

L. Danielsson

The title of Professor Alfvén's communication was semantics and I have a purely semantic objection, and that is the distinction between observation and experiment. Experimental work is mainly observation.

H. Persson

You stress that theoretical terms should not be too frequently used when observations are described, since this may give rise to misunderstandings about what is actually the cause of the observed fact. Instead you recommend the use of a purely observational language. I agree that theoretical concepts should not be misused; on the other hand one always wants to relate observations to some over-all picture, and this requires the use of theoretical words. I therefore suggest that a third language—that of theoretical conjectures—is introduced to meet those two requirements. For instance, instead of calling a certain object X "cometary" (theoretical, unless X is a comet), or describing it using a purely observational term, for instance referring to the density, one should use a word that stresses a) that X has some connection with comets, and b) that this connection is not yet complete or even certain. Suggestions: "pseudo-cometary", "cometacious", "cometarious"—or rather a better word that imagination may invent.

V. Vanýsek

If we could change these terms "cometary" etc. and use only the observational point of view we have a "soft" and "hard" component in meteorites, because we observe aerodynamical effects in the atmosphere. But I personally think that these old terms are good for us.

D. Lal

This long discussion shows that there is a semantic problem.

B. A. Lindblad

I like to make a minor change here, if permitted, and introduce the word "meteoroids" instead. I agree completely with Hannes Alfvén that if you use the term cometary meteoroids and asteroidal meteoroids when you only mean low-density meteoroids and high-density meteoroids it is confusing.

However, most people are implying that the low-density stuff comes from comets and that the high-density meteoroids could come from the asteroid belt, i.e. they are implying an evolutionary process.

P. M. Millman

I should add one further word resulting from personal conversation with Sir Harrie. This research on particles in the upper atmosphere will require great ingenuity, and here is a chance for the half-space nations to use ingenuity rather than money to get their results.

E. Anders

At the risk of antagonizing my friends in the life sciences, I want to express extreme pessimism about biological material in particles in the upper atmosphere. Ten years ago I wrote a little paper about the possibility of looking on the Moon for biological material from outside our solar system (Anders, 1961). In my calculations I assumed, much too optimistically, that it is possible to eject biological material from a planet of the size of the Earth and then have it drift off into space so that some particles leave the planetary system. Estimating the number of life-bearing planets in the Galaxy as 10^{11}, I came up with the result that of 10^{12} biological particles falling on the Moon, one might come from outside the solar system, the others being largely of terrestrial origin. One also has to remember that these particles travel in space for a long time and get sterilized by radiation, so really the chance of finding viable biological particles of extraterrestrial origin is quite small.

From a practical standpoint, let us remember that people have worked very diligently on dust collection in the upper atmosphere and still have not come up with any convincing evidence for extra-terrestrial particles. The terrestrial background is very high. Now, we know that maybe 10^4 or 10^5 tons of extraterrestrial material comes in every year, largely as dust. If people have had trouble finding a single authentic particle of this dust, how are they going to identify the 2.5 spores that may arrive from Arcturus or wherever per century?

E. Anders, 1961, Science, *133*, 1115.

H. Alfvén

I interpreted this remark as a check whether this material was of terrestrial origin. Is that right?

P. M. Millman

I think so. I do not think that Sir Harrie was emphasizing the search for extraterrestrial biological material at all. The experiment should be organized so that you could determine composition, crystal structure and the presence of biological material.

In answer to Dr Anders' comment I quite agree that nobody has yet positively identified space dust. I am quite sure, with the quantity coming in, that many of the particles collected probably have a connection with space, but the identification of any single particle is another matter. Repeating what Fred Whipple said some years ago, I would like to see just one particle that definitely came from space without any shadow of doubt.

H. Alfvén

In this region of the world, around the year 1900, Professor Svante Arrhenius discussed this problem and even if it is not believed very much today it was fascinating and many people got interested in it.

S. K. Runcorn

I was going to ask the same question whether the work of Parkin is regarded now as a kind of hopeless quest?

G. Arrhenius

I have had some experience by following closely Parkin's interesting work with inorganic particles suspected possibly to be of extraterrestrial origin; Parkin found a wide assortment of shapes and sizes of particles, and microprobe analysis demonstrated also a wide variety of compositions. The major difficulty lies in deciding where these particles come from, if they are terrestrial contamination and if any could be extraterrestrial, and in such a case which ones.

P. Pellas

Dr Maurette has proposed an intelligent way to identify extraterrestrial small particles. If they are extraterrestrial they are irradiated by the solar wind. When they travel through the Earth's atmosphere, they are fused, and the amorphous skin produced by the solar wind during the fusion can produce some annealed phases which can be detected by electron microscopy. In short, you will obtain the solar wind signature.

Discussion after a Summary Paper by G. Arrhenius and E. Anders

(The papers by G. Arrhenius and E. Anders also include the contents of their Summary Paper.)

D. Lal

It is generally believed that the low energy cosmic ray particle record seen in single grains of meteorites refers to particles of solar origin. Whereas this may be true, there is no *a priori* reason for this assertion. Within the solar plasma modulation zone, the time averaged flux of $\leqslant 1$ BeV/nucleon protons and heavy nuclei is mostly contributed by solar particles, a result based on recent observations during solar cycles 19 and 20.

In the case of meteorites, some of the evidence already at hand indicates the presence of an appreciable flux of medium energy particles (few hundred MeV/n), in excess to that observed near Earth, for both protons and heavy nuclei. Thus if meteorite grains were irradiated at distances where solar modulation was small, the low energy particles may belong to the continuous galactic stream. This discussion just points out the necessity of being careful in assigning an origin to particles we know so little about—their history in time and space.

H. Alfvén

We should remember what we discussed this morning, namely that if we have a sunspot and a prominence around this, we get solar flares, as Dr Öhman pointed out, by a disruption of the current. If instead of a sunspot with a rotation in it, we have a central body (be it the Sun or a planet) with a magnetic field and a plasma cloud in the neighbourhood, we must have currents in order to put the plasma into a partial corotation. This means that the current system may be similar to what Dr Öhman showed us this morning, including that they may flare. One should expect the same spectrum from a solar flare as from a flare in this situation. I think that if Dr Lal observed an irradiation which could be interpreted as due to the present solar radiation, we could very well think that this instead is produced by a similar phenomenon at the time of formation. The solar activity today is probably a very unstable phenomenon, which is due to the present condition of the Sun, and the Sun, when it generated the planets, may have had a different structure, so we should not take for granted that we could have anything like solar flares in the Sun, but similar phenomena may have occurred in the current system around the Sun.

The more of the hetegonic processes we can refer to present day phenomena in space, the more we are on a somewhat secure ground in our interpretation.

Fossil Magnetic Fields

By S. K. Runcorn

In the problem of the origin of the solar system, it is evident that in addition to gravitational forces one has to consider electromagnetic forces. We have, as a result of Professor Alfvén's work, gradually realized that large-scale motions in the cosmos cannot be discussed without taking magnetic forces into account. One would like, of course, to be able to say something about the strengths and nature of the magnetic fields in the solar system in the formative stage. There seems to be a possibility that one could apply the principles of paleomagnetism to this problem, that is to make use of the fossil magnetization which is retained by small magnetic particles, either of iron or of iron oxides for the periods of time that we are interested in. I think I should, perhaps, emphasize here that the theory of the retention of magnetization by small particles is now very well understood as a result of the work of Professor Néel, and it is no longer surprising that, for example, it is possible to recover from the magnetization of terrestrial sandstones the magnetic field of even as long ago as 3 000 million years. That naturally brings me to the question of the natural permanent magnetization found in meteorites. Can one use this natural remanent magnetization to infer something about the magnetic fields in the early history of the meteorite and the solar system, and possibly say something about the origin of meteorites? Unfortunately this subject has not been very well studied. We are trying to begin work again on this in Newcastle, so I was rather hesitant to say anything at all about the subject in its present rudimentary stage, but I will try to summarize what is known. One of the difficulties was that in the development of paleomagnetism, it was very hard to study specimens of rather irregular shape for technical reasons that I won't go into, and you can imagine that custodians of meteorites were very reluctant to let us do what se wanted on these specimens, and that undoubtedly has been a difficulty which it is now possible to get over. The second difficulty that I must refer to is that when measurements were first made on meteorites, people did not understand the importance of distinguishing between the original magnetization and magnetizations of a secondary kind. The first measurement as far as I know was made by Lovering who took iron meteorites, some four of them, and found that they possessed natural remanent magnetization. But he did not apply any of the demagnetizing techniques, so it was impossible for him to know what is now clear; that the magnetization he was looking at was one picked up in the Earth's field since the fall occurred. Later on a carbonaceous chondrite and three chondrites were studied by Stacey, Lovering and Parry in Australia and they found that all these specimens possessed a natural remanent magnetization. They then demagnetized their specimens. There are various kinds of demagnetizing techniques. One consists of heating the specimen and observing the decrease of magnetization with temperature. Another is to use alternating magnetic fields and to measure the magnetization after the specimen has been subjected to successively higher values of the alternating field. The Australian workers found that the carbonaceous chondrite had no natural remanent magnetization above about 200°C whereas the chondrites did possess a natural remanent magnetization which could be split into two parts—a part which disappeared at rather low temperatures and was evidently spurious and had been picked up since the fall, and an original magnetization which was present right up to the Curie point of the iron particles which were responsible for the magnetic properties. A technique which is often useful is to draw the direction of the natural remanent magnetization on a stereographic projection, then if an original magnetization and a secondary magnetization are present, the natural remanent magnetization lies on a great circle joining these two directions of the separate components. As you heat the specimen up the less stable magnetization disappears and the remanent magnetization of the specimen moves along this great circle. It becomes clear when one does this demagnetizing experiment what the direction and intensity of the original magnetization is. A little bit later, Weaving in London studied a section of the Brewster meteorite, and he was able to show, again, that the magnetization of the meteorite consisted of a secondary magnetization and an original magnetization, and he was able, for example, to show by doing laboratory experiments in cooling the meteorite from high temperature in a magnetic field, that the thermal remanent magnetization produced by exposing the cooling sample to the earth's magnetic field disappears in a very different way from the isothermal remanent magnetization and so this is another way of distinguishing whether you are looking at a magnetization which is original or which has been picked up since the return to Earth. I should emphasise of course that the heating in the passage through the atmosphere mainly affects the magnetic properties in a very thin skin, as of course everyone knows. When one looks through the data that has been collected on the handful of specimens that have been studied, then it is clear—take for example some Russian work—that about half of the chondritic meteorites do possess a magnetization which appears to be of the type of thermo-remanent magnetization resulting from cooling. But we now know that one can produce a strong, stable magnetization not only by cooling from above the Curie point in a field, but also by chemical changes when particles grow—for example they will reach the critical size at which stability is possible, and then if there is a magnetic field present, a remanent magnetization will be found. So, when one looks at the literature and sees various attempts, for example by Guskova and Pochtarev, to determine the field, in which the magnetization was produced (and the values that were found were all around, say, a tenth of an oersted) I think one must, in the light of more recent knowledge, view this with considerable scepticism, because it has been just recently demonstrated by Cox and Butler in the USA that radiation may alter the coercive force—this was done particularly in connection with the problem of the magnetization of lunar samples—but the principle is, of course, that the radiation can produce dislocations in the solid material, which can alter the coercive force. So at the present time I don't think one should say that we know the process by which the permanent magnetization of meteorites could have been acquired, and in particular the values of the field which is responsible for the natural remanent magnetization cannot yet be determined. When people wrote about this subject some years ago, they argued that the presence of natural remanent magnetization demonstrated that the meteorites

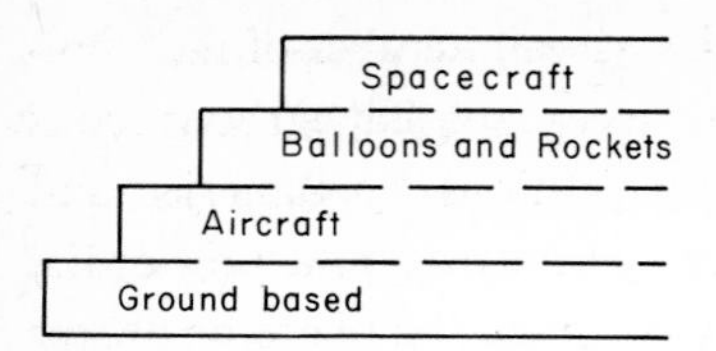

tasks? For the Imaging Photopolarimeter task on the Pioneer missions I found a good (partial) solution by writing the proposal with an industrial partner. The NASA-Ames Research Center kindly agreed to separate contracts: to the industry for the instrument, and to the university for our scientific work. This arrangement was made in lieu of the usual one where the Principal Investigator has the whole contract and must then handle the administration of subcontracts. The arrangement is working to everyone's satisfaction.

The Stairway Rule appears to favor meteoritic research (Anders 1971) and to emphasize groundbased telescopic observations of comets, asteroids and satellites. Even with active and optimistic planning for future missions, it looks as if we shall have quite a few years to concentrate on groundbased studies anyhow. Examples of urgent groundbased observations and equipment are given in my other paper in this book and in the NASA SP-267 book (see References).

For the study of complex interrelations as found in the solar system, theoreticians and observers should get together more often. The present Symposium has been most successful. The International Astronomical Union is considering to add physical studies of minor planets to those of comets already being pursued by one of its commissions. NASA might wish to consider an institute of space sciences with special emphasis on comets and asteroids.

As for the planning of the first missions to asteroids and comets, we shall have to consider the present climate of financial restriction in the various space programs. Our proposed comet/asteroid missions will require severe competition with those to planets and for stellar and other investigations. A painful compromise we have to find between the wish for a sample-return mission and the feasibility of a fast flyby. From the film shown by Dr Newell it appears realistic to plan for a cometary rendezvous that also gives a flyby of an asteroid. A rather modest spacecraft may be employed that is spin stabilized and has a simple scan imager (Gehrels et al. 1971) on board. For description of various techniques as well as of possible scientific studies on asteroid missions see NASA SP-267, and Roberts (1971) for comet missions.

References

Anders, E., 1971, NASA SP-267.
Gehrels, T., Suomi, V. E. and Krauss, R. J., 1971, in "Space Research XI" (ed. A. C. Stickland). Akad. Verlag, Berlin.

Jager, C. de, 1971, Report of COSPAR IAU Symposium, in "Space Research XI", (ed. A. C. Stickland). Akad. Verlag, Berlin.
NASA SP-267, 1971, "Physical Studies of Minor Planets" (ed. T. Gehrels), National Aeronautics and Space Administration, Special Publication 267. US Government Printing Office, Washington, D.C. 20546.
Roberts, D. L., 1971, "Proceedings of the Cometary Science Working Group" (ed. D. L. Roberts). IIT Research Institute (Chicago, Ill.), November 1971; also to be published in NASA-SP series US Government Printing Office, Washington, D.C. 20546.

Discussion

E. Anders

I would like to draw attention to an exceedingly fruitful line of asteroid research, based on Prof. Gehrels' own work (Gehrels 1970). Hapke (1971) has shown that the asteroids can be divided into 4 groups on the basis of Gehrels' UBV color measurements. These groups show a definite correlation with semimajor axis (Fig. 1); nearly all the asteroids in the inner part of the belt belong to Group I, while Groups II and III are found mainly in the middle and outer part. Hapke has also made some very preliminary attempts to determine the UBV colors of meteorites, and it seems that all the more reduced meteorites (H- and E-chondrites; aubrites) fall into Group II, while the more oxidized L-chondrites fall into Group I. The data are not yet good enough and complete enough for any definitive assignments, but it appears that the asteroids become more reduced with increasing distance from the Sun. The larger asteroids (Ceres, Vesta) differ from the smaller ones; presumably their surfaces are differentiated. Icarus, which Arnold and I declared to be a cometary object in

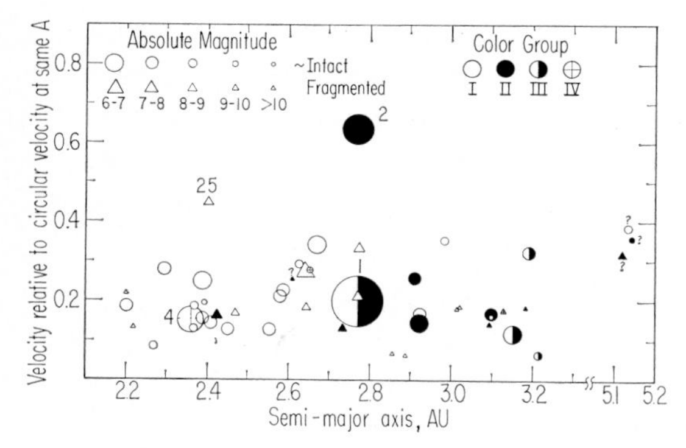

Fig. 1. Asteroid colors (Hapke 1971; Gehrels 1970) as a function of semimajor axis, inclination, eccentricity, size, and degree of fragmentation. The most clear-cut correlation is found with semimajor axis. Nearly all asteroids in the inner half of the belt belong to color Group I, while those in the center and outer half tend to belong to Groups II and III. No significant correlation is found between color and degree of fragmentation. Asteroids were defined as "fragmented" if their lightcurves had $\Delta g > 0.4$, or, in the case of families, if the largest asteroid comprised $< 60\%$ of the mass of the family.

1965, falls well outside the 4 asteroidal color groups. This strengthens the case for its non-asteroidal nature.

It seems that ground-based color and spectroscopic work (Chapman and McCord 1971) can provide valuable clues to the composition of asteroids. Such information can be exceedingly useful in the selection of targets for a possible asteroid mission.

Gehrels, T., 1970, Surface and Interiors of Planets and Satellites (ed. Dollfus) p. 319
Hapke, 1971, NASA SP-267.

Z. Kopal
I would like to ask whether or not, in connection with asteroid missions, consideration is being given to a mission to the Martian satellites?

H. Alfvén
Oh, yes. Fred Singer has argued for this for quite a long time.

Z. Kopal
And are the missions under study?

F. L. Whipple
The importance of photometric observations of asteroids appears of critical importance, both for understanding their true nature and in selecting the ones most valuable scientifically for space probe studies.

H. Alfvén
Summarizing what we have heard so far, there has been a general demand for more emphasis on the study of small bodies. This, of course, is not representative for the scientific community as a whole, because we constitute a highly selective sample of that. But I think that the interest for small bodies is rapidly growing, and this is, indeed, a homage to a grand old man of the field: Fred Whipple, who for a very long time has stood up in defense of the little man i.e. the small bodies in space. Dr Kopal added Martian satellites to the list of the small bodies, and I agree with him, with the exception that I should like to cancel "Martian", I should like to add satellites on the whole, because—from the hetegonic point of view—the most important data one can gain from the Grand Tour may be a detailed study of the outer satellite systems. In that way one gets independent information about how secondary bodies are formed around primary bodies. Even if the mass of the primary body in these cases differs from the solar mass by several orders of magnitude, still the structure of the satellite systems is so similar.

Planning of Space Experiments

By H. E. Newell

As the group enters a general discussion of important questions relevant to understanding how the solar system evolved, there are many related questions as to how space research can best help in trying to answer those questions. Following are some that the group might want to discuss.

1. Which are the most important new missions to try to introduce into the program? I would include here repeats of earlier missions as new ones.
2. In designing new spacecraft and mission profiles, engineers have most often found themselves having to start before the scientists have fully made up their minds as to what should be the primary objectives. Yet it would be much better if the engineers could know in advance what are the most important features to design into their systems. So an important question is: For each class of missions, what is the single most important objective that should guide the spacecraft and mission design?
3. What are the relative priorities of comets, asteroids, Deimos and Phobos? Other satellites?
4. On the path from plasma to planet, what are the key questions for each major phase, and what are the most significant experiments or observations that should be considered?
5. What are the specific combined efforts of ground based programs and space missions that should be fostered?

Discussion

H. Alfvén
We thank Doctor Newell for this, I think, very important survey of the problems. You put five questions, which we should answer. As we have only five minutes left, I don't think we could give a very detailed answer to anyone of them, but I trust that you meant that this would be what we should consider for a year or two, write a number of papers on and discuss at a number of committee meetings. I think that we have time for a few brief comments.

Studies of Other Planetary Systems with Space Techniques

By Z. Kopal

I think the proceedings of our meeting would be scarcely complete if we didn't mention, at this closing stage of the conference, space programs which one should keep in mind in an effort to discover other solar systems in space.

We have been concerned throughout this week with problems largely close on hand, concerned with the system we know best, which we can explore at close range. However, I think one of the points of NASA's charter calls for an exploration of the possibility of an existence of other solar systems outside our own. And perhaps, I may be allowed to spend two or three minutes to mention space missions which should enable us to advance the solution of this problem. First, in reference to Dr Gehrels' "escalation diagram", let me stress that the contribution of ground-based astronomy to quest of other solar systems in space is virtually completed. A law of diminishing returns would be severely operative not only because of limitation of the instruments, but also because of limitation imposed by the atmosphere, plus the fact that the surface of the Earth does not provide a sufficiently steady platform for astrometric studies of highest accuracy. It is obviously not a task for intermediate means of space research, but calls for major space missions. And, for this reason, I should perhaps address my remarks mainly to the representatives of the space super powers on either side of the aisle.

There are, in principle, two ways in which one could go about detecting the existence of other solar systems in space. One is astrometric, the other is photometric. The photometric approach was already pioneered by the Russians; for I think it was academician Fessenkov who some fifteen years ago first pointed out that an indicator of the existence of a solar system, which may precede the discovery of actual planets, is the presence of a "zodiacal light" or dust cloud, which we have reasons to believe may be present in every planetary system. The amount of light of the central star, scattered by such a dust cloud exceeds widely the amount of light scattered by the planets themselves. Therefore, the presence of such a cloud could indicate that planets may be present in there as well. Moreover, its orientation would indicate to us the approximate orientation of the "invariable plane" of the respective system. This quest for photometric evidence of such a cloud represents essentially a photometric problem, but one which cannot be tackled from the Earth. In order to do so with any chance of success, one has to move far away from the Earth, and possibly far away above the ecliptic. Lyman Spitzer some years ago discussed the possibilities of accomplishing such a mission, and I think the methods he had in mind could very well be simplified today, but still require a deep-space mission in a direction perpendicular to the ecliptic, in order to get away from the interplanetary haze that may interfere with its success.

The second method, the actual astrometric discovery of the presence of bodies with small mass in the proximity of a star, would call for the use of large telescopes under the stablest conditions, far away from the Earth, preferably on the Moon. If one considers the pros and cons of the use of a telescope in orbit versus the one on the Moon, I think the use of the Moon as a "stable platform" would prove to possess overwhelming advantages. It is not a work, which one could easily do with an orbiting telescope. The use of the Moon has inestimable advantages of stability for such a work. In fact, we should be very grateful to Nature for having created such a wonderful space platform in so close a proximity to us. It is well outside the terrestrial exosphere as well as magnetosphere with as black and refraction-free a sky as we can get so close to the plane of the ecliptic. It is also seismically very stable, far more so than the Earth. This is an important point for success of this work, which would require a combination of observations made over extended intervals of time.

One problem arises in this connection which astronomers will have to tackle before the engineers will get down to the design of the respective space mission; and that is to investigate, with the requisite degree of precision, the extent of the stability of the lunar platform in the gravitational field of the Earth and the Sun (or, to a lesser extent, of other planets). The motions of the Moon about its centre of gravity—the "physical librations" of our satellite—are much larger and more complicated than is the case for the Earth, but should be much more stable and more easily predictable than for any circum-terrestrial orbiting space station.

A mathematical analysis of such librations represent the type of work in which Professor Gyldén or Professor Bohlin—distinguished predecessors of the present director of the Stockholm Observatory—were the past masters. May we hope that, even at the present time, mathematicians of the requisite calibre may be attracted to the problem, and provide its solution in advance of the space missions that will require its knowledge? I take it from earlier remarks, Dr Newell, that missions to the Moon equipped for astrometric work are being contemplated for the 1980's?

Discussion

H. E. Newell

Yes, that is correct. The studies of the use of a base on the moon are under way, and they are being conducted at many places, both within NASA and outside of NASA. The office of Manned Space Flight, for example, is spending a lot of time supporting such studies, and I would say, the possibility of a lunar base exists for some time in the 1980's.

Z. Kopal

I think we all agree that it is bound to come. Whatever we can do—individually or collectively—to bring it about sooner, will provide also many exciting results in the field of astrometry. I'd like to stress to end up my remarks—

because I see academician Petrov would like to comment on this—that it is not only the spectroscopists, photometrists or investigators of polarization, who have every reason to go into space. It is positional astronomers as much as anybody else, because by getting rid of limitations of positional work on the Earth, we should gain a factor at least a hundred or more in precision—something that we could never hope to do from the surface of the Earth; and I believe we may get some hope in this respect from academician Petrov.

Comments on Space Programs

By G. I. Petrov

Planning space exploration is a very difficult problem since any experiment which utilizes space technology is always very expensive and requires the efforts of many men. Attempts at developing systems and algorithms for producing optimum investigation programs so far have not led to satisfactory results, I do not have much hope that it might be done in the future.

Considerations which I will present further on the most economical ways of revealing of laws of origin of planetary systems are my own opinion only and do not represent a viewpoint of any organization.

Astronomical observatories on orbital stations, free from the atmosphere interference, can increase resolution as much as by an order of magnitude compared to the best instruments installed on Earth.

Professor Kopal presented in his lecture rather interesting observation results of planetary system origination near other stars.

Even greater possibilities are opened by radio-interferometric measurements with bases larger than on earth, at centimeter wavelengths, for example, 1.35 cm, 18.6 cm, and other. These measurements can assure resolution up to 10^{-4} arcseconds and provide a sufficiently detailed picture of the processes occurring during origination of planetary systems near sufficiently remote objects.

The same means will provide information about the far planets of our system and will allow us to advance substantially in studying other astrophysical objects.

The possibilities of using automatic stations on the Moon I have outlined in my lecture.

Of particular importance is the study of Jupiter and the system of its satellites.

In the Soviet Union there are also supporters of the "Grand Tour" mission; but in my opinion, being a very effective mission from the viewpoint of ballistics, it is at the same time an expensive experiment that is not very effective from the viewpoint of obtaining new information. I am more inclined for Jupiter fly-by with a turn perpendicular to the plane of the ecliptic. This experiment will allow to determine the structure of magnetic fields in our planetary system and the structure of all kinds of corpuscular flows from the Sun.

For studying the system of Jupiter and its satellites it is preferable to have an artificial Jupiter satellite and, what is especially desirable, a penetrating probe into the atmosphere of Jupiter.

Landing on an asteroid with the purpose of delivering back to Earth primordial samples is a very interesting problem but, at the same time, it is a very difficult one, due first of all, to the quite insufficient accuracy of our knowledge of the ephemeris of these very small bodies.

Therefore, a spacecraft must have a selfguidance system and very large reserves in energetics. This will require a large amount of work. I believe that some amount of primordial substance from micrometeorites can be obtained on Earth more cheaply if in the Earth laboratories a method would be developed of catching these particles by soft braking without explosion. Some preliminary work undertaken in this direction yielded hopeful results.

The construction and insertion into an orbit, even essentially distant from the Earth, of a trap, with a size of the order of meters (which could substantially simplify the solution of the braking problem and ensure the delivery to Earth of a container with caught substance) can be accomplished in a shorter period of time and at less cost compared to a spacecraft visiting an asteroid.

General Discussion

P. M. Millman

I have a brief comment on Dr Newell's remarks in connection with the programs of upper atmosphere research by means of rockets and balloons. It is true that a lot of work has now been done, and we hope that this activity will continue. I also hope that in the future more attention will be paid to devising new techniques that are promising, and that the experimenters will not feel that they have to come up with a significant result at the first or second trial. I am sure that the early rocket experiments were hampered by the desire to produce a flux for meteoritic material or to derive other similar results, right away. Not enough attention was paid to calibration and other instrumentation problems. Ten years ago we started in our section in Ottawa some experiments in connection with micrometeorite detection from rockets, and I told the personnel concerned: "I don't care whether you have any results at all for five or even ten years. I want you to learn about instrumentation on upper air rockets—what it does, how you calibrate it, what errors you may run into." We have not conducted this program with the idea of coming up with a flux of any kind. I think, as of now, more programs should be planned in this way. It requires a difficult type of philosophy at a university, because a graduate student can't come up with a thesis in one or two years. But I think in the long run we will get the kind of things Sir Harrie Massey was thinking of.

There is one particular type of program that I have been very impressed with. It is the Pegasus experiment. I think this is an example of a carefully planned program with a new technique and great possibilities, and I feel that it has given us some of the most important results we have yet attained. That is the type of thinking that should go into the upper air rockets—long-term, careful planning.

B. A. Lindblad

In discussing the comet intercept and comet rendez-vous missions as well as missions of type "Grand Tour" where we pass near to some of the minor Jovian satellites, I think it is very important to remember that for some of these objects we do not have too good ephemerides. My question is: Is it not necessary to immediately start a large, ground-based program of observations of the interesting comets and of the minor Jovian satellites, so that we in time for the launches have improved orbits? Is NASA planning to support ground-based studies of these questions?

H. E. Newell

The answer very simply is: Yes. We have been doing this sort of thing. The Jet Propulsion Laboratory has been doing some of it, NASA has supported the new construction and use of planetary telescopes, and we are in discussion now with some people on the possibility of additional planetary telescopes. This is the sort of thing, that I was referring to, when I said: What are the important combinations of space and ground missions, that we are considering?

G. Arrhenius

May I ask a brief question, mainly directed to Dr Newell. It is clear from the discussion of Dr Newell's and Dr Petrov's presentations that much revolves around the strategy of selection, with the extreme choices being sample return or flyby. There is an enormous difference in the type and quality of information that one can extract in the two cases. Dr Gehrels put it in the perspective of time, with only our children and grandchildren reaping the fruits of our experimental labor. Dr Newell projected the question of choice into terms of cost, as did also Dr Petrov. I wonder if it is possible to give some idea about the magnitude of cost difference or time difference, as you see them now, in the case of, say, a mission to an asteroid or a comet.

H. E. Newell

Yes, I didn't come prepared to talk about cost, so I have to talk just in terms of general orders of magnitude. A Pioneer spacecraft costs ten million dollars or so—you can make it cost more by instrumenting it more fully, and with more complex equipment, but that is the order of magnitude for Pioneer—say ten to twenty millions of dollars. The Mariner spacecraft costs from fifty to a hundred million dollars, depending on how complex you make it. The Grand Tour spacecraft, for the four of them that we are talking about, including the operations for the testing and so on, will cost us about three quarters of a billion dollars. That's why there has to be a careful strategy with the use of the Grand Tour spacecraft, because if you just use such a spacecraft for a single-minded mission, then it gets to be very expensive, as Dr Petrov says. On the other hand, if you use it the way I described yesterday to make twenty-seven encounters with sixteen different objects, then the cost per encounter comes down to around thirty million dollars, which is far less than using a Mariner for such an encounter. And furthermore, since the program is extended over a decade, the total cost is distributed over a decade, and that means the yearly cost is not very great. In that sense, it is not a very expensive project. Did I answer your question with these sums?

T. Gehrels
Would you have an estimate for a sample return mission?

H. E. Newell
Oh yes, you asked about sample-return. For sample return missions, then the costs will go up by a factor of an order of ten. That is because of the guidance and control needed after you get there, permitting one to land and then to get back to the Earth, and then you have to have recovery methods. You add about four new phases in the operation, that you have to pay for.

G. Arrhenius
It is important then to note that the information return from a sampling goes up by more than an order of magnitude.

H. E. Newell
Yes, you have to ask yourself now: What information return am I getting per dollar that I invest?

T. Gehrels
In addition, we must, of course, face the question if the total cost of the mission is not too high to obtain approval.

H. Alfvén
Could we have the prices in rubels also?

G. I. Petrov

In my opinion it is nearly the same.

D. Lal

In this connection, if I understood right, academician Petrov said that there was a way of bringing back samples of any matter impacting the instruments. Can flyby objects be sampled?

G. I. Petrov

Yes, some work has been started on the design of an instrument with which micrometeorites might be collected in space and brought back to Earth without being destroyed.

H. Alfvén

What we experience now is a new age, a new epoch in astronomy. Astronomy started with the planetary system. Like a shock-wave it has extended further and further out, but now a second wave is coming, which concentrates the interest again on the planetary system. This is a different type of astronomy—much less speculative, much more connected with experiments. We should remember what Sir Harrie Massey pointed out yesterday after the lectures; namely that if somebody had said all this fifteen years ago he would have been considered to be completely out of contact with reality—a man who speculated and talked nonsense. Fifteen years ago no one could talk about the possibility that an American or Russian is driving a car on the moon.

F. L. Whipple

I beg your pardon. I'll show you a couple of volumes written in 1952.

H. Alfvén

Yes. Very, very few people in any case. I think that it is somewhat of a shame to the scientific community that it is the engineers and not the scientists who have been the main driving force in space research. When listening to the discussion, we should remember this. Many people say: Well, we must be realistic, we should not ask for too much, nothing which is not already in the plans, can be realized in the near future. This is an unfortunate attitude. After all, the era of inventions and new ideas has not ended. There will be a number of new devices, by which we can get new information. We have recently seen the invention of solar-electric propulsion engines which make it much easier to go to comets and asteroids, and make other trips. It is not impossible at all, that we very soon will experience a new revolution of the same kind as was started fourteen years ago. Is it only fourteen years ago? That is a very short time. What may happen before the end of the next fourteen-year period we do not know, but I hope that we will meet again long before that to discuss, how to use the new techniques, which I am sure will emerge. And with these concluding remarks the 21st Nobel Symposium "From Plasma to Planet" should proceed to the final point on the program, the closing dinner.